Organelles in Eukaryotic Cells

Molecular Structure and Interactions

Organelles in Eukaryotic Cells

Molecular Structure and Interactions

Edited by

Joseph M. Tager
University of Amsterdam
Amsterdam, The Netherlands

Angelo Azzi
University of Bern
Bern, Switzerland

Sergio Papa and
Ferruccio Guerrieri
University of Bari
Bari, Italy

Plenum Press • *New York and London*

Library of Congress Cataloging-in-Publication Data

Federation of European Biochemical Societies Advanced Course on
  Organelles in Eukaryotic Cells: Molecular Structure and Interactions
  (1988 : Bari, Italy)
    Organelles in eukaryotic cells : molecular structure and
  interactions / edited by Joseph M. Tager ... [et al.].
        p.   cm.
    Proceedings of the Federation of European Biochemical Societies
  Advanced Course on Organelles in Eukaryotic Cells: Molecular
  Structure and Interactions, held May 16-27, 1988 in Bari, Italy"-
  -T.p. verso.
    ISBN 0-306-43388-5
    1. Cell organelles--Congresses. 2. Eukaryotic cells--Congresses.
  I. Tager, J. M. II. Federation of European Biochemical Societies.
  III. Title.
  QH573.F43 1988
  574.87'34--dc20   89-25545
                                                            CIP

Proceedings of the Federation of European Biochemical Societies
Advanced Course on Organelles in Eukaryotic Cells: Molecular
Structure and Interactions, held May 16-27, 1988, in Bari, Italy

© 1989 Plenum Press, New York
A Division of Plenum Publishing Corporation
233 Spring Street, New York, N.Y. 10013

Printed in the United States of America

PREFACE

Every year, the Federation of European Biochemical Societies sponsors a series of Advanced Courses designed to acquaint postgraduate students and young postdoctoral fellows with theoretical and practical aspects of topics of current interest in biochemistry, particularly within areas in which significant advances are being made. This volume contains the Proceedings of FEBS Advanced Course No. 88-02 held in Bari, Italy on the topic "Organelles of Eukaryotic Cells: Molecular Structure and Interactions."

It was a deliberate decision of the organizers not to restrict FEBS Advanced Course 88-02 to a discussion of a single organelle or a single aspect but to cover a broad area. One of the objectives of the course was to compare different organelles in order to allow the participants to discern recurrent themes which would illustrate that a basic unity exists in spite of the diversity. A second objective of the course was to acquaint the participants with the latest experimental approaches being used by investigators to study different organelles; this would illustrate that methodologies developed for studying the biogenesis of the structure-function relationships in one organelle can often be applied fruitfully to investigate such aspects in other organelles. A third objective was to impress upon the participants that a study of the interaction between different organelles is intrinsic to understanding their physiological functions.

This volume is divided into five sections. Part I is entitled "Structure and Organization of Intracellular Organelles." The two contributions on mitochondrial respiratory chain complexes and the F_o-F_1 ATP synthase illustrate the general principles involved in studying structure-function relationships in membrane-bound components of organelles. The third contribution deals with peroxisomes; the essential role of these organelles in mammalian physiology is illustrated graphically by the recent discovery of lethal genetic diseases in man in which peroxisomal functions are impaired.

Membrane flow through the vacuolar system has become a major area of research during the last few years. The two contributions in Part II deal with this subject. In one, vesicular traffic between elements of the vacuolar system is discussed. In the other attention is paid not only to protein transport but also to the equally important question of the movement of lipids during membrane flow.

Part III is devoted to a consideration of the interactions between organelles during metabolism. The specific subjects covered include Control Analysis, intra- and intercellular interactions during ureogenesis and pH homeostasis, and the mechanism of action, both at the cellular and at the molecular level, of a bacterial toxin.

Genetic interactions during the biogenesis of different organelles is covered by the five contributions in Part IV. The two contributions on mitochondrial biogenesis discuss the organization and evolution of the mitochondrial genome and, on the other hand, the expression of the nuclear genes coding for mitochondrial respiratory chain complexes which, in mammals, clearly is tissue specific.

Finally, Part V is a unique feature of this volume. It is a methodological section, and it contains detailed protocols for studying various aspects of structure-function relationships in complexes of the mitochondrial inner membrane, the biogenesis of mitochondria, the regulation of lipid and nitrogen metabolism in the liver, and membrane functions in chromaffin granules.

J.M. Tager
A. Azzi
S. Papa
F. Guerrieri

CONTENTS

THE MITOCHONDRIAL RESPIRATORY CHAIN

Angelo Azzi, Michele Müller and Néstor Labonia

Institut für Biochemie und Molekularbiologie der Universität

Bühlstrasse 28, CH-3012 Bern, Switzerland

Electron flow through the respiratory chain of mitochondria from NADH to dioxygen is associated with the synthesis of ATP. This process is essential for life as it appears when a block of the respiratory chain is induced by deadly poisons, such as hydrogen sulfide or cyanide, or by the defective synthesis of some proteins, an event which is at the basis of many respiratory chain diseases (1).

The concepts that electrons flow to oxygen and the role of the oxidase in the mitochondrial chain were already clear to Keilin (2) and Wartburg (3). The contribution which was given during the last half a century of research has been in the understanding the role of the single components of the respiratory chain relative to the complex transport function. A second major point of interest has been the clarification of the mechanism by which the electron energy fall in the respiratory chain is conserved and transformed into usable forms of energy. Among these transformations a primary importance has been given to the detail mechanism of ATP synthesis. Finally the question of the control of these processes has been considered, at both biogenetic and postsynthetic levels.

THE MOLECULAR STRUCTURE OF THE RESPIRATORY CHAIN

The classical separation into four complexes of the mitochondrial respiratory chain obtained by Hatefi laboratory (4) has survived the challenge of time. Complex I, III and IV represent the three enzyme units through which the electrons donated by NADH have to pass in an obligatory sequence in order to reach oxygen (Fig. 1). The electron transport units may not be physically organized as a chain of complexes, entertaining stable interactions. They may rather interact with each other as freely colliding molecules resulting in a kinetic chain of events (5)

COMPLEX I

Bovine heart complex I (NADH : ubiquinone reductase) (M_r 610,000) contains 25 polypeptides, one molecule of FMN, 24 non-heme irons and acid-labile sulfides, and 2-4 ubiquinone molecules. It has been resolved in three sub-complexes: the flavoprotein complex contains 3 subunits, FMN, 6 Fe and the NADH binding site; the iron-sulfur protein complex contains 10 irons associated with 6 subunits; the hydrophobic protein complex contains 16 subunits and 6 irons (6).

A low resolution three-dimensional structure of NADH: ubiquinone reductase from *Neurospora crassa* mitochondria has been determined by electron microscopy of membrane crystals (7).

The structure shows that NADH: ubiquinone reductase extends 15 nm across the membrane, projecting 9 nm from one membrane side and 1 nm from the opposite side. Only about one-third of the total protein mass is located in the membrane.

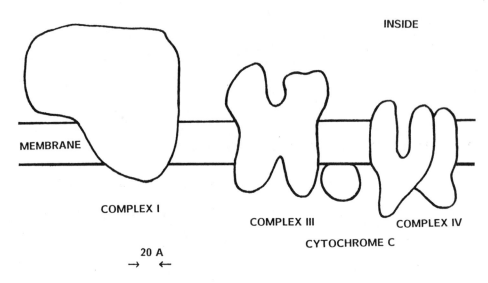

Figure 1. Model of the three electron transport complexes. The dimensions and shape were obtained from the electron microscopic analysis of two dimensional crystals from (11).

At the level of complex I the first coupling site of the electron transfer chain is present and it is linked to a proton translocation which is associated with the electron flow (8). However, the complication and the dimensions of the bovine heart complex I are such that, until now, no more detailed information is available as to the role of its composing polypeptides, the mechanism by which the proton translocation occurs and by what controls the system is regulated. Clinical and biochemical data are frequently

TABLE I

THE POLYPEPTIDE COMPOSITION OF COMPLEX I.
(Modified from Morgan et al (8))

Fraction	Number of subunits	Mass (kDa)	FeS cluster
Hydrophobic Protein (HP)	16	42,39,35, 26,25,23.5, 22,22,21,20,18, 16,13.5,8.5	2FeS 4FeS
Iron sulfur Protein (IP)	6	75	2FeS 2FeS
		49, 30, 13	2FeS 4FeS
		18, 15	
Flavoprotein (FP)	3	51 24, 9	4FeS 2FeS

reported for patients with mitochondrial (encephalo-) myopathies associated with complex I alterations. The symptoms can range from exercise-induced muscle weakness to signs of multisystem disease (9).

COMPLEX III

Bovine heart complex III (ubiquinol : cytochrome c reductase) receives electrons from complex I, *via* ubiquinone, and transfers them to complex IV, *via* cytochrome c. It contains 11 subunits, three of which have a known catalytic function (10, 11).

TABLE II

The polypeptide components of complex III and their function (from von Jagow et al (10)

Subunit	Name	Features	M_r
I	Core I	unknown	47
II	Core II	unknown	45
III	Cytochrome b	2 hemes	42.7
IV	Iron-sulfur Protein	Fe_2S_2	21.6
V	Cytochrome c_1	heme	27.3
VI	QP-C	binds ubiquinone	13.4
VII	--	associated with core proteins	9.5
VIII	Hinge protein	associated with cytochrome c_1	9.2
IX	DCCD binding protein	H^+-translocation ?	8.0
X	--	associated with cytochrome c_1	7.2
XI	ISP	associated	6.4

The four redox centers of the catalytic subunits are three hemes and one iron-sulfur center. While the redox centers and three subunits are constantly present throughout evolution, the remaining subunits are not always present. The number of subunit can be as low as three in bacteria, 4-5 in chloroplasts, 10 in fungi and 11 in mammals. All catalytic and most (except for the two core proteins) non-catalytic subunits have been sequenced.

The coupling between electron flow and proton translocation at the level of complex III, the second coupling site in the respiratory chain is still a matter of discussion.

The available experimental data suggest that complex III functions according to a cyclic electron flow mechanism, in which ubiquinone plays a central role as H^+ translocator across the mitochondrial membrane. Variations on this theme have been proposed, such as a double ubiquinone cycle, a b-cycle and a ubiquinone gated H^+ pump. In all cases the experimental data of the electrogenic translocation of $2H^+$ per pair of electrons crossing complex III has to be taken into consideration.

COMPLEX IV

Complex IV, cytochrome c oxidase, has been one of the most intensively studied mitochondrial enzymes in the last few years. The mammalian enzyme is rather complicated, since it contains 13 subunits (12), two hemes and two or three coppers as recently suggested by Buse's group (13). The enzymes of lower forms of life are much simpler, raging from the two subunits of the oxidase-c complex of *Thermus thermophylus* to the 3 subunits of *Paracoccus denitrificans*, to the 5 polypeptides of higher plants, the 7-8 subunits of fungi and protozoa. As for complex III the reason for the different complexity of the different enzymes is far from being understood. It appears moreover that the mammalian enzyme can exist in several tissue specific isozyme forms (14).

The study of evolutionary conserved residues which may be considered candidates to the role of metal ligands (15) has brought to the conclusion that subunit II cannot ligate heme iron with imidazol nitrogens as indicated to be necessary by

TABLE III

Subunit composition of complex IV (modified from Kadenbach et al (12))

Subunit	Feature	M_r
I	Contains heme and at least one Cu	56,993
II	Contains on Cu and binds cytochrome c	26,049
III	Important in H^+ pump and dimerization	29,918
IV	Necessary for the enzyme assembly	17,153
Va	Redox state affects antibody reactivity	12,436
Vb	not necessary for H^+ and e^- transport	10,670
VIa	not necessary for H^+ and e^- transport	9,419
VIb	not necessary for H^+ and e^- transport	10,068
VIc	differs in tissue specific isozymes	8,480
VIIa	differs in tissue specific isozymes	6,244
VIIb	differs in tissue specific isozymes	6,000
VIIc	differs in tissue specific isozymes	5,541
VIII	differs in tissue specific isozymes	4,962

precise spectroscopic studies. Recent finding on the subunit II gene of cytochrome c oxidase from *Trypanosoma brucei, Crithidia fasciculata and Leishmania tarentolae* (16) have in fact indicated that the only possible histidine (after the binding of copper to a cluster of residues typical of copper proteins) is not conserved in this species leaving subunit I as the only polypeptide responsible for heme binding (Fig. 2). Direct proof for this conclusion has been afforded by serial proteolytic degradation of *P. denitrificans* oxidase by chymotrypsin and *Staphylococcus aureus V8* protease resulting in a single polypeptide (subunit I) active cytochrome c oxidase (3).

What are the subunits responsible for cytochrome c binding?

Cytochrome c was modified by introducing a 3-nitrophenylazido group at lysine 13 (20). The cytochrome c modified at lysine 13 in the presence of ultraviolet light formed a covalent complex with cytochrome c oxidase (21) through specific covalent interaction of the lysine 13 derivative of cytochrome c with subunit II (22). It appears probable that also other subunits are involved in the delimitation of cytochrome c binding site, such as subunit III and some low molecular weight component in the mammalian enzyme. The recent finding that the single polypeptide enzyme resulting from digestion of the subunit II of *P. denitrificans* oxidase is capable of oxidizing reduced cytochrome c with affinity higher than that of the two subunits complex suggests that subunit I may play an important role in binding not only of one substrate, oxygen, but also the other, cytochrome c. It seems at least possible that, through subunit II, the contact is realized necessary to the passage of electrons between cytochrome c and one of the cytochrome c oxidase coppers. Cytochrome c oxidase monomers are active.

Gel filtration chromatography on Ultrogel AcA 34 in the presence of dodecylmaltoside (25-29) separated the monomeric and dimeric forms of bovine cytochrome c oxidase, both of them containing all 12-13 subunits described for this enzyme. Sucrose density centrifugation analysis and analytical centrifugation gave sedimentation coefficients of 15.5 and 9.6 S for the dimer and the monomer respectively. With the Stokes radius (measured by laser light-scattering to be 7 nm for the dimeric detergent-lipid-protein complex) and the partial specific volumes of the detergent and the enzyme (determined densitometrically), the molecular weights of

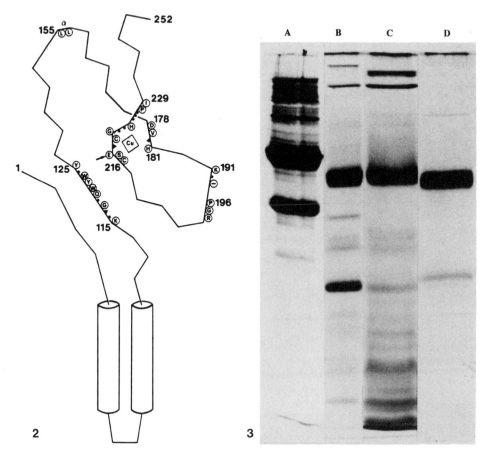

Figure 2. Model of oxidase subunit II. Numbers or symbols correspond to evolutionary conserved residues of the bovine heart enzyme sequence. Circles: aminoacids or charges; triangles: hydrophobic residues;dots: residues with no apparent function. The two cylinders represent alpha helices spanning the membrane. Arrows represent negatively charged amino acids which cytochrome c protected against modification by water soluble carbodiimides (15).

Figure 3. SDS-polyacrylamide gel electrophoresis of native, chymotrypsin treated and *S.aureus* V8 protease digested cytochrome c oxidase . Silver stained gels; Lane A: native enzyme; lane B: Chymotrypsin treated enzyme; lane C: digestion of subunit IIc by *S. aureus* V8 protease; Lane 4: one-subunit enzyme after HPLC ion-exchange chromatography; the band was migrating at the level of ovalbumin, which has an apparent molecular weight of 43,000 (46,47).

400,000 for the protein moiety of the dimer and 170,000-200,000 for the monomer were calculated. The monomer/dimer equilibrium was found to be dependent (under very low ionic strength conditions) on the protein concentration: Monomers were predominant at low (10^{-9} M) and dimers (plus aggregates) at high (> 5 X 10^{-6} M) enzyme concentrations. The dimeric enzyme showed a biphasic activity and monomers gave monophasic kinetics as a function of cytochrome c concentration (measured spectrophotometrically and analyzed by Eadie-Hofstee plots), compatible with a homotropic negative cooperative mechanism for the dimer of cytochrome c oxidase (30). From observations that in bidimensional crystals the bovine heart oxidase is dimeric the conclusion was inferred that the enzyme is only active in such a form. The conclusions reported above indicate that in detergent dispersed cytochrome oxidase not only dimers but also monomers are efficient catalysts, leaving the question open as to the activity of the enzyme as a proton pump, due to the lack of information on the state of aggregation of the enzyme in the phospholipid membrane (31).

The enzyme from *Paracoccus denitrificans* was always found monomeric, at whatever salt and protein concentration in dodecylmaltoside and Triton X-100 which may point to an intrinsic difficulty of this enzyme to form aggregates (26,29): still the enzyme reconstituted in lipid vesicles was able to pump, although with a limited efficiency.

Vectorial proton translocation is an intrinsic function of the oxidase and can be inhibited by hydrophobic carbodiimides.

For a quantitative analysis of the proton pump associated with bovine cytochrome c oxidase (32-35) we have synthesized fluoresceinphosphatidyl-ethanolamine which, after reconstitution with cytochrome c oxidase into phospholipid vesicles, has been used as a specific and sensitive indicator of the intravesicular pH (36-39).

It was observed that cytochrome c oxidase catalyzed the abstraction of almost 2 protons from the intravesicular medium per molecule of ferrocytochrome c oxidized while in the external medium approximately 1.0 proton per molecule appeared. That cytochrome c oxidase behaves as a proton pump seems to be strongly supported by the above data. It was also shown that apparent anomalies in the kinetics of H^+ re-entry (36) resulted from insufficient transmembrane charge equilibration during oxidase turnover (37) and were fully normalized by appropriately adjusting K^+ and valinomycin concentrations.

Dicyclohexylcarbodiimide (DCCD) was found to bind to cytochrome c oxidase with a parallel inhibition of ferrocytochrome c-induced H^+ translocation (38,39). The maximal overall stoichiometries of DCCD molecules bound per cytochrome c oxidase was 1 for the reconstituted enzyme. The DCCD- binding site was identified to be in a non polar sequence (the hydrophilic carbodiimide, 1-ethyl-(3-dimethylaminopropyl)-carbodiimide affected to a minor extent H^+ translocation) located in the subunit III of the oxidase (40, 41-43). It appears, from these observations that subunit III of cytochrome c oxidase plays an important role in H^+ translocation by the enzyme. An evolutionary conserved glutamic acid residue was found to be the specific target of DCCD.

A possible role of subunit III may be that of protecting the enzyme from a loss of efficiency of the proton pump when the enzyme is subjected to multiple turnovers. A subunit III depleted enzyme has in fact a high efficiency only when it is allowed to translocate protons for a very limited number of turnovers. It is evident that the extractions of protons becomes more and more difficult, after each turnover, due to the small inner volume and the decrease in the concentration of the transported substrate, the proton itself. What subunit III may do is to increase by creating the appropriate environment (through the evolutionary conserved glutamate 90, whose inhibition blocks the proton pump?), the pK value of a protonatable group associated with the proton pump. Such a conclusion is also supported by the observation that the midpoint potential of the heme a becomes pH insensitive after subunit III extraction.

Cytochrome c oxidase activity can be modulated by free fatty acids.

The system of cytochrome c oxidase reconstituted in phospholipid vesicles has been used to study the regulation of the enzyme by free fatty acids (44). The proton pump of the redox complex was found not to be affected, even at concentrations of 10 μM by the presence of palmitic acid (Na salt). The H^+/e^- ratio remained at values close to 0.9 in the presence or absence of the fatty acid. The proton permeability of the vesicles was studied under three conditions: 1. passive re-entry of protons following the active H^+ extrusion 2. passive re-entry of H^+ following an acid pulse. 3. passive entry of H^+ driven by a diffusion potential (negative inside). In all cases, under the experimental conditions employed, no increase of the passive H^+ permeability was produced by concentrations of palmitic acid known to uncouple oxidative phosphorylation (45). Also the permeability for potassium ions was not affected by palmitic acid. Despite the lack of proton permeability changes and of molecular uncoupling of the pump (slippage) the respiratory control index was constantly diminished in the presence of the fatty acid due to an increased basal respiration rate (in the absence of uncoupler and/or valinomycin). The diminution of respiratory control was always partial, despite the high concentrations employed, as if the fatty acid had an effect only on one of two possible components of the respiratory control. Besides the control exerted by the membrane potential opposing the intraenzyme electron flow, a conformational transition may follow the potential onset which would bring the enzyme into a less active state. This second event may be prevented by free fatty acid by making the enzyme conformation insensitive to the potential.

Another possibility of control of the respiratory chain may be related to the finding that cytochrome c oxidase binds ATP at physiological concentrations at the cytosolic side of the enzyme. The binding and release of ATP result in changes of the K_m for cytochrome c (12).

ACKNOWLEDGEMENTS

Part of the work reviewed in this article has been supported by the Swiss National Science Foundation.

REFERENCES

1. J. Bioenerg. Biomembr. 20, 291-391
2. Keilin, D. (1925) Proc. R. Soc. Lond. 98, 312-320
3. Wartburg, O., and Negelein, E. (1929) Biochem. Z. 214, 64-69
4. Hatefi, J. (1976) in Enzymes of Biological Membranes (A. Martonosi, Ed.), vol 4, Plenum Press, New York & London, pp. 3-41
5. Hackenbrook, C.R. (1981) Trends Bioch. Sci. 6, 151-154
6. Ohnishi, T., Meinhardt, S.W., Yagi, T., and Oshima, T. (1987) in Advances in Membrane Biochemistry and Bioenergetics (Ch. H. Kim, H. Tedeschi, J.J. Diwan and J.C. Salerno, Eds.), Plenum Press, New York, pp. 237-248
7. Leonard, K., Haiker, H., and Weiss, H. (1987) J. Mol. Biol. 194, 277-86
8. Morgan-Hughes, J.A., Schapira, A.H.V., Cooper, J.M., and Clark, J.B. (1988) J. Bioenerg. Biomembr. 20, 365-382
9. Schapira, A.H., Cooper, J.M., Morgan-Hughes, J.A., Patel, S.D., Cleeter, M.J., Ragan, C.I., and Clark, J.B. (1988) Lancet 8584, 500-503
10. von Jagow, G., Link, T.A., and Ohnishi, T. (1986) J. Bioenerg. Biomembr. 18, 157-179
11. Capaldi, R.A., Halphen, D.G., Zhang, Y.-Z., and Yanamura, W. (1988) J. Bioenerg. Biomembr. 20, 291-365
12. Kadenbach, B., Kuhn-Nentwig, L., and Büge, U. (1987) Current Topics in Bioenerg. 15, 113-161
13. Steffens, G., and Buse, G. (1986) 4th Eur. Bioenerg. Conference, Vol. 4, p. 191
14. Kuhn-Nentwig, L., and Kadenbach, B. (1985) Eur. J. Biochem. 149, 147-158
15. Millett, F., de Jong, C., Paulson, L., and Capaldi, R.A. (1983) Biochemistry 22, 546-552

16. Benne, R., Van Den Burg, J., Brakenhoff, J.P.J., Sloof, P. Van Boom, J.H., and Tromp, M.C. (1986) Cell 46, 819-826
17. deVrij, W., Poolman, B., Konings, W.N., and Azzi, A. (1986) Methods Enzymol. 126, 159-173
18. Gennis, R.B., Casey, R.P., Azzi, A., and Ludwig, B. (1982) Eur. J. Biochem. 125, 189-195
19. DeVrij, W., Azzi, A., and Konings, W.N. (1983) Eur. J. Biochem. 131, 97-103
20. Bisson, R., Gutweniger, H., and Azzi, A. (1978) FEBS Lett. 92, 219-222
21. Bisson, R., Azzi, A., Gutweniger, H., Colonna, R., Montecucco, C., and Zanotti, A. (1978) J. Biol. Chem. 253, 1874-1880
22. Bisson, R., Gutweniger, H., Montecucco, C., Colonna, R., Zanotti, A., and Azzi, A. (1977) FEBS Lett. 81, 147-150
23. Bisson, R., Montecucco, C., Gutweniger, H., and Azzi, A. (1979) Biochem. Soc. Trans. 7, 156-159
24. Bisson, R., Montecucco, C., Gutweniger, H., and Azzi, A. (1979) J. Biol. Chem. 254, 9962-9965
25. Bolli, Nałęcz, K.A., and Azzi, A. (1985) Arch. Biochem. Biophys. 240, 102-116
26. Nałęcz, K.A., Bolli, R., Ludwig, B., and Azzi, A. (1985) Biochim. Biophys. Acta 808, 259-272
27. Bolli, R., Nałęcz, K.A., and Azzi, A. (1985) Biochimie 67, 119-128
28. Nałęcz, K.A., Bolli, R., and Azzi, A. (1983) Biochem. Biophys. Res. Commun. 114, 822-828
29. Bolli, R., Nałęcz, K.A., and Azzi, A. (1986) J. Bioenerg. Biomembr. 18, 277-284
30. Nałęcz, K.A., Bolli, R., and Azzi, A. (1986) Methods Enzymol. 126, 45-64
31. Casey, R.P., O'Shea, P.S., Chappell, J.B., and Azzi, A. (1984) Biochim. Biophys. Acta 765, 30-37
32. Azzi, A. (1980) Biochim. Biophys. Acta 594, 231-252
33. Müller, M., Thelen, M., O'Shea, P., and Azzi, A. (1986) Methods Enzymol. 126, 78-87
34. Casey, R.P., Broger, C., Thelen, M., and Azzi, A. (1981) J. Bioenerg. Biomembr. 13, 219-228
35. Azzi, A. (1984) Experientia 49, 901-906
36. Casey, R.P., and Azzi, A. (1983) FEBS Lett. 154, 237-242
37. Thelen, M., O'Shea, P.S., Petrone, G., and Azzi. A. (1985) J. Biol. Chem. 260, 3626-3631
38. Azzi, A., Müller, M., O'Shea, P., and Thelen, M. (1985) J. Inorg. Biochem. 23, 341-347
39. Thelen, M., O'Shea, P.S., and Azzi, A. (1985) Biochem. J. 227, 163-167
40. Casey, R.P., Thelen, M., and Azzi, A. (1980) J. Biol. Chem. 255, 3994-4000
41. Casey, R.P., Broger, C., and Azzi, A. (1981) Biochim. Biophys. Acta 638, 86-93
42. Azzi, A., Casey, R.P. and Nałęcz, M.J. (1984) Biochim. Biophys. Acta 768, 209-226
43. Nałęcz, M.J., Casey, R.P., and Azzi, A (1986) Methods Enzymol. 125, 86-100
44. Casey, R.P., Ariano, B.H., and Azzi, A. (1982) Eur. J. Biochem. 122, 313-318
45. Labonia, N., Müller, M., and Azzi, A. (1988) Biochem. J. 254, 139-145
46. M. Müller, B. Schläpfer, A. Azzi (1988) Proc. Natl. Acad. Sci. USA 85, 6647-6651.
47. M. Müller, B. Schläpfer, A. Azzi (1988) Biochemistry, 27, 7546-7551

THE F_0F_1 H^+-ATP SYNTHASE OF MITOCHONDRIA

Sergio Papa

Institute of Medical Biochemistry and Chemistry

University of Bari, Bari, Italy

The H^+-ATP synthase complex of coupling membranes (mitochondrial membrane, plasma membrane of bacteria, thylakoid membrane of chloroplasts) (E.C.3.6.1.34) is a key enzyme in eukaryotes and prokaryotes, being responsible for ATP synthesis supported by the protonmotive force (PMF) generated in these membranes by redox or photoredox systems (1,2). To give an example of its role in humans, an adult with a daily energy need of around 3.000 Kcal utilizes $\approx$200 Kg of ATP. About 90% of this amount of ATP turning over in our cells – they are practically impermeable to ATP (and ADP)– is, under normal conditions, synthesized from ADP and Pi by the H^+-ATP synthase of mitochondria.

The H^+-ATP synthase or H^+-ATPase catalyzes reaction (1)

$$(1) \quad ATP + H_2O + nH_N^+ \longleftrightarrow ADP + POH + nH_P^+$$

associated with H^+ transport across the coupling membrane and poised in cells against a phosphate potential ($\Delta G_p'$) of $\approx$12-16 Kcal/mol and a PMF of $\approx$230 mV (with a predictable H^+/ATP ratio of 3 for H^+ pumped from the inner (N) to the outer (P) aqueous phase (see Table 1 from ref.3).

The H^+-ATP synthase of coupling membranes is structurally and functionally made up of three parts: (i) the catalytic sector or F_1 (knob) universally consisting of 5 subunits; (ii) the H^+-translocating or membrane-integral sector (F_0)(consisting of a variable number of subunits); (iii) the stalk (coupling sector or gate) made up of some F_1 and F_0 components. This tripartite structure was first revealed by electron microscopy of negatively stained vesicles of the inner mitochondrial membrane (14). Recently the same structure has also been obtained by image analysis of electron micrographs of rapidly frozen, unstained samples of E.Coli F_0F_1 H^+-ATP synthase (Fig.1, from ref.5), this indicating that the knob-on-stalks structure represents a general feature of the enzyme (see however,6).

In prokaryotes F_0,F_1 H^+-ATP synthase functions, under physiological conditions, as ATP synthase (oxidative phosphorylation) or ATP hydrolase generating a PMF (from ATP produced in fermentative phosphorylation)which can be used to drive transport of nutrients and inorganic ions.

Table 1. Phosphorylation Potential and Protonmotive Force in State 4
         Mitochondria
         Temperature 25°C

$$\Delta G'_p = \Delta G'_o + 1.36 \log \frac{[ATP]}{[ADP][Pi]}$$

$$\Delta \tilde{\mu} H = \Delta \psi - 59 \; \Delta pH$$

	pH	Mg^{2+}	$\Delta G'_o$	$\Delta G'_p$	$\Delta \tilde{\mu} H^+$ (mV) (necessary) H^+/ATP =2	$\Delta \tilde{\mu} H^+$ (measured) H^+/ATP =3
			(Kcal/mol)			
Mitchell, 1966	7.0	+	7.0	12.4	270	180
Cockrell et al., 1966	7.8	−	9.6	15.6	338	225
Slater, 1969	7.8	−	9.4	15.8	343	228
Rosing and Slater, 1973	7.7	−	8.7	16.1	350	233
Mitchell and Moyle, 1969						$\sim$230

(From Papa, 1976, ref.3. See this for the other references quoted).

In eukaryotes F_oF_1 is made to function strictly as ATP synthase by various factors like pH activation (7,8) and thiol modulation in chloroplasts (8) and the ATP-ase-inhibitor protein in mitochondria (9,10).

Protein subunits and genes

The F_1 peripheral sector (knob) of the synthase has both in prokaryotes and eukaryotes an invariable composition consisting of $3\alpha, 3\beta, 1\gamma, 1\delta$ and 1ϵ subunits (see Table 2) (1). The genes of these subunits have been identified in prokaryote genomas (Fig.2) (11,12) and nuclei of eukaryotes (see Table 2) (13-15) and their primary structure and interspecies homologies have been described (13-15).

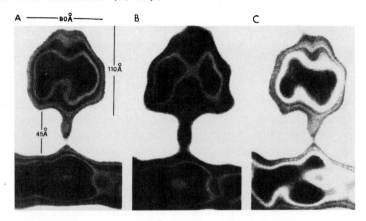

Fig.1. Averages of side views of purified F_o, F_1 H^+-ATP synthase from
       E.Coli reconstituted in phospholipid membranes. (Reproduced
       from Gogol et al., 1987 ref.5).

Table 2. Subunits of bovine-heart F_oF_1 H^+-ATP Synthase. (For additional information and references see Text)

Denomination and number copies	Number of residues	Apparent Mr from gels (Daltons)	Genes (N,nuclear) (M,mitochondrial)	Calculated Mr Daltons	N-terminus	Function
F_1:						
3-α	509	53000	N gene	55164	(E)KTGTA...	Catalalytic or regulatory ADN binding domain
3-β	480	50000	N gene	51595	(QA)SPSP...	Catalytic ADN binding domain
1-γ	272	33000	N gene	30141	ATLKDI...	Unknown
1-δ	146	17000	N gene	15065	(A)EAAAA...	Unknown coupling and gate?
1-ε	50	7500	N gene	5652	VAYWRQ...	Unknown
1F$_1$	84	10000	N gene	9582	GSESGD...	pH, ΔpH regulation of catalysis and H^+ conduction
F_o:						
PVP (F_oI)	214	24000-27000	N gene	24668	PVPPLP...	Binding of F_1 to F_o Modulation of H^+ conduction Sensitivity of H^+ conduction to oligomycin
OSCP	190	21000-25000	More N pseudogenes	20968	FAKLVR...	Connection of F_1 to F_o sensitivity of catalysis to oligomycin
ATPase-6	226	19000-23000	M gene	24816	f-SFITPV...	H^+ conduction?, Oligomycin sensitivity, Assembly
d	160	19000-26000	N gene	18600	Ac-AGRKLA...	Unknown
F6	76	9000	N gene	9071	NKELDP...	Binding of F_1 to F_o
A6L	66	8000	M gene	7965	f-MPQLDT...	Assembly,(Pi binding)?
(?6)-C	6	8000- 9000	Two N genes encoding different presequences	8000	DIDTAA...	Proton conduction

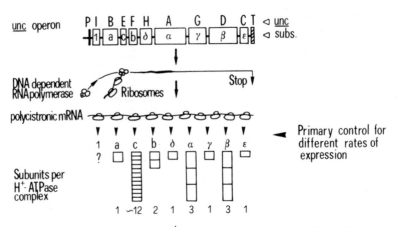

Fig.2. The unc operon genes of H^+-ATP synthase of E.Coli and its expression (Reproduced from McCarthy, ref.12).

The F_0 intrinsic membrane sector, responsible for proton translocation, consists in E.Coli of three subunits a, b and c. Their genes are located in the unc or atp operon (Fig.2) (11,12) and they have been characterized as essential components for a proper assembly and function of F_0 in the ATP synthase complex (16-18). The subunit composition of F_0 of eukaryotic ATP synthase appears to be more complex (1,2,15) (Table 2.). Even preparations of ATP synthase from bovine heart mitochondria exhibit a range of subunit complexities (19-21,23,24). In addition to the five F_1 subunits and the ATPase-inhibitor protein(9), calculated Mr from the gene sequence 9582 (22), the following proteins have been identified as components of bovine-heart ATP-synthase: (1) A nuclear encoded protein, calculated Mr 24,668 (apparent Mr from gels 24,000-27,000) (20,21,23,24); (2) the oligomycin-sensitivity-conferral protein (OSCP) (27), encoded by nuclear pseudogenes (22), calculated Mr 20,968 (apparent Mr gels 21,000-25,000) (20, 24,28); (3) a mitochondrial encoded protein (ATPase-6) (29) calculated Mr 24,816 (apparent Mr from gels 19,000-23,000 (29); (4) a nuclear-encoded protein, calculated Mr 18,600 (apparent Mr from gels 19,000-26,000) (23, 26); (5) subunit F_6 (30), nuclear encoded (22), calculated Mr 9.071 (apparent Mr from gels 9,000); (6) subunit A6L, mitochondrial-encoded (29), calculated Mr 7,965 (apparent Mr from gels 8,000); (7) subunit c (DCCD-binding protein) (31), encoded separately by two nuclear genes (15), calculated Mr 8,000 (apparent Mr from gels 8,000-9,000) (32,33).

Besides these seven subunits certain preparations of bovine-heart ATP synthase contain a protein with apparent Mr of 31,000 (uncoupler-binding protein) (34) and one with apparent Mr, in its monomeric form, of 13,000 (Factor B) (35).

From comparison of the aminoacid sequences and hydrophobic profiles it appears that subunit c, ATPase 6 and OSCP of the bovine enzyme are equivalent to subunit c, a, and F_1-δ of the E.Coli enzyme respectively. On the other hand subunits A6L, F_6 and the nuclear-encoded subunit Mr 18,600 (subunit d in Walker's terminology [23])of the bovine enzyme do not have specific counterparts in E.Coli.

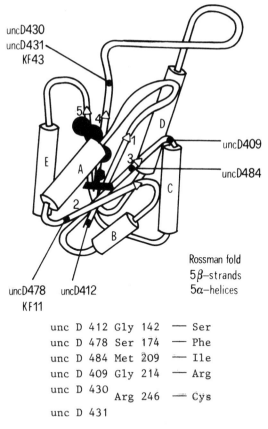

unc D 412 Gly 142 — Ser
unc D 478 Ser 174 — Phe
unc D 484 Met 209 — Ile
unc D 409 Gly 214 — Arg
unc D 430
            Arg 246 — Cys
unc D 431

Fig.3. Proposed tertiary structure of the catalytic and nucleotide-
binding domain of β-subunit with some missense unc D mutants
of E.Coli affecting catalysis. (Reproduced from Parsonage et al.
ref.41). For details see Text and ref.s 41,42.

The catalytic process in F_1

Work carried out in various laboratories (2,36,37), has provided definite progress in the elucidation of the catalytic process in F_1 (38,39).
Three rapidly exchangeable nucleotide binding sites have been identified
per F_1 unit, which are directly involved in the catalytic process (2,40).
There are additional tighter nucleotide binding sites (regulatory sites)(38).

The following pieces of evidence indicate that the segment extending
from residue 140 to 320 in the β subunit contains the catalytic nucleotide
binding domain of F_1 (Fig.3) (2,41,42). β-Subunits exhibit extensive sequence homology among species (2,14,39). Polyclonal antibodies against
β-subunits exhibit high cross-reactivity among species. Nucleotide affi-
nity-labelling analogues, when bound to intact F_1 react predominantly with
residues in β-subunits. (Ile-290, Tyr-297,Lys-301, Leu-328, Ile-330, Tyr-
331, Asp-338, Tyr-354 (43). Chemical modifiers which inhibit catalytic
activity of F_1 react with residues in β-subunits (2,44).

Missense mutations in β-subunit which depress catalysis without per-
turbing assembly involve 10 residues in the segment 137-246 (41). These
mutations impaire the positive catalytic co-operativity. The segment 140-

13

320 apparently contains two conserved nucleotide binding sequences and a catalytic region (Gly-142, Ser-174, Met-209, Gly-214, Arg-246) (41,42).

The α-subunit of E.Coli contains a nucleotide binding domain in the segment 160-340 analogue to that in the β-subunit (45). Mutation in the α subunit provides genetic evidence that a segment at the C terminus of the nucleotide binding domain (between Ser-347 and Ser-375) is involved in β-α-β co-operative interaction essential for catalysis (2,45).

Evidence has been produced showing that the catalytic process involves two (or three) co-operating alternating sites in β subunits (or at the α-β interfacie) (2,38,39). The various sequential steps can be summarized as follows:

1. $\beta_1(\alpha_1) \xrightleftharpoons{ADP+Pi} \beta_1(\alpha_1) \underset{ADP}{\overset{Pi}{\rightleftharpoons}} \beta_1(\alpha_1) = ATP+H_2O$

bound ATP is formed at a single site from bound ADP + Pi, without involvement of ∿P or PMF, K ∿1, small ΔG (unisite catalysis) (46).

2. $\beta_2(\alpha_2) \xrightleftharpoons{ADP+Pi} \beta_2(\alpha_2) \underset{Pi}{\overset{ADP}{\diagup}}$

binding of ADP+Pi at a second site enhances binding affinity of ADP+Pi and lowers binding affinity of ATP at the first site.

3. $\beta_1(\alpha_1) = ATP + nH_P^+ \underset{}{\overset{\beta_2(\alpha_2)ADP+Pi}{\rightleftharpoons}} \beta_1(\alpha_1) + ATP + nH_N^+$

PMF induces release of ATP from $\beta_1(\alpha_1)$ and thus promotes catalysis of ATP synthesis (47).

4. $\beta_2(\alpha_2) \underset{Pi}{\overset{ADP}{\diagup}} \rightleftharpoons \beta_2(\alpha_2) \underset{Pi}{\overset{ADP}{\diagup}}$

the second low affinity site is converted to high affinity site.

5. $\beta_2(\alpha_2) \underset{Pi}{\overset{ADP}{\diagup}} \rightleftharpoons \beta_2(\alpha_2) = ATP$

the second site alternates with the first and a second ATP molecule is formed.

Proton conduction in F_o

Subunit c is a constant hydrophobic small component of F_o in prokaryotic and eukaryotic ATP-synthase. It has conserved high homology during evolution and seems to represent an essential component of the transmembrane proton channel in F_o (48). It is present in multiple copies and probably forms, in its oligomeric state (hexamer or dodecamer), a proton conducting channel (Fig.4) (49). It is now clear that subunit c is not the only component of the proton conducting device of F_o. In fact the polar residues in subunit c conceived to contribute a hydrogen-bonded proton-conducting wire (49) are not highly conserved and, in the case of the E. Coli protein, are not enough to provide a membrane spanning path (see Fig. 5) (50). Evidence has been produced showing that also subunit a (51) (see also 52) and possibly also subunit b (53) contribute to the H^+ conduction pathway in E.Coli F_o.

Recent work by our groups (24-26) has provided direct evidence showing that the nuclear encoded protein of Mr 25,000 of bovine ATP synthase (23) is an intrinsic component of F_o, is covered by F_1 and involved in proton conduction by F_o and its sensitivity to oligomycin, a specific inhibitor

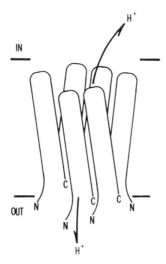

Fig.4. Possible oligomeric arrangement of subunits c to form a bundle with a central proton conducting device (see Papa et al.ref.49). According to Cox et al.(52) a ring of subunits c may be filled with membrane spanning segments of subunits a and b.

of proton conduction by F_O in the mitochondrial enzyme.

Fig.6 presents the SDS-PAGE pattern of the H^+-ATPase isolated from submitochondrial particles of beef-heart (ESMP) showing between the γ and δ subunits of F_1, four protein bands with apparent Mr of 31,000, 27,000, 25,000 and 23,000. The band of ≈ 25,000 consists, in fact, of a closely-spaced doublet (see also 20).

SDS-PAGE electrophoresis of the purified F_O (Fig.7), obtained by extensive removal with urea of F_1 from ATP synthase complex (25) showed the

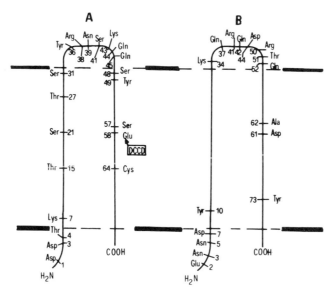

Fig.5. Polar residues in subunit c of bovine heart (A) and E.Coli (B) F_O (see ref.s 48 and 50).

15

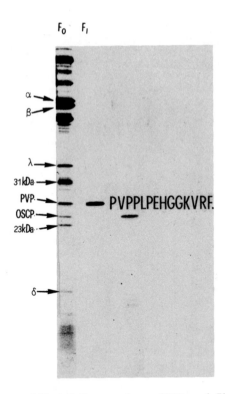

Fig.6. SDS-PAGE of purified F_oF_1 complex, OSCP and PVP protein.
On SDS slab-gels the following protein samples were analyzed:
15 μg protein of F_oF_1 1.2 μg protein of OSCP and approximately
0,5 μg of PVP protein extracted from purified F_o(25).After electrophoresis, proteins were detected by staining with Coomassie-
blue. The letters listed to the right of the figure indicate
the amino terminal sequence of PVP protein (from Papa et al.
ref.26).

five bands described above, detected both by Coomassie-blue staining or
by reaction with the fluorescent hydrophobic maleimide derivative, N-(dime-
thylamino-4-methyl-coumarinyl)-maleimide (DACM). After electrophoretic separation the protein bands were electroeluted in glycerol-H_2O (25). The
purified proteins, whose homogeneity was checked electrophoretically, were
used for production of polyclonal antibodies (24), or subjected to automated Edman degradation for determination of the aminoacid sequence at the
N-terminus (25).
    Aminoacid sequencing of the protein of apparent Mr 27,000 revealed
an aminoacid sequence at the N-terminus consisting of PVPPLPEHGGKVRF (25)
which is exactly that determined on the protein of apparent Mr of 24,000
isolated by Walker et al.(23) and corresponds to the nucleotide sequence
of the cognate cDNA (23). The protein was denominated PVP from its first
three residues at the N-terminus. It can also be noted that the cDNA coding for the PVP subunit has one codon for cysteine and the protein isolated here reacts with the thiol reagent DACM (see Fig.7). Thus the protein of apparent Mr 27,000 isolated in our laboratory from bovine F_o, is
the same as the one described by Walker et al., as subunit b (23). The

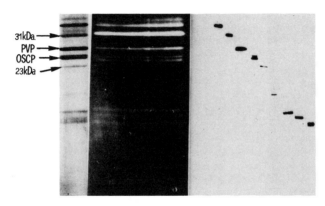

31kDa

PVP

OSCP

23kDa

Fig.7. Electrophoretic separation of the subunits of purified F_O.
Preparative electrophoresis was run as described by Zanotti et
al. (53). After electrophoresis the bands were identified by
Coomassie-blue staining of a small strip cut from the gel or
by reaction with   fluorescent DACM. An aliquot of the single
bands electroeluted from the gel was used for electrophoretic
identification of the individual protein fractions. Redrawn from
Zanotti et al.(54).

difference in the apparent Mr estimated from electrophoretic mobility has
to do with differences in the conditions of SDS-PAGE.

The band of apparent Mr 25,000, which is the lower component of the
doublet, corresponds to OSCP as revealed by comparison of its electrophoretic mobility with that of samples of OSCP isolated in other laboratories
(21,24). Aminoacid sequencing of the band of apparent Mr of 26,000 (upper
component of the doublet) revealed that its N-terminus is blocked (26).
This band, which also reacted with DACM may correspond to subunit d of
Walker et al. (23), which has the N-terminal alanine blocked by an acyl
group and its cDNA has one codon for cysteine.

The subunit of apparent Mr 23,000 did not react with DACM (see Fig.7).
It may correspond to the protein encoded by the ATP-6 gene which, in fact,
has no codon for cysteine (29).

Rabbit immunization with the 25,000-27,000 Da protein fractions from
F_O(Fig.7) gave antisera (Fig.8, strips 3-5), which detected three different antigens (A_1,A_2, and A_3) in ESMP (24). As indicated by mobilities of
F_1 subunits(α,δ) recognized on the same blot by anti-F_1 serum (Fig.8, slot
1), the apparent molecular mass of the A_1-A_3 antigens were within the range
27,000-23,000. None of the A_1,A_3 antigens were recognized by the anti-F_1
serum. The majority of the isolated SDS-treated F_O proteins below 25,000
(see Fig.7), including the 23,000 Da protein of F_O were not immunogenic.
A high immunogenicity of SDS-treated 25,000-27,000 Da proteins of F_O is consistent with the reactivity of the antiserum raised against whole ATP synthase complex (complex V), (Fig.8, slot 2) which detects both A_1 and A_2
antigens. However no reactivity with the A_3 antigen was observed with the
serum against complex V (Fig.8, slot 2).

Analysis of the reactivity of the antisera prepared against the isolated fractions with preparations of H^+-ATP synthase, F_O and isolated fractions showed that the A_1 antigen is the PVP protein and A_2 antigen is OSCP

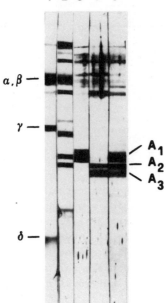

Fig.8. Immunoblot pattern of bovine heart ESMP decorated using various sera against components of H^+-ATPase. 300 μg ESMP protein was subjected as a single sample to electrophoresis on a slab SDS gel and transferred electrophoretically to nitrocellulose.The nitrocellulose strips were subsequently immunodecorated using different sera against $F_1(1)$, isolated F_oF_1 (2) and isolated F_o protein fractions containing mainly PVP protein (3), OSCP protein (4) and a mixture containing both OSCP and PVP protein (5) respectively. The position of antigens A_1,A_2,A_3 (detected by sera in 2-5) are indicated (From Houstek et al.,ref.24).

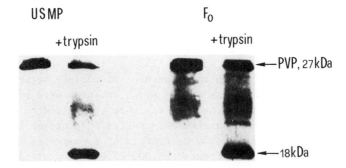

Fig.9. Immunoblot analysis of PVP content in USMP and F_o,isolated from USMP, before and after trypsin digestion.F_o was directly isolated from USMP using CHAPS as detergent. Trypsin (50 μg/mg protein) digestion of USMP was carried out for 20 min as described by Houstek et al.(24). Immunoblot analysis of USMP (20 μg) and F_o (20 μg)was carried out with specific serum anti-PVP subunit as described in(24). 18 kDa indicates the specific products of trypsin digestion of PVP protein (from Papa et al.,ref.26).

Reactivity of the PVP protein with a monospecific antiserum against a cop-
per-o-phenathroline reactive polypeptide (21) showed the identity of the
two proteins (24).

Topology of the PVP protein was studied by immunological analysis of
its accessibility to exogenous proteases. Incubation with trypsin of particles devoid of F_1 by urea(USMP)resulted in extensive digestion of the PVP
protein (Fig.9) (25,26). In contrast no digestion was observed in F_1 containing particles (24,26). The decrease of the immunoreactive band of PVP

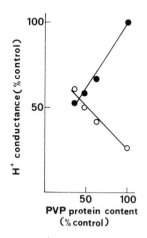

Fig.10. Relationship between H^+ conduction in USMP and the content of
PVP protein. For USMP preparation, immunoblot analysis,trypsin
treatment and evaluation of PVP content see legend to Fig.9 and
ref.s 24 and 25. H^+ conduction by F_o was measured electrochemi-
cally by analysis of anaerobic release of respiratory proton
gradient as described by Zanotti et al.(54). The reciprocal
value of $t\frac{1}{2}$ of this process in untreated USMP was taken as
100%. ●—● control; o—o + oligomycin 2 µg/mg protein (From
Zanotti et al.ref.25).

was accompanied by the appearance of an immunoreactive band of 18,000 Da.
This proteolytic fragment could not be removed by washing and centrifugation of USMP indicating that it remained bound to the membrane. The fragment was co-purified with F_o. Aminoacid sequencing and labelling with
thiol reagents showed that the fragment resulted from proteolytic removal
at the C-terminal region of not more than 17 residues (total residues in
PVP protein amount to 214 (55).

Fig.10 shows that the progressive digestion of the PVP subunit to the
18,000 fragment, effected by incubation of USMP with increasing concentrations of trypsin resulted in a progressive inhibition of passive proton
conduction in the particles and loss of sensitivity of this process to
the inhibitory action of oligomycin (25). Incorporation of F_o, isolated

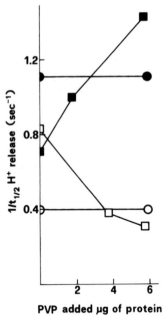

Fig.11. Reconstitution of H^+ conduction and oligomycin sensitivity of F_o extracted from trypsin treated USMP by addition of isolated PVP protein. F_o liposomes were prepared by dialysis method as described by Zanotti et al.(25).Where indicated,purified PVP (see legend to Fig.8)was added to F_o-liposomes suspension and incubated for 10 min before valinomycin addition.
Symbols: ●——●, F_o extracted from untreated USMP; ■——■ , F_o extracted from USMP treated with trypsin (50 µg/mg protein), ○——○ F_o extracted from untreated USMP + oligomycin (2 µg/mg protein); ☐——☐ , F_o extracted from USMP treated with trypsin (50 µg/mg protein) + oligomycin (2 µg/mg protein).

from trypsinized USMP, with about 75% of the PVP subunit digested, showed a rate of passive H^+ conduction about 40% lower than that exhibited by liposomes reconstituted with untreated F_o. Proton conduction in the vesicles reconstituted with F_o from trypsinized USMP was, in addition, completely insensitive to oligomycin which in contrast markedly inhibited proton conduction in liposomes reconstituted with F_o isolated from untreated USMP (Fig.11). Reconstitution of the liposomes with F_o isolated from trypsinized USMP with increasing concentration of the intact PVP subunit, isolated from untreated F_o, resulted in a progressive and full recovery of the proton conducting activity of F_o and its inhibition by oligomycin (25). It may, in fact, be noted that at the higher concentrations of added PVP subunit the proton conductivity of vesicles with trypsinized F_o was even higher than that exhibited by vesicles with untreated F_o, but this high proton conduction was still suppressed by oligomycin. It thus appears that the isolated native PVP subunit is able to replace its 18,000 inactive fragment in F_o and restore in this way normal proton conductivity.

These observations provide evidence that the PVP protein (F_oI) (25) encoded by a nuclear gene (23)., is an intrinsic component of mitochondrial F_o involved in proton conduction. This subunit appears, furthermore, to

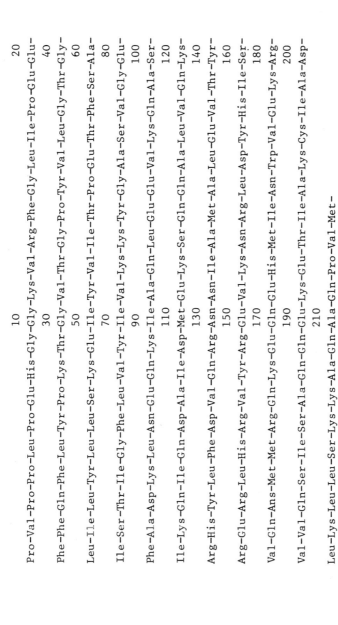

```
                                                        10                                          20
          Pro-Val-Pro-Leu-Pro-Glu-His-Gly-Lys-Val-Arg-Phe-Gly-Leu-Ile-Pro-Glu-Glu-
                                                        30                                          40
          Phe-Phe-Gln-Phe-Leu-Tyr-Pro-Lys-Thr-Gly-Val-Thr-Gly-Pro-Tyr-Val-Leu-Gly-Thr-Gly-
                                                        50                                          60
          Leu-Ile-Leu-Tyr-Leu-Leu-Ser-Lys-Glu-Ile-Tyr-Val-Ile-Thr-Pro-Glu-Thr-Phe-Ser-Ala-
                                                        70                                          80
          Ile-Ser-Thr-Ile-Gly-Phe-Leu-Val-Tyr-Ile-Val-Lys-Lys-Tyr-Gly-Ala-Ser-Val-Gly-Glu-
                                                        90                                         100
          Phe-Ala-Asp-Lys-Leu-Asn-Glu-Gln-Lys-Ile-Ala-Gln-Leu-Glu-Glu-Val-Lys-Gln-Ala-Ser-
                                                       110                                         120
          Ile-Lys-Gln-Ile-Gln-Asp-Ala-Ile-Asp-Met-Glu-Lys-Ser-Gln-Gln-Ala-Leu-Val-Gln-Lys-
                                                       130                                         140
          Arg-His-Tyr-Leu-Phe-Asp-Val-Gln-Arg-Asn-Asn-Ile-Ala-Met-Ala-Leu-Glu-Val-Thr-Tyr-
                                                       150                                         160
          Arg-Glu-Arg-Leu-His-Arg-Val-Tyr-Arg-Glu-Val-Lys-Asn-Arg-Leu-Asp-Tyr-His-Ile-Ser-
                                                       170                                         180
          Val-Gln-Ans-Met-Met-Arg-Gln-Lys-Glu-Gln-Glu-His-Met-Ile-Asn-Trp-Val-Glu-Lys-Arg-
                                                       190                                         200
          Val-Val-Gln-Ser-Ile-Ser-Ala-Gln-Gln-Glu-Lys-Glu-Thr-Ile-Ala-Lys-Cys-Ile-Ala-Asp-
                                                       210
          Leu-Lys-Leu-Ser-Lys-Lys-Ala-Gln-Ala-Gln-Pro-Val-Met-
```

Fig.12. Aminoacid sequence of the PVP protein from bovine heart ATP synthase as deduced from sequence analysis of the cDNA (23) and partial aminoacid analysis (ref.s 23 and 25).

be involved in the inhibition by oligomycin of proton conduction in mito-chondrial F_o. Thus the PVP protein is the second subunit of mitochondrial F_o, in addition to the 8,000 Da, DCCD-binding protein (31), which is shown to be involved in proton conduction in mitochondrial ATP synthase.

It is shown that the tail of this protein extending from carboxyl-terminal Met-214 to Lys-202, which has a relative abundance of polar resi-dues (Fig.13), protrudes out of the M surface of the membrane, is covered by F_1 and modulates proton translocation by transmembrane segments of the proteins constituting the channel. The loss of oligomycin sensitivity of proton conduction in F_o caused by proteolytic cleavage of the peripheral

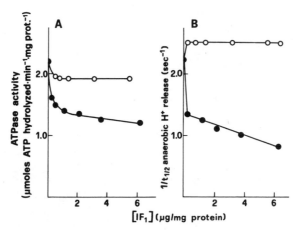

Fig.13. Effect of diethylpyrocarbonate treatment of purified IF_1 on its inhibitory effect on ATPase activity and anaerobic release of respiratory proton gradient in submitochondrial particles deplet-ed of IF_1 (Sephadex particles). Submitochondrial particles and IF_1 preparations, treatment of IF_1 with EFA and measurements of ATPase activity and passive H^+ conductivity were carried out as described in ref.61. (From Guerrieri et al.ref.61). •——• native IF_1; o——o IF_1 modified by diethylpyrocarbonate.

carboxy-terminal region of PVP indicates that this probably affects the conformation of the proton channel in the membrane. Oligomycin, which is a bulky hydrophobic reagent, has in fact been shown to cover a number of residues in subunit c exposed towards the hydrophobic lipid core on the membrane (56).

The regulatory protein in mitochondrial ATP synthase

In the H^+-ATP synthase of mitochondria a specific and important regu-latory role is played by a protein of Mr 10 kDa which is absent in the enzyme of bacteria and chloroplasts. This component, denominated ATPase inhibitor protein (9,10) has an amphiphatic nature and binds specifically by its hydrophobic moiety to the β subunit thus inhibiting the catalytic activity (10). The binding of IF_1 to the ATPase complex is suppressed by respiratory $\Delta\mu H^+$, this allowing ATP synthesis to proceed in the respiring steady-state (10,57,58). In the absence of $\Delta\mu H^+$ IF_1 binds to the complex thus preventing futile ATP hydrolysis. The content of IF_1 varies from tissue to tissue in animals and in the heart it is inversely correlated with beating frequency (59). It has been found in our laboratory that IF_1 exerts also an inhibitory action on passive proton conduction through the F_0 moiety and that this action has the same titer (Fig.13) and pH dependence (Fig.14) or the effect exerted on the catalytic process (60,61). The inhibitory activity decreases with pH, showing half-maximal activity around 7.0. This suggests a critical role of hystidine residues and in fact chemical modification of one out of the five hystidine residues present in the

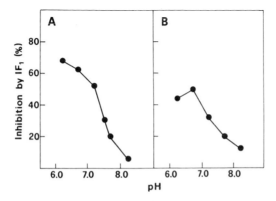

Fig.14. pH dependence of the inhibition of IF_1 on ATPase activity (A) and anaerobic release of respiratory proton gradient (B) in Sephadex particles. The experiments were carried out as described in ref.59.(From Guerrieri et al.ref.61).

inhibitor protein (10) prevents the activity of the isolated protein when reconstituted with the enzyme (Fig.13). The dual inhibitory effects exerted by IF_1 on both catalytic activity and proton conduction qualify this protein as a component of the gate. It is possible that the protein, in addition to its binding site, on the β subunit of F_1 (10), also binds to component(s) of the membrane sector (see also 62,63). The pH dependence of the inhibitor suggests that IF_1 will be without significant effect on the enzyme under normal aerobic conditions where the cytosolic pH is around neutrality and that of the matrix space is around 8. However under conditions leading to cellular ischemia, like an intensive work performance by the skeletal muscle, circulatory insufficiency or artery occlusion, respiratory activity may stop with drop of the transmembrane $\Delta\mu H^+$ and interruption of oxidative phosphorylation. Glycolysis will be activated as a consequence of the decrease in cytosolic phosphate potential with substantial lowering of pH well below neutrality. This acidification will promote inhibition of the H^+-ATPase, which otherwise, in the absence of the pressure exterted by $\Delta\mu H^+$ will rapidly consume cellular ATP. Thus the inhibitory protein may represent an extremely important specific defence mechanism for animals and in particular humans (64,65).

An increase in the content of the inhibitor protein has been observed in tumor cells (66) where it may serve to prevent hydrolysis of the glycolytic ATP by the mitochondrial enzyme. In fact in tumor cells there is a decreased utilization of respiratory substrates and Pi by mitochondria (67), with consequent depression of PMF and tendency of the ATP synthase to hydrolyse ATP.

This role,that the ATPase inhibitor protein may play in the cell provides a nice example of the functions that supernumerary subunits, characteristics of the protonmotive energy transfer systems of eukaryotes, can serve as a response to enhanced functional complexity and versatility acquired by these systems during evolution.

ACKNOWLEDGEMENTS
This work was supported by C.N.R. Grants n.88.00376.11 and n.88.00794.44

References

1. S.Papa, K.Altendorf, L.Ernster and L.Packer "H^+-ATPase (ATPsynthase): Structure, Function, Biogenesis. The F_oF_1 complex of coupling membranes". ICSU Press, Miami - Adriatica Editrice, Bari.
2. A.E.Senior, Physiological Reviews 68: 177 (1988)
3. S.Papa, Biochim.Biophys.Acta 456: 39 (1976)
4. H. Fernandez-Moran, T.Oda, P.V.Blair and D.E.Green, J.Cell Biol.22:63 (1964).
5. E.P.Gogol, U.Lücken and R.A.Capaldi, FEBS Lett. 219:274 (1987).
6. W.W.Wainio, J.Ultrastructure Res. 93:138 (1985).
7. W. Junge, Eur.J.Biochem. 14:582 (1970)
8. J.D.Mills, in ref.1, p.340
9. M.E.Pullman and G.C.Monroy, J.Biol.Chem. 238:3762 (1963).
10. D.A.Harris, in ref.1, p.387
11. F.Gibson, Proc.Roy.Soc.Lon. B 315:1 (1982)
12. J.E.G. Mc Carthy, J.Bioenerg.Biomembr. 20:19 (1988).
13. J.E.Walker, V.L.J. Tybulewicz, G.Falk, N.J.Gay and A.Hampe, in ref.1 p.1
14. J.E.Walker, I.M.Fearnley, N.J.Gay, B.W.Gibson, F.D.Northrop,S.J.Powell, M.J.Runswick, M.Saraste and W.L.J.Tybulewicz, J.Mol.Biol. 184:677(1985).
15. J.E.Walker, A.L.Cozens, M.r.Dyer, I.M.Fearnley, S.J.Powell and M.J. Runswick, in "Bioenergetics:Structure and Function of Energy Transducing Systems", T.Ozawa and S.Papa, eds., Japan Sci.Soc.Press, Tokyo - Springer-Verlag, p.167 (1987).
16. E.Schneider and K.Altendorf, Proc.Nat.Acad.Sci.81:7279 (1984).
17. E.Schneider and K.Altendorf, EMBO J. 4:515 (1985).
18. P.Friedl, J.Hoppe, R.P.Gunsalus, O.Michelsen, K. von Meyenburg and H.U. Schairer EMBO J. 2:99 (1983).
19. D.L.Stigall, Y.M.Galante and Y.Hatefi, J.Biol.Chem.259:956 (1978)
20. C.Montecucco, F.Dabbeni-Sala, P.Friedl and Y.M.Galante, Eur.J.Biochem. 132:189 (1983)

21. K.Torok and S. Joshi, Eur.J.Biochem.153:155 (1985)

22. J.E.Walker, N.J.Gay, S.J.Powell, M.Kostina and M.R.Dyer, Biochemistry 26:8613 (1987).

23. J.E.Walker, M.J.Runswick and L.Poulter, J.Mol.Biol. 197:89 (1987).

24. J.Houstek, J.Kopecky, F.Zanotti, F.Guerrieri, E.Jirillo, G.Capozza and S.Papa, Eur.J.Biochem. 173:1 (1988).

25. F.Zanotti, F.Guerrieri, G.Capozza, J.Houstek, S.Ronchi and S.Papa, FEBS Lett. 237:9 (1988).

26. S.Papa, F.Guerrieri, F.Zanotti, J.Houstek, J.Kopecky, G.Capozza, and S.Ronchi, in: "Molecular Basis of Biomembrane Transport", F.Palmieri, and E.Quagliariello, eds., Elsevier, Amsterdam, p.29 (1988).

27. D.H.Mc Lennan and A. Tzagaloff, Biochemistry 7:1603 (1968).

28. A.Depuis, G.Zaccai and M.Satre, Biochemistry 22:5951 (1983).

29. I.M.Fearnley and J.E.Walker, EMBO J., 5:2003 (1986).

30. B.I.Kanner, M.Serrano, M.A.Kandrach and E.Racker, Biochem.Biophys.Res. Commun.69:1050 (1976).

31. W.Sebald and J.Hoppe, Curr.Top.Bioenerg. 12:1 (1981).

32. H.Sigrist, K.Sigrist-Nelson and C.Gitler, Biochem.Biophys.Res.Commun. 74:178 (1977).

33. J.Kopecky, F.Guerrieri and S.Papa, Eur.J.Biochem. 131:17 (1983).

34. W.C.Hanstein and Y.Hatefi, J.Biol.Chem.249:1356 (1974).

35. D.R.Sanadi, Biochim.Biophys.Acta 683:39 (1982).

36. M.J.J.Gresser, J.A.Myers and P.D.Boyer, J.Biol.Chem. 257:12030 (1982).

37. H.S.Penefsky, J.Biol.Chem.260:13728 (1985).

38. R.L.Cross, Annu.Rev.Biochem. 50:681 (1981).

39. X. Ysern, L.M.Amzel and P.L.Pedersen, J.Bioenerg.Biomembr. 20:423 (1988).

40. F.A.S., Kironde and R.L.Cross, J.Biol.Chem. 262:3488 (1987)

41. D.Parsonage, T.M. Duncan, S.Wilke-Mounts, F.A.S.Kironde, L.Hatch and A.L.Senior, J.Biol.Chem. 262:6301 (1987).

42. T.M.Duncan, D.Parsonage and A.E.Senior, FEBS Lett. 208:1 (1986).

43. P.V.Vignais, and J.Lunardi, Annu.Rev.Biochem.54:977 (1985).

44. W.W.Andrews, F.C.Ill and W.S.Allison, J.Biol.Chem. 259:14378 (1984).

45. M.B.Maggio, J.Pagan, D.Parsonage, L.Hatch and A.L.Senior, J.Biol.Chem. 262:8981 (1987).

46. C.R.Grubmeyer, L.Cross, and H.S.Penefsky, J.Biol.Chem. 257:12092 (1982).

47. H.S.Penefsky, J.Biol.Chem. 260:13735 (1985).

48. J.Hoppe and W.Sebald, Biochim.Biophys.Acta 768:1 (1984).

49. S.Papa, F.Guerrieri, F.Zanotti and R.Scarfò, in:"Biological Membranes" L.Bolis, E.J.M.Helmerich and H.Passow, eds., Alan R.Liss, New York p.187 (1984).

50. K.H.Altendorf, U.Hammel, G.Deckers, H.H.Kilts and R.Schmid, in:"Function and Molecular Aspects of Biomembrane Transport", E.Quagliariello et al., eds.,Elsevier-North Holland,Amsterdam, p.53 (1979).

51. K.von Meyenburg, B.B.Jørgensen, O.Michelsen, L.Sørensen and J.E.G. Mc Carthy, EMBO J., 4:2357 (1985).

52. G.B.Cox, A.L.Fimmel,F.Gibson, L.Hatch, Biochim.Biophys.Acta 849:62 (1986).

53. K. Steffens, J.Hoppe and K.Altendorf, Eur.J.Biochem. 170:627 (1988).

54. F.Zanotti, F.Guerrieri, Y.W.Che, R.Scarfò and S.Papa, Eur.J.Biochem. 164:517 (1987).

55. S.Papa, F.Guerrieri, F.Zanotti, J.Houstek, G.Capozza and S.Ronchi, manuscript in preparation.

56. J.Hoppe, D.Gatti, H.Weber and W.Sebald, Eur.J.Biochem.155:259 (1986).
57. G.Lippe, M.C.Sorgato and D.A.Harris, Biochim.Biophys.Acta,933-1 (1988).
58. G.Lippe, M.C.Sorgato, and D.A.Harris, Biochim.Biophys.Acta 933:12 (1988).
59. W.Rouslin and M.M.Pullman, J.Med.Cell.Biol. 19:661 (1987).
60. F.Guerrieri, R.Scarfò, F.Zanotti, Y.W.Che and S.Papa, FEBS Lett. 213: 67 (1987).
61. F.Guerrieri, F.Zanotti, Y.W.Che, R.Scarfò and S.Papa, Biochim.Biophis. Acta 892:284 (1987).
62. T. Hashimoto, Y.Yoshida and K.Tagawa, J.Biochem. 95:131 (1984).
63. T.Hashimoto, Y.Yoshida, and K.Tagawa, J.Biochem. 94:715 (1983).
64. W.Rouslin, The J.Biol.Chem. 262:3472 (1987).
65. S.Papa, in: "The Molecular Biology of Human Diseases", J.W.Gorrod, O. Albano and S.Papa, eds., Ellis Horwood Ltd., Chichester, England, Vol.I (1989).
66. K.Luciakova and S.Kuzela, FEBS Lett. 177:85 (1985).
67. F.Capuano, G.Paradies, N.Capitanio and S.Papa, in"Membranes in Tumor Growth", T.Galeotti et al.eds., Elsevier Biomedical, Amsterdam, p.345 (1982).

FUNCTIONS, BIOGENESIS AND PATHOLOGY OF PEROXISOMES IN MAN

Erik A.C. Wiemer[1], Stanley Brul[1], Abraham Bout[1], Anneke Strijland[1], Judith C. Heikoop[1], Rob Benne[1], Ronald J.A. Wanders[2], Andries Westerveld[3] and Joseph M. Tager[1]

[1]E.C. Slater Institue for Biochemical Research and Departments of [2]Pediatrics and [3]Human Genetics, University of Amsterdam, Academic Medical Centre, Amsterdam The Netherlands

INTRODUCTION

Peroxisomes (microbodies; microperoxisomes), were first described morphologically by Rhodin (1) in 1954 and characterized biochemically by De Duve and coworkers (2) in the early 1960's. They are defined as intracellular organelles bounded by a single membrane and containing catalase and H_2O_2-producing oxidases. This definition emphasizes the oxidative functions of peroxisomes. However, the peroxisome is a highly versatile organelle. While oxidative processes are certainly a characteristic feature of the metabolic pathways in which peroxisomes participate other types of reactions, too, occur in peroxisomes.

In mammals, including man, peroxisomes are present in virtually all cells except mature erythrocytes. For several years after the discovery of peroxisomes their function was not clear; they were even referred to as fossil organelles (see ref. 3). Now, however, it is evident that peroxisomal enzymes are required for many metabolic transformations in the cell. In fact, additional functions of peroxiosmes presumably remain to be discovered.

The importance of peroxisomes in human physiology has been given particular emphasis by the recent discovery of a group of genetic diseases in man in which peroxisomal functions are impaired (see refs. 4-9 for reviews). The peroxisomal diseases are usually fatal, the patients often dying in early infancy. The prototype of this group of genetic diseases is the cerebro-hepato-renal syndrome of Zellweger.

An important milestone in the history of the peroxisomal diseases was the discovery in 1973 by Goldfischer and colleagues (10) that morphologically distinguishable peroxisomes are virtually absent in liver and kidney of patients with the Zellweger syndrome. This key observation was subsequently confirmed in several different laboratories.

Later studies showed that morphologically distinguishable peroxisomes are also deficient in several other disease (4-9). Thus we have the remarkable situation of a group of genetic diseases in man in which a whole organelle is deficient. These are diseases of peroxisomal biogenesis, or, more accurately, diseases of peroxisome assembly (see ref. 9).

The discussion in this section will be restricted to a consideration of the functions of peroxisomes in mammalian species, particularly in man. In assigning functions to peroxisomes two problems arise.

Firstly, certain enzymes and enzyme systems are present not only in peroxisomes but also in other intracellular compartments. The question arises then of whether the enzymes present in different compartments have distinct functions and metabolize different compounds or whether they provide alternative routes for the metabolism of the same compounds. Examples of both possibilities can be given. As will be discussed below the fatty acid β-oxidation system in peroxisomes is capable of oxidizing fatty acids which cannot be catabolized by the mitochondrial β-oxidation system. On the other hand dicarboxylic acids can apparently be catabolized both in peroxisomes and in mitochondria (at least in rat liver (11,12)).

Secondly, the intracellular distribution of an enzyme may differ in different species. Thus alanine: glyoxylate aminotransferase is a mitochondrial enzyme in the dog and the cat, a peroxisomal enzyme in man, and is present in equal amounts in both compartments in the rat (13,14). As will be discussed below, the catabolism of L-pipecolic acid is a peroxisomal process in the Cynomolgus monkey (15) and man (16), but occurs in mitochondria in the rabbit (15).

Apart from direct evidence for a role of peroxisomes in metabolic pathways obtained from cell fractionation studies, presumptive evidence indicating that peroxisomes must be essential for certain metabolic pathways is provided by studies of the biochemical abnormalities encountered in diseases like the Zellweger syndrome where peroxisomes are deficient.

In man, peroxisomes play an essential role in the following metabolic pathways.
1. The biosynthesis of ether phospholipids
2. The oxidation of very-long-chain fatty acids and mono- and poly-unsaturated fatty acids.
3. The biosynthesis of bile acids
4. The degradation of phytanic acid
5. The oxidation of L-pipecolic acid
6. The catabolism of glyoxylate

The synthesis of ether phospholipids

Ether phospholipids are analogues of diacylglycerophospholipids in which the acyl group at position 1 of the glycerol backbone is replaced by an alkyl group in ether linkage. Two enzyme acting consecutively are responsible for the introduction of the ether bond in ether phospholipids, acyl-CoA: dihydroxyacetonephosphate acyltransferase (DHAP-AT) and alkyldihydroxyacetonephosphate synthase (alkylDHAP synthase); both are peroxisomal enzymes (for a review see ref. 17).

In man and other mammals, the most common ether phospholipids are the plasmalogens (1-0-alk-1-enyl-2-acylphosphoglycerides), which are major constituents of total phospholipids in the membranes of a number of tissues. Plasmalogens can scavenge reactive oxygen species (18) and may therefore play a role in protecting cells from the damaging effects of oxidative stress. An ether phospholipid with a well-defined function is platelet activating factor.

Peroxisomal β-oxidation

In 1976 Lazarow and De Duve (19) demonstrated that rat-liver peroxisomes are able to bring about the β-oxidation of fatty acids. The peroxisomal β-oxidation system is outlined in Fig. 1.

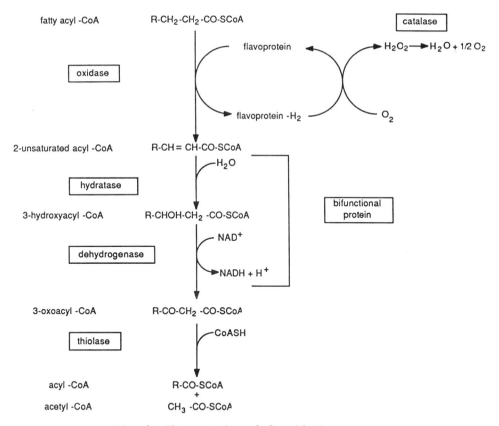

Fig. 1 The peroxisomal β-oxidation system.

The initial substrate for β-oxidation in peroxisomes is, as in mitochondria, an acyl-CoA ester. During each cycle of peroxisomal β-oxidation the
acyl-CoA is shortened by 2 C atoms and acetyl-CoA is formed. The consecu-
tive steps of dehydrogenation, hydration, dehydrogenation and thiolytic
cleavage are catalysed by an acyl-CoA oxidase, a bifunctional protein
with enoyl-CoA hydratase and 3-hydroxyacyl-CoA dehydrogenase activity and
a 3-oxoacyl-CoA thiolase, respectively. These enzyme proteins are quite
distinct from the corresponding enzymes in mitochondria (for reviews see
refs. 20,21).

Bremer (22) in a note published in 1977 was the first to suggest
that the primary function of the peroxisomal β-oxidation system might be
the chain-shortening of fatty acids in order to prepare them, as it were,
for β-oxidation to completion in the mitochondria The correctness of this
prophetic suggestion has subsequently been amply verified. Peroxisomes
are particularly equipped to oxidize fatty acids which the mitochondrial
β-oxidation system cannot take care of, e.g. saturated very-long-chain
fatty acids (C24:0, C26:0). It was first shown by Singh et al. (23) that
very-long-chain fatty acids are oxidized in peroxisomes. This is due to
the fact that the activation of very long chain fatty acids is catalysed
by a very-long-chain acyl-CoA synthetase (which is present in peroxisomes
and microsomes but not in mitochondria) and not by long-chain acyl-CoA
synthetase (which is present in all three compartments)(24-28).

Acyl-CoA oxidase is poorly reactive with acyl-CoA esters of chain
length 8C atoms or less (29,30). Thus the major products of the peroxisomal β-oxidation of fatty acids are octanoyl-CoA and acetyl-CoA, which
must be transported out of the peroxisomes, perhaps to be catabolized
further in the mitochondria. Significantly, peroxisomes contain carnitine

octanoyltransferase and carnitine acetyltransferase, which may be involved in carnitine-dependent transport from the peroxisomes to the cytosol of the octanoyl and acetyl moieties. Although no carnitine dependency of peroxisomal fatty acid oxidation can be demonstrated in isolated organelles (31), this could be an isolation artifact. However, it remains to be established whether the permeability properties of peroxisomes in situ in the intact cell differ from those of the isolated organelle or not.

Compounds catabolized via β-oxidation in peroxisomes

Peroxisomes are able to oxidize, in addition to very-long-chain fatty acids, also polyunsaturated fatty acids, monounsaturated long-chain fatty acids like erucic acid, dicarboxylic acids, prostaglandins and xenobiotics with an acyl side chain (see ref. 8 for a review). One may speculate that a general function of peroxisomes is to bring about the catabolism by β-oxidation of a variety of compounds which, if allowed to accumulate, would be toxic.

It is not clear in all cases whether β-oxidation of a particular compound occurs exclusively in the peroxisomes. If a compound accumulates in patients with the Zellweger syndrome this is presumptive evidence for an exclusive peroxisomal localization of its catabolism. This appears to be the case for e.g. trihydroxycoprostanoic acid (32). However, damage to mitochondria is sometimes a feature of the Zellweger syndrome (10), presumably arising as a secondary effect due to the impairment of peroxisomal functions; such damage could conceivably account for the accumulation of certain compounds.

In those cases where the catabolism of a particular compound can take place both in peroxisomes and in mitochondria information on the relative contribution of each organelle to its catabolism is urgently required. We are at present generating Chinese hamster ovary cell mutants with impairments in peroxisomal functions in order to investigate this problem.

The Biosynthesis of Bile Acids

The final steps in the biosynthesis of bile acids from cholesterol involve the conversion of trihydroxycholestanoic acid to cholic acid and of dihydroxycholestanoic acid to chenodeoxycholic acid. This conversion is an oxidative chain shortening and occurs in the peroxisomes, propionyl-CoA being formed instead of acetyl-CoA (33).

Since trihydroxycholestanoic acid and dihydroxycholestanoic acid accumulate in patients with an isolated deficiency of peroxisomal 3-oxo-acyl-CoA thiolase (34,35) or bifunctional protein (36) but not in patients with an isolated deficiency of acyl-CoA oxidase (37), one may conclude that the bifunctional protein and peroxiosmal thiolase are required for chain-shortening both of fatty acids and of bile acid intermedicates but that a separate oxidase is involved. Direct evidence for separate acyl-CoA oxidases for fatty acids and for intermediates of bile acid synthesis respectively has recently been brought forward (38,39).

Pedersen and Gustafson (40) suggested in 1980 that the accumulation of trihydroxycholestanoic acid and dihydroxycholestanoic acid in patients with the Zellweger syndrome might be involved in the pathogenesis of the clinical signs and systems in the patients.

Of interest is the observation that the activation of trihydroxycholestanoic acid occurs in the endoplasmic reticulum (41) so that trihydroxycholestanoyl-CoA must be transported from the endoplasmic reticulum to the peroxisomes. On the other hand very-long-chain acyl-CoA formed in the endoplasmic reticulum is apparently not available for the peroxisomal β-oxidation system. This is indicated by the fact that in patients with X-linked Adrenoleukodystrophy, in which the peroxisomal very-long-chain acyl-CoA synthetase is deficient, very-long-chain fatty acids cannot be oxidized even though there is no deficiency of the endoplasmic reticulum very-long-chain acyl-CoA synthetase (42,43).

The Catabolism of Phytanic Acid

Since phytanic acid accumulates in body fluids of patients with the Zellweger syndrome (5-8), it can be inferred that at least one step in the catabolism of the compound must occur in peroxisomes. The initial step in the catabolism of phytanic acid involves α-oxidation and decarboxylation so that pristanic acid is formed; in rat liver this step occurs in the mitochondria (44). Pristanic acid is broken down further by β-oxidation.

The Catabolism of L-Pipecolic Acid

One of the features of the Zellweger syndrome and related disorders is that L-pipecolic acid accumulates in body fluids of the patients. Mihalik and Rhead (15) have recently shown that L-pipecolic acid is converted to α-aminoadipic acid in peroxisomes in kidney of the Cynomolgus monkey. Wanders et al. (16) have demonstrated that human liver homogenates contain L-pipecolic acid oxidase activity localized in peroxisomes and that there is a deficiency of this activity in liver from patients with the Zellweger syndrome.
The situation is different in the rabbit. As shown by Mihalik and Rhead (15), a mitochondrial dehydrogenase is responsible for the oxidation of L-pipecolic acid in this species.
In an earlier study on the intracellular distribution of pipecolic acid catabolism DL-pipecolic acid was used as substrate (45) so that the contribution of peroxisomal D-amino acid oxidase to oxidation of the racemic mixture complicated the picture and led to unequivocal results (see ref. 46).

The Catabolism of Glyoxylate

In 1979 Noguchi and Takada (13) discovered that alanine: glyoxylate aminotransferase, an enzyme required for the catabolism of glyoxylate, is localized in peroxisomes in human liver. In contrast, the enzyme is entirely mitochondrial in dog and cat liver and is distributed equally between both organelles in rodent liver (for a review see ref. 14). Noguchi and Takada (13) predicted that the enzyme would be found to be deficient in patients with primary Hyperoxaluria type 1, a genetic disease in which an impairment in glyoxylate catabolism leads to accumulation of glyoxylate and oxalate. That the enzyme is, indeed, deficient in this disease was subsequently demonstrated by Danpure and Jennings (47).

Other Metabolic Pathways that Have Been Demonstrated in Peroxisomes

A small percentage of the total 3-hydroxy-3-methylglutaryl-CoA reductase activity in rat liver is present in peroxisomes (48) and evidence has been brought forward indicating that isolated peroxisomes are able to synthesize cholesterol (4a) and dolichol (50). The quantitative significance of the peroxisomal pathways for the synthesis of these compounds is not clear.
Finally, peroxisomes contain polyamine oxidase activity (51).

Table 1. Classification of Peroxisomal Disorders

Disorders	Morphologically distinguishable peroxisomes[a]	Enzyme defects
A. Generalized loss of peroxisomal functions		
Cerebro-hepato-renal (Zellweger) Syndrome	Absent	
Neonatal Adrenoleukodystrophy	Absent	
Infantile Refsum disease	Absent	
Hyperpipecolic Acidaemia	Absent	
B. Loss of multiple peroxisomal functions		
Rhizomelic Chondrodysplasia punctata	Present	DHAP-AT; Alkyl-DHAP synthase; phytanic acid oxidase; pro-thiolase protease[b]
Zellweger-like Syndrome	Present	Peroxisomal β-oxidation enzyme proteins; DHAP-AT
C. Loss of a single peroxisomal function		
X-linked Adrenoleukodystrophy	Present	Peroxisomal very-long-chain acyl-CoA synthetase
Thiolase deficiency (pseudo-Zelweger Syndrome)	Present	Peroxisomal thiolase
Acyl-CoA oxidase deficiency (Pseudo Neonatal Adrenoleuko-dystrophy)	Present	Acyl-CoA oxidase
Bifunctional protein deficiency	Present	Bifuctional protein
Adult Refsum disease[c]	Present	Phytanic acid α-hydroxylase
Hyperoxaluria type 1	Present	Alanine:gly-oxylate amino-transferase
Acatalasaemia	Present	Catalase

[a]Defined as organelles staining for catalase in an enzyme histochemical test

[b]Abbreviations: DHAP-AT, acyl-CoA: dihydroxyacetone phosphate acyltransferase; alkylDHAP synthase, alkyldihydroxyacetonephophate synthase

[c]It is not clear whether or not this disease should be classified as a peroxisomal disorder since the intracellular localization of phytanic acid α-hydroxylase, the deficient enzyme, has not yet been established unequivocally in man.

PATHOLOGY OF PEROXISOMES

In 1973 Goldfischer et al. (10) reported that peroxisomes, defined as subcellular organelles staining for catalase in an enzyme histochemical test, are absent in liver and kidney of patients with the cerebro-hepato-renal (Zellweger) syndrome. At first the importance of this observation was not appreciated and it was only later that other investigators, too, saw a connection between the absence of peroxisomes and the pathogenesis of the Zellweger syndrome. Specific suggestions were that the symptoms of the disease might be caused by the accumulation of intermediates in the biosynthesis of bile acids (40) or by the deficiency of plasmalogens (3).

The Zellweger syndrome is the prototype of what has come to be recognized as a distinct group of inherited diseases in man in which peroxisomal functions are impaired (for recent reviews see refs. 4-9). A classification of the peroxisomal disease is given in Table 1.

BIOGENESIS OF PEROXISOMES AND DISORDERS OF PEROXISOME ASSEMBLY

All peroxisomal proteins studied so far, including integral membrane proteins, are synthesized on free ribosomes and are imported post-translationally into peroxisomes (see refs. 52-54 for reviews). It is now generally agreed that peroxisomes, like mitochondria, arise by division of preexisting peroxisomes. The signal required for the import of some (but not all) proteins into peroxisomes has recently been identified (55); it consists of the tripeptide sequence Ser-Lys-Leu and is usually, but not always, present at the C terminus.

In contrast to mitochondrial proteins peroxisomal proteins are usually synthesized in their final size and thus do not possess a cleavable signal sequence (52-54). There are, however, at least two exceptions, i.e. acyl-CoA oxidase (56) and 3-oxoacyl-CoA thiolase (56,57).

Biosynthesis and Import into Peroxisomes of Acyl-CoA Oxidase and 3-Oxoacyl-CoA Thiolase

We have investigated the biosynthesis and import into peroxisomes of acyl-CoA oxidase and thiolase in isolated rat hepatocytes. In agreement with Hashimoto and coworkers (56) we found that rat-liver acyl-CoA oxidase is synthesized as a precursor molecule with an M_r of 72 kDa which is cleaved to a 52 kDa form (Fig. 2). The other product of the cleavage, a 20 kDa polypeptide, is present in isolated enzyme preparations but cannot be visualised is such labelling experiments due to its low methionine content (56). The half life of the 72 kDa precursor molecule was estimated by densitometric scanning to be >3h.

Fig. 2 clearly depicts that most of the 72 kDa and almost all of the 52 kDa form were associated with the particulate fraction, even after a pulse of 30 min with no chase. Import of the respective molecular forms of acyl-CoA oxidase into a putative peroxisomal compartment was determined by treatment of the particulate fraction with a mixture of proteases. Fig. 2 indicates that already at the end of the pulse period part of the 72 kDa form was protease-resistent. Addition of Triton X-100 (final concentration 0.1% v/v) before the incubation with proteases caused complete degradation of the 72 kDa form. No information could be obtained with respect to the possible import of the 52 kDa form, because the protease treatment used gave rise to three proteolytic intermediates with M_r values in the range of 52 kDa. We conclude that proteolytic cleavage of the precursor takes place, at least in past, within the peroxisomes.

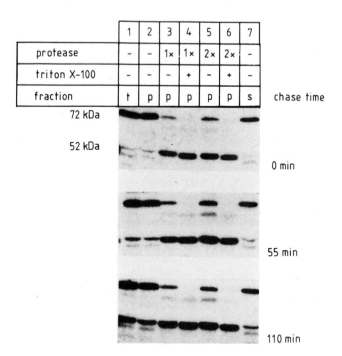

	1	2	3	4	5	6	7	
protease	–	–	1×	1×	2×	2×	–	
triton X-100	–	–	–	+	–	+	–	
fraction	t	p	p	p	p	p	s	chase time

Fig. 2. Biosynthesis, proteolytic processing and intracellular transport of acyl-CoA oxidase in isolated rat hepatocytes. Isolated hepatocytes from a rat treated with di(ethylheptyl)phthalate were radiolabelled with [^{35}S]methionine for 30 min, after which the cells were chased in the presence of an excess amount of unlabelled methionine. At intervals samples of the cell suspension were taken and incubated with digitonin (0.2 mM) for 2 min at 0°C. Subsequently the cell suspension was fractionated into a soluble and a particulate fraction by means of centrifugation. The particulate fraction was incubated with a protease mixture either in the presence or in the absence of 0.1% (v/v) Triton X-100. The samples were then supplemented with a mixture of protease inhibitors, sodium dodecyl sulphate (SDS) and Triton X-100 and stored at −20°C. β-Oxidation enzyme proteins were isolated by immunoprecipitation and visualised by SDS-polyacrylamine gel electrophoresis and fluorography. Abbreviations: p, particulate fraction; s, supernatant fraction; t, total cell suspension.

Pulse-chase experiments showed, in agreement with Hashimoto and coworkers (56) and Fujiki et al. (57), that 3-oxoacyl-CoA thiolase is synthesized as a 44 kDa precursor which is converted within a 2 h chase period to the 41 kDa mature protein. Both the precursor and the mature form were associated in part with the particulate fraction and were partly protease-resistant (Fig. 3). Densitometric scanning indicated that the ratio 41 kDa form: 44 kDa form was higher in the particulate fraction than in the soluble fraction (4.3 and 2.8, respectively, in the experiment of Fig. 3). This implies that proteolytic processing of the precursor of thiolase occurs after association with the particulate fraction. Our data lead us to conclude that at least part, if not all, of the proteolytic processing of 3-oxoacyl-CoA thiolase takes place within the peroxisomal compartment.

We have also studied the biosynthesis of acyl-CoA oxidase and 3-oxoacyl-CoA thiolase in cultured human skin fibroblasts (58). In cells from control subject the 72 kDa precursor of acyl-CoA oxidase formed during a 1 h pulse period is completely converted to the 52 kDa mature protein

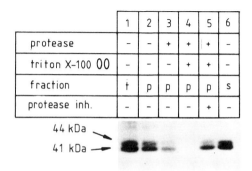

	1	2	3	4	5	6
protease	–	–	+	+	+	–
triton X-100 00	–	–	–	+	+	–
fraction	t	p	p	p	p	s
protease inh.	–	–	–	–	+	–

44 kDa
41 kDa

Fig. 3. Biosynthesis, proteolytic processing and intracellular transport of 3-oxoacyl-CoA thiolase in isolated rat hepatocytes. For experimental details and abbreviations see legend to Fig. 2.

within 24 h. This conversion does not take place in fibroblasts from patients with the Zellweger syndrome and the 72 kDa precursor protein gradually disappears. Similarly, there is no processing of the precursor of 3-oxoacyl-CoA thiolase in the mutant fibroblasts and the precursor of the enzyme is rapidly broken down.

Two conclusions may be drawn. Firstly, functional peroxisomes are required for the proteolytic processing of the precursors of both acyl-CoA oxidase and 3-oxoacyl-CoA thiolase. Secondly, in the absence of functional peroxisomes the precursors of the two enzymes appear to be unstable and are rapidly degraded, presumably in the cytosol (or in the lysosomes). In accordance with these conclusions are the results of immunoblotting experiments showing that the peroxisomal β-oxidation enzyme proteins are deficient in the liver of patients with the Zellweger syndrome and related disorders (59-61).

A corollary to these two conclusions is that peroxisomes contain specific proteases responsible for the proteolytic processing of the precursor forms of acyl-CoA oxidase and 3-oxoacyl-CoA thiolase, respectively (see below).

The basic defect in the diseases in category A of Table 1. Clearly the basic defect in the diseases in category A of Table 1 is a deficiency of a protein required for the assembly of peroxisomes. This protein could be a subunit of a membrane protein, for instance a receptor required for the import of proteins into peroxisomes. It could also be a soluble protein required for the biogenesis of peroxisomes, for instance an "unfoldase" (see ref. 62). Another possibility is that the mutation affects the structure of a peroxisomal protein destined for import in such a way that the mutant protein forms an abortive complex with a receptor or an unfoldase.

The basic defect in the diseases in category B of Table 1: The rhizomelic form of Chondrodysplasia punctata. The biochemical abnormalities observed so far in the rhizomelic form of Chondrodysplasia punctata are: an impairment in the ability to synthesize plasmalogens due to a deficiency of acyl-CoA:dihydroxyacetonephosphate acyltransferase (DHAP-AT) and alkylDHAP synthase, an accumulation of phytanic acid due to a deficiency of phytanic acid oxidase, and an inability to convert precursor 3-oxoacyl-CoA thiolase to the mature protein, presumably due to the deficiency of a prothiolase protease (63-65). Since the maturation of acyl-CoA oxidase is not impaired in rhizomelic Chondrodysplasia punctata there is probably a separate protease which is responsible for the proteolytic processing of the precursor of acyl-CoA oxidase.

The basic defect in this disease could be a mutation leading to a defective protein, perhaps a receptor, specifically required for the import into peroxisomes of DHAP-AT, alkyl-DHAP synthase, prothiolase protease and a component of phytanic acid oxidase. Another possibility is that one of the four enzymes is mutated and forms an abortive complex with a specific receptor.

Comparison between the diseases in category B and certain lysosomal storage diseases. There are certain similarities between two of the lysosomal storage diseases and the diseases of peroxisomal assembly in category B. The two diseases are Mucolipidosis II (MLII; I-cell disease) and Mucolipidosis III (MLIII; pseudo-Hurler polydystrophy). In both there is a deficiency of UDP-N-acetylglucosamine: lysosomal enzyme precursor N-acetylglucosaminephosphate transferase, one of the two enzymes required for the generation of mannose-6-phosphate groups in the oligosaccharide chains of the lysosomal enzyme precursors (66). The mannose-6-phosphate groups are required for targeting of soluble lysosomal enzymes to the lysosomes. In MLII and MLIII the enzymes are secreted instead of being transported to their appropriate destination. MLII and MLIII can thus be considered diseases of lysosomal assembly.

The basic defect in the Zellweger-like syndrome. There have recently been two reports on patients with a Zellweger-like syndrome (67,68). In both patients the results of immunoblotting experiments indicated that all three peroxisomal β-oxidation enzyme proteins were deficient in liver. This was reflected in an accumulation of very-long-chain fatty acids and intermediates of bile acid synthesis. Furthermore, DHAP-AT was deficient. All these findings are characteristic of diseases in category A of Table 1. However, in both patients catalase-containing particles were present in liver.

Unfortunately no fibroblasts were available. Thus it was not possible to confirm that peroxisomes are present by using an independent, biochemcial method (percentage of particle-bound catalase in cells). It is conceivable that the peroxisomes in the two patients were able to import catalase but not the β-oxidation enzyme proteins and DHAP-AT.

Another possibility is that the mutation was such that catalase was bound to the membranes of ghost-like structures without being imported.

The basic defect in the diseases in category C of Table 1: Hyperoxaluria Type 1. Danpure et al. (69) have recently described an interesting mutation in two patients with Hyperoxaluria type 1. In these two patients there is a decrease in the amount of alanine: glyoxylate aminotransferase in the liver as indicated by immunoblotting. However, fractionation studies indicated that the residual protein is present in the mitochondria rather than in the peroxisomes. Thus we have a trafficking defect in these two patients. These findings raise the interesting possibility that in the rat, where alanine:glyoxylate aminotransferase is equally distributed between mitochondria and peroxisomes, the proteins in both organelles might be encoded for by the same gene.

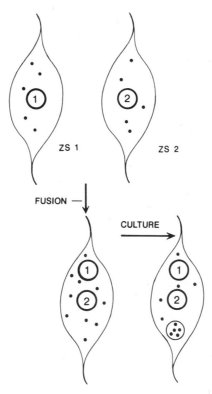

Fig. 4 Diagrammatic representation of complementation analysis following somatic cell fusion of two cell lines deficient in peroxisomes, using the intracellular localization of catalase as an index of peroxisome assembly. In the parental cell lines catalase (represented by the black dots) is present in the cytosol. After fusion the percentage of particle-bound catalase in the heterokaryons is measured at intervals by means of digitonin titration (73). If the cell lines are complementary assembly of peroxisomes will occur in the heterokaryons and catalase will be imported into the peroxisomes, i.e. the enzyme will become particle-bound.

We have used complementation analysis after somatic cell fusion of cultured skin fibroblasts to investigate the genetic relationship between diseases of peroxisome assembly (70). The principle of complementation analysis is as follows. If two mutant cell lines deficient in peroxisomes are cultured together they can be induced to fuse with the aid of poly-ethyleneglycol or inactivated Sendai virus; multinucleate cells are obtained. The heterokaryons will contain the genetic information from both parental cell lines. If the mutations affect different genes the heterokaryons will contain the wild type alleles derived from the respective parental cell lines. Thus if peroxisome assembly occurs, i.e. if the cell lines complement each other, this is an indication that the mutations affect different genes. However, lack of complementation does not necessarily mean that we are dealing with mutations in the same gene; since peroxisomes arise by division of pre-existing peroxisomes lack of complementation may be due to the absence of pre-existing peroxisome-like structures in the parental cell lines.

DHAP-AT activity, which is deficient in the diseases in category A of Table 1 and in rhizomelic Chondrodysplasia punctata, can be used as an index of peroxisome assembly. In those diseases in which peroxisomes are deficient and catalase is present in the cytosol (category A) the appearance of particle-bound catalase can be used as an index of peroxisome assembly and hence of complementation (Fig. 4).

The cell lines we have studied from patients with diseases in category A fall into five complementation groups; a cell line from a patient with rhizomelic Chondrodysplasia punctata falls into a separate complementation group (Table 2). Roscher et al. (71) have recently described six complementation groups and Poll-Thé et al. (72) two complementation groups in cell lines from patients with diseases in category A.

Examination of Table 2 indicates that patients with a well-defined clinical phenotype such as the Zellweger syndrome fall into different complementation groups. Thus there is genetic heterogeneity in the

Table 2. Complementation groups in diseases with a partial or generalized impairment of peroxisomal functions.

Complementation group	Phenotypes[a]
1	RCDP
2[b]	ZS1, IRD1, HPA
3	ZS2
4	NALD
5	ZS3
6	IRD2

[a] Abbreviations: HPA, Hyperpipecolic Acidaemia; IRD, infantile Refsum disease; NALD, neonatal Adrenoleukodystrophy; ZS, Zellweger syndrome.
[b] Equivalent to group 1 of Roscher et al. (71).

Zellweger syndrome. The same applies to infantile Refsum disease (Table 2; ref. 71) and neonatal Adrenoleukodystrophy (71).

The following conclusions can be drawn.
1.   Several genes are required for the assembly of a functional peroxisome.
2.   Mutations in a single gene can lead to different clinical phenotypes.
3.   Mutations in different genes can lead to the same clinical phenotype.

Kinetics of Peroxisome Assembly in Heterokaryons Obtained after Fusion of Complementary Cell Lines Deficient in Peroxisomes

We have measured the kinetics of peroxisome assembly in heterokaryons of complementary cell lines by measuring the rate of incorporation of catalase, initially present in the cytosol, into particles (see Fig. 4). In some combinations of cell lines assembly was rapid and insensitive to cycloheximide. Thus the components required for peroxisome assembly must have been present in the parental cell lines. Furthermore, at least one of the parental cell lines must have contained peroxisomal ghost-like structures.

ZS2 x HPA

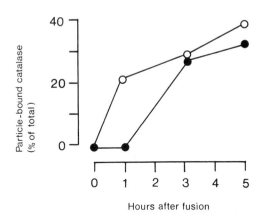

Fig. 5 Kinetics of the assembly of peroxisomes in heterokaryons obtained after fusion of complementary cell lines deficient in peroxisomes: effect of leupeptin. Confluent fibroblasts were cultured in HAM F10 medium containing 10% foetal calf serum prior to fusion with the aid of polyethyleneglycol. Leupeptin (50 μg/ml) was either absent (•) or present (o) 1 day prior to fusion and after fusion. Particule-bound catalase was assessed at inverals in the heterokaryons as described in [73]. The cell lines fused were ZS2 (complementation group 3) and HPA (complementation group 2).

In complementary cell lines which when fused give slow, cyclohexi-mide-sensitive assembly of peroxisomes there is a small but reproducible lag after fusion before assembly of peroxisomes commences (see Fig. 1 of ref. 74). This lag can be circumvented by culturing the parental cell lines in the presence of leupeptin, a protease inhibitor, prior to fusion. A typical experiment is shown in Fig. 5. This effect of leupeptin is reproducible. We conclude that the components required for the assembly of peroxisomes are unstable in some cell lines but that they can be stabilized by inhibiting proteolysis.

The Presence of Peroxisomal Ghosts in Cell Lines Deficient in Functional Peroxisomes

In rat-liver peroxisomes several integral membrane proteins, defined as proteins in membranes resistent to extraction at an alkaline pH, have been identified ranging in M_r from 69 kDa to 15 kDa (75-77). Three of these proteins, with M_r's of 69 kDa, 36 kDa and 22 kDa, respectively, are immunologically distinct. All three are synthesized at their final size on free polyribosomes and are posttranslationally inserted into the peroxisomal membrane (78,79).

As discussed above, in cells deficient in peroxisomes some peroxisomal proteins appear to be unstable and are degraded rapidly after synthesis; this is the case with the peroxisomal β-oxidation enzyme proteins (59-61). Other peroxisomal proteins, like catalase, are stable and biologically active in the absence of peroxisomes but are localized in the cytosol (73,80). The peroxisomal integral membrane proteins (PMP) appear to be present in normal amounts in liver of some patients with the Zellweger syndrome and related disorders (81) but to be slightly to markedly deficient in the liver of other patients (82,83). A complicating factor in assessing the results of such immunoblotting experiments is that proteolytic degradation of the proteins may occur artifactually during storage and extraction of autopsy or biopsy material.

Santos and coworkers (84,85) have reported that peroxisomal integral membrane proteins with M_r's of 140, 69, 53 and 22 kDa are present both in control fibroblasts and fibroblasts from Zellweger patients. These authors presented morphological evidence for the presence of peroxisomal ghost-like structures containing peroxisomal integral membrane proteins in cells from Zellweger syndrome patients.

We have carried out analogous studies (83). When cryosections of cells were examined by electron microscopy using immunogold labelling, control cells contained small structures consisting of a single membrane enclosing a homogeneous matrix; the membranes reacted with anti-(69 kDa PMP) and the matrix with anti-(catalase). The mutant cell lines contained spherical or ellipsoidal structures the membranes of which reacted with anti-(69 kDa PMP); no labelling was observed with anti-(catalase).

The subcellular localization in fibroblasts of catalase and the 69 kDa PMP was also studied by indirect immunofluorescence. A characteristic punctate fluorescence was seen in control cells incubated with either anti-(catalase) or with anti-(69 kDa PMP). Incubation of all mutant cell lines with anti-(catalase) resulted in a diffuse fluorescence, whereas with anti-(69 kDa PMP) fluorescent particles were visualized which, in some cell lines, were larger and fewer in number than in control cells. Fluorescent particles were observed in all of the cell lines studied, which belonged to complementationg groups 2-6 (see Table 2). Thus peroxisomal ghosts (empty vesicles) are present in all of these cell lines.

We are at present generating Chinese hamster ovary mutant cell lines deficient in peroxisomes. One of our aims is to obtain cell lines deficient in peroxisomal ghosts so that the role of such ghosts in the assembly of peroxisomes can be evaluated.

CONCLUSIONS

Peroxisomes play an essential role in cellular metabolism. Characteristically the metabolic pathways in which peroxisomes participate are oxidative pathways. In particular, β-oxidation in peroxisomes is responsible for the catabolism of a range of compounds which cannot be degraded in mitochondria. These compounds include very-long-chain fatty acids and mono-unsaturated long-chain fatty acids. In addition peroxisomes participate in the oxidative catabolism of pipecolic acid and phytanic acid, in the conversion of cholesterol to bile acids and in the biosynthesis of ether phospholipids.

Peroxisomes arise by division of pre-existing peroxisomes. Peroxisomal proteins, including integral membrane proteins, are synthesized on free ribosomes in the cytosol and are imported posttranslationally into peroxisomes.

A group of genetic diseases in man is now recognised in which peroxisomal functions are impaired. In some of those diseases there is a single enzyme defect. For instance, X-linked Adrenoleukodystrophy is due to a deficiency of peroxisomal very-long-chain fatty acyl-CoA synthetase. In other peroxisomal diseases there is a deficiency of multiple peroxisomal functions and in some cases morphologically distinguishable peroxisomes are absent. These are diseases of peroxisome assembly.

Complementation analysis following somatic cell fusion has indicated that there is genetic heterogeneity within the disorders of peroxisome assembly and that several genes are required for the assembly of functional peroxisomes.

The kinetics of the assembly of peroxisomes can be studied in heterokaryons formed by fusion of complementary fibroblast cell lines. After fusion of certain combinations of cell lines assembly of peroxisomes in the heterokaryons is rapid and cycloheximide insensitive, indicating that the components required for assembly of peroxisomes are stable in the parental cell lines, in at least one of which peroxisomal ghosts must have been present. In combinations of cell lines which give rise to heterokaryons in which assembly of peroxisomes is slow, due to a requirement for protein synthesis, the rate of assembly can be increased by culturing the parental cell lines in the presence of the protease inhibitor leupeptin, prior to fusion.

Peroxisome-deficient cell lines contain peroxisomal ghost-like structures (empty vesicles).

ACKNOWLEDGEMENTS

The studies by the authors described in this review were supported by a Programme Grant from the Netherlands Organization for Scientific Research (NWO), under the auspices of the Netherlands Foundation for Medical and Health Research (Medigon).

REFERENCES

1. J. Rhodin. Correlation of ultrastructural organization and function in normal and experimentally changed proximal convoluted tubule cells of the mouse kidney. Ph.D. thesis. Aktiebolaget Godoil, Stockholm, pp. 1-76 (1954)
2. H. Beaufay, P. Jacques, P. Baudhuin, O.Z. Sellinger, J. Berthet and C. De Duve. Tissue fractionation studies. 18. Resolution of mitochondrial fractions from rat liver into three distinct populations of particles by means of density equilibration in various gradients. Biochem. J. 92: 184-205 (1964)

3. P. Borst. Animal peroxisomes (microbodies), lipid biosynthesis and the Zellweger syndrome. Trends Biochem. Sci. 8: 269-272 (1983)

4. S. Goldfischer and J.K. Reddy. Peroxisomes (microbodies) in cell pathology. Int. Rev. Exp. Path. 26: 45-84 (1984)

5. R.B.H. Schutgens, H.S.A. Heymans, R.J.A. Wanders, H. van den Bosch and J.M. Tager. Peroxisomal disorders: a newly recognized group of genetic diseases. Eur. J. Pediatr. 144: 430-440 (1986)

6. H.W. Moser, S. Naidu, A.J. Kumar and A.E. Rosenbaum. The Adrenoleukodystrophies. CRC Critical Reviews on Neurobiology 3: 29-83 (1987)

7. H. Zellweger. The cerebro-hepato-renal (Zellweger) syndrome and other peroxisomal disorders. Developm. Med. Child Neurol. 29: 821-829 (1987)

8. R.J.A. Wanders, H.S.A. Heymans, R.B.H. Schutgens, P.G. Barth, H. van den Bosch and J.M. Tager. Peroxisomal disorders in neurology. J. Neurol. Sci. 88: 1-39 (1988)

9. P.B. Lazarow and H.W. Moser. Disorders of peroxisome biogenesis. In: The Metabolic Basis of Inherited Disease, 6th Edition (C.R. Scriver, A.L. Baudet, W.S. Sly and D. Valle, eds.), McGraw-Hill, New York, in press (1989)

10. S. Goldfischer, C.L. Moore, A.B. Johnson, A.J. Spiro, M.P. Valsamis, H.K. Wisniewski, R.H. Ritch, W.T. Norton, I. Rapin and L.M. Gartner. Peroxisomal and mitochondrial defects in the cerebro-hepato-renal syndrome. Science 182: 62-64 (1973)

11. S. Kolvraa and N. Gregersen. In vitro studies in the oxidation of medium-chain dicarboxylic acids in rat liver. Biochim. Biophys. Acta 876: 515-525 (1986)

12. J. Vamecq and J.P. Draye. Interactions between the α- and β-oxidations of fatty acids. J. Biochem. 102: 225-234 (1987)

13. T. Noguchi and Y. Tokada. Peroxisomal localization of alanine:glyoxylate aminotransferase in human liver. Arch. Biochem. Biophys. 196: 645-647 (1979)

14. T. Noguchi. Amino acid metabolism in animal peroxisomes. In: Peroxisomes in Biology and Medicine (H.D. Fahimi and H. Sies, eds.), Springer, Heidelberg, pp. 234-243 (1987)

15. S.J. Mihalik and W.J. Rhead. L-Pipecolic acid oxidation in the rabbit and Cynomolgus monkey. Evidence for differing organellar locations and cofactor requirements in each species. J. Biol. Chem. 264: 2509-2517 (1989)

16. R.J.A. Wanders, G.J. Romeyn, C.W.T. van Roermund, R.B.H. Schutgens, H. van den Bosch and J.M. Tager. Identification of L-pipecolic oxidase in human liver and its deficiency in the Zellweger syndrome. Biochem. Biophys. Res. Commun., 154: 33-38 (1989)

17. A.K. Hajra and J.E. Bishop. Glycerolipid biosynthesis in peroxisomes via the acyl dihydroxyacetone phosphate pathway. Ann. N.Y. Acad. Sci 386: 170-182 (1982)

18. O.H. Morand, R.A. Zoeller and C.R.H. Raetz. Disappearance of plasmalogens from membranes of animal cells subjected to photosensitized oxidation. J. Biol. Chem. 263: (1988)

19. P.B. Lazarow and C. de Duve. A fatty acid acyl-CoA oxidizing system in rat liver peroxisomes: enhancement by clofibrate, a hypolipidemic drug. Proc. Natl. Acad. Sci. USA 73: 2043-2046 (1976)

20. T. Hashimoto. Individual peroxisomal β-oxidation enzymes. Ann. N.Y. Acad. Sci. 386: 5-12 (1982)

21. T. Hashimoto. Comparison of enzymes of lipid β-oxidation in peroxisomes and mitochondria. In: Peroxisomes in Biology and Medicine (H.D. Fahimi and H. Sies, eds.), Springer, Heidelberg, pp. 97-104 (1987)

22. J. Bremer. Carnitine and its role in fatty acid metabolism. Trends Biochem. Sci. 2: 207-209 (1977)

23. I. Singh, A.B. Moser, S. Goldfischer and H.W. Moser. Lignoceric acid

is oxidized in the peroxisome: implications for the Zellweger cerebro-hepato-renal syndrome and adrenoleukodystrophy. Proc. Natl. Acad. Sci. USA 81: 4203-4207 (1984)

24. A. Bhusnan, R.P. Singh and I. Singh. Characterization of rat brain microsomal acyl-coenzyme A ligase: diferent enzymes for the synthesis of palimtoyl-CoA and lignoceroyl-CoA. Arch. Biochem. Biophys. 246: 374-380 (1986)

25. R.J.A. Wanders, C.W.T. van Roermund, M.J.A. Van Wijland, R.B.H. Schutgens, J. Heikoop, H. Van den Bosch, A.W. Schram and J.M. Tager. Peroxisomal fatty acid β-oxidation in relation to the accumulation of very long chain fatty acids in peroxisomal disorders. J. Clin. Invest. 80: 1778-1783 (1987)

26. H. Singh, N. Derwas and A. Poulos. Very long chain fatty acid β-oxidation by rat liver mitochondria and peroxisomes. Arch. Biochem. Biophys. 359: 382-390 (1987)

27. S. Miyazawa, T. Hashimoto and S. Yokota. Identity of long-chain acyl-CoA synthetase in microsomes, mitochondria and peroxisomes in rat liver. J. Biochem. 98: 723-733 (1985)

28. I. Singh, A. Bhushan, N.K. Relan and T. Hashimoto. Acyl-CoA ligases from rat brain microsomes: an immunochemical study. Biochim. Biophys. Acta 963: 509-514 (1985)

29. T. Osumi, T. Hashimoto and N. Ui. Purification and properties of acyl-CoA oxidase from rat liver. J. Biochem. 87: 1735-1746 (1980)

30. P.B. Lazarow. Rat liver peroxisomes catalyse the β-oxidation of fatty acids. J. Biol. Chem. 253: 1522-1528 (1978)

31. G.P. Mannaerts, L.I. De Beer, J. Thomas and R.J. De Schepper. Mitochondrial and peroxisomal fatty acid oxidation in liver homogenates and isolated hepatocytes from control and clofibrate-treated rats. J. Biol. Chem. 254: 4585-4595 (1979)

32. S. Goldfischer and H.J. Sobel. Peroxisomes and bile-acid synthesis. Gasteroenterology 81: 196-197 (1981)

33. F. Kase, I. Björkhem, and J.I. Pedersen. Formation of cholic acid from 3α, 7α, 12α-trihydroxy-5β-cholestanoic acid by rat-liver peroxisomes. J. Lipid. Res. 24: 1560-1567 (1983)

34. S. Goldfischer, J. Collins, I. Rapin, P. Neumann, W. Neglia, A.J. Spiro, T. Ishii, F. Roels, J. Vamecq and R. Van Hoof. Pseudo-Zellweger syndrome: deficiencies in several peroxisomal oxidative activities. J. Pediatr. 180: 25-32 (1986)

35. A.W. Schram, S. Goldfischer, C.W.T. Van Roermund, E.M. Brouwer-Kelder, J. Collins, T. Hashimoto, H.S.A. Heymans, H. Van den Bosch, R.B.H. Schutgens, J.M. Tager and R.J.A. Wanders. Human peroxisomal 3-oxo-acyl-CoA thiolase deficiency. Proc. Natl. Acad. Sci. USA 84: 2494-2497 (1987)

36. P.A. Watkins, W.W. Chen, C.J. Harris, G. Hoefler, S. Hoefler, D.C. Blake, A. Balfe, R.I. Kelley, A.B. Moser, M.E. Beard and H.W. Moser. Peroxisomal bifunctional enzyme deficiency. J. Clin. Invest. 83: 771-777 (1989)

37. B.T. Poll-Thé, F. Roels, H. Ogier, J. Scotto, J. Vamecq, R.B.H. Schutgens, C.W.T. Van Roermund, M.J.A. Van Wijland, A.W. Schram, J.M. Tager and J.M. Saudubray. A new peroxisomal disorder with enlarged peroxisomes and a specific deficiency of acyl-CoA oxidase (pseudo neonatal adrenoleukodystrophy). Am. J. Hum. Genet. 42: 422-434 (1988)

38. M. Casteels, L. Schepers, J. Van Eldere, H. Eyssen and G.P. Mannaerts. Inhibition of 3α, 7α, 12α-trihydroxy-5β-cholestanoic acid oxidation and of bile acid secretion in rat liver by fatty acids. J. Biol. Chem. 263: 4654-4661 (1988)

39. J.I. Pedersen, E. Hvattum, T. Flatabo and I. Björkhem. Clofibrate does not induce peroxisomal 3α, 7α, 12α-trihydroxy-5β-cholestanoyl coenzyme A oxidation in rat liver: evidence that this reaction is catalyzed by an enzyme system different from that of peroxisomal

acyl-CoA oxidation. Biochem. Int. 17: 163-169 (1988)

40. J.I. Pedersen and J. Gustafson. Conversion of 3α, 7α, 12α-trihydroxy-5β-cholestanoic acid into cholic acid by rat liver peroxisomes. FEBS Lett. 121: 345-348 (1980)

41. L. Schepers, M. Casteels, K. Verheyden, G. Parmentier, S. Asselberghs, H.J. Eyssen and G.P. Mannaerts. Subcellular distribution and characteristics of trihydroxycoprostanoyl-CoA synthetase in rat liver. Biochem. J. 257: 221-229 (1989)

42. O. Lazo, M. Contreras, A. Bhushan, W. Stanley and I. Singh. Adrenoleukodystrophy: impaired oxidation of fatty acids due to peroxisomal lignoceroyl-CoA deficiency. Arch. Biochem. Biophys. 270: 722-728 (1989)

43. R.J.A. Wanders, C.W.T. van Roermund, M.J.A. van Wijland, R.B.H. Schutgens, H. van den Bosch and J.M. Tager. Direct demonstration that the deficient oxidation of very long chain fatty acids in X-linked adrenoleukodystrophy is due to an impaired ability of peroxisomes to activate very long chain fatty acids. Biochem. Biophys. Res. Commun. 153: 618-624 (1988)

44. O.H. Skjeldal and O. Stokke. The subcellular localization of phytanic acid oxidase in rat liver. Biochim. Biophys. Acta. 921: 38-42.

45. J.M.F. Trijbels, L.A.H. Monnens, G. Melis, M. Van den Broek-van Essen and M. Bruckwilder. Localization of pipecolic acid metabolism in rat liver peroxisomes: probable explanation for hyperpipecolataemia in Zellweger syndrome. J. Inher. Metab. Dis. 10: 128-134 (1987)

46. K. Zaar, S. Angermüller, A. Völkl and H.D. Fahimi. Pipecolic acid is oxidized by renal and hepatic peroxisomes: Implications for Zellweger's cerebro-hepato-renal syndrome (CHRS). Exp. Cell Res. 164:267-271 (1986)

47. C.J. Danpure and P.R. Jennings. Peroxisomal alanine: glyoxylate aminotransferase deficiency in primary hyperoxaluria type I. FEBS Lett. 201: 20-24 (1986)

48. G.A. Keller, M.C. Barton, D.J. Shapiro and S.J. Singer. 3-Hydroxy-3-methylglutaryl coenzyme A reductase is present in peroxidomes in normal rat liver cells. Proc. Natl. Acad. Sci. USA 82: 770-774 (1985)

49. S.L. Thompson, R. Burrow, R.J. Laub and S.K. Krisans. Cholesterol synthesis in rat liver peroxisomes. J. Biol. Chem. 262: 17420-17425 (1987)

50. E.L. Appelkvist and G. Dallner. Dolichol metabolism and peroxisomes. In: H.D. Fahimi and H. Sies (Eds.) Peroxisomes in Biology and Medicine. Springer, Heidelberg pp. 53-66

51. M.E. Beard, R. Baker, P. Conomos, D. Pugatch and E. Holtzman. Oxidation of oxalate and polyamines by rat liver peroxisomes. J. Histochem. Cytochem. 33: 460-464 (1985)

52. P.B. Lazarow and Y. Fujiki. Biogenesis of peroxisomes. Ann. Rev. Cell Biol. 1: 489-530 (1985)

53. P. Borst. Review. How proteins get into microbodies (peroxisomes, glyoxysomes, glycosomes). Biochim. Biophys. Acta 866: 179-203 (1986)

54. P. Borst. Peroxisome biogenesis revisited. Biochim. Biophys. Acta. 100: 1-13 (1989)

55. S.J. Gould, G.A. Keller and S. Subramani. Identification of a peroxisomal targeting signal at the carboxy terminus of firefly luciferase. J. Cell Biol. 105: 2923-2931 (1987)

56. S. Miura, M. Mori, M. Takiguchi, M. Tatibana, S. Furuta, S. Miyazawa and T. Hashimoto. Biosynthesis and intracellular transport of enzymes of peroxisomal β-oxidation. J. Biol. Chem. 259: 6397-6402 (1984)

57. Y. Fujiki, R.A. Rachubinski, R.M. Mortensen and P.B. Lazarow. Synthesis of 3-ketoacyl-CoA thiolase of rat liver peroxisomes on free polyribosomes as a larger precursor. Biochem. J. 226: 697-704 (1985)

58. A.W. Schram, A. Strijland, T. Hashimoto, R.J.A. Wanders, R.B.H. Schutgens, H. Van Den Bosch and J.M. Tager. Biosynthesis and

maturation of peroxisomal β-oxidation enzymes in fibroblasts in relation to the Zellweger syndrome and infantile Refsum disease. Proc. Natl. Acad. Sci. USA 83: 6156-6158 (1986)

59. J.M. Tager, W.A. Ten Harmsen Van Der Beek, R.J.A. Wanders, T. Hashimoto, H.S.A. Heymans, H. Van Den Bosch, R.B.H. Schutgens and A.W. Schram. Peroxisomal β-oxidation enzyme proteins in the Zellweger syndrome. Biochem. Biophys. Res. Commun. 126: 1269-1275 (1985)

60. Y. Suzuki, T. Orii, and T. Hashimoto. Biosynthesis of peroxisomal β-oxidation enzymes in infants with Zellweger syndrome. J. Inher. Metab. Dis. 9: 292-296 (1986)

61. W.W. Chen, P.A. Watkins, T. Osumi, T. Hashimoto and H.W. Moser. Peroxisomal β-oxidation enzyme proteins in adrenoleukodystrophy: distinction between X-linked adrenoleukodystrophy and neonatal adrenoleukodystrophy. Proc. Natl. Acad. Sci. USA 84: 1425-1428 (1987)

62. N. Pfanner, F.-U. Hartl and W. Neupert. Review. Import of proteins into mitochondria: a multistep process. Eur. J. Biochem. 175: 205-212 (1988)

63. R.B.H. Schutgens, H.S.A. Heymans, R.J.A. Wanders, J.W.E. Oorthuys, J.M. Tager, G. Schrakamp, H. Van den Bosch and F.A. Beemer. Multiple peroxisomal enzyme deficiencies in rhizomelic chondrodysplasia punctata. Comparison with Zellweger syndrome, Conradi-Hünermann syndrome and X-linked dominant type of Chondrodysplasia punctata. In: Advances in Clinical Enzymology. Vol. 6 Enzymes-Tools and Targets (D.M. Goldberg, D.W. Moss, E. Schmidt and F.W. Schmidt, eds) Karger, Basel pp. 57-65 (1988)

64. G.H. Hoefler, S. Hoefler, P.A. Watkins, W.W. Chen, A. Moser, V. Baldwin, B. McGillivary, J. Charro, J.M. Friedman, L. Rutledge, T. Hashimoto and H.W. Moser. Biochemical abnormalities in rhizomelic chondrodysplasia punctata. J. Pediatr. 112: 726-733 (1985)

65. J. Heikoop, R.J.A. Wanders, R.B.H. Schutgens, A.W. Schram and J.M. Tager. Maturation of peroxisomal thiolase protein in rhizomelic chondrodysplasia punctata (RCDP). Abstracts Int. Congr. Cell Biol. p. 341 (1988)

66. T.Mueller, N.K. Honey, L.E. Little, A.L. Miller and T.S. Shows. Mucolipidosis II and III. The genetic relationships between two disorders of lysosomal enzyme biosynthesis. J. Clin. Invest. 72: 1016-1023 (1983)

67. Y. Suzuki, N. Shimazawa, T. Orii, N. Igarashi, N. Kono, A. Matsui, Y. Inoue, S. Yokota and T. Hashimoto. Zellweger-like syndrome with detectable hepatic peroxisomes: A variant fom of peroxisomal disorder. J. Pediatr. 113: 841-845 (1988)

68. M. Paterneau-Jouas, F. Taillard, A. Gansmuller, R. Schutgens, J. Mikol, M.-S. Aigrot and C. Sereni. Clinical, biochemical, pathological "Zellweger-like" disorder with morphologically normal peroxisomes. in: Lipid Storage disorders. Biological and Medical Aspects (R. Salvayre, L. Douste-Blazy, and S. Gatt, eds) p. 805-807 (1989)

69. C.J. Danpure, P.J. Cooper, P.J. Wise, and P.R. Jennings. An enzyme trafficking defect in two patients with primary hyperoxaluria type 1: Peroxisomal alanine/glyoxylate aminotransferase rerouted to mitochondria. J. Cell. Biol. 108: 1345-1352 (1989)

70. S. Brul, A. Westerveld, A. Strijland, R.J.A. Wanders, A.W. Schram, H.S.A. Heymans, R.B.H. Schutgens, H. Van Den Bosch and J.M. Tager. Genetic heterogeneity in the cerebro-hepato-renal (Zellweger) syndrome and other inherited disorders with a generalized impairment of peroxisomal functions: A study using complementation analysis. J. Clin. Invest. 81: 1710-1715 (1988)

71. A.A. Roscher, S. Hoefler, G. Hoefler, E. Paschke, F. Paltauf, A. Moser and H. Moser. Genetic and phenotypic heterogeneity in disorders of peroxisome biogenesis. A complementation study involving cell

lines from 19 patients. Pediatr. Res. In press (1989)

72. B.T. Poll-Thé, O.H. Skjeldal, O. Stokke, A. Poulos, F. Demaugre and J.M. Saudubray. Phytanic acid alpha-oxidation and complementation analysis of classical Refsum disease and peroxisomal disorders. Human. Genet. 81: 175-181 (1989)

73. R.J.A. Wanders, M. Kos, B. Roest, A.J. Meijer, G. Schrakamp, H.S.A. Heymans, W.H.H. Tegelaers, H. Van Den Bosch, R.B.H. Schutgens and J.M. Tager. Activity of peroxisomal enzymes and intracellular distribution of catalase in Zellweger syndrome. Biochem. Biophys. Res. Commun. 123: 19054-1061 (1984)

74. S. Brul, A.E.C. Wiemer, A. Westerveld, A. Strijland, R.J.A. Wanders, A.W. Schram, H.S.A. Heymans, R.B.H. Schutgens, H. Van Den Bosch and J.M. Tager. Kinetics of the assembly of peroxisomes after fusion of complementary cell lines from patients with the cerebro-hepato-renal (Zellweger) syndrome and related disorders. Biochem. Biophys. Res. Commun. 152: 1083-1089 (1988)

75. Y. Fujiki, S. Fowler, H. Shio, A.L. Hubbard and P.B. Lazarow. Polypeptide and phospholipid composition of the membrane of rat liver peroxisomes: Comparison with endoplasmic reticulum and mitochondrial membranes. J. Cell Biol. 93: 103-110 (1982)

76. F.-U. Hartl, W.W. Just, A. Köster and H. Schimassek. Improved isolation and purification of rat liver peroxisomes by combined rate zonal and equilibrium density centrifugation. Arch. Biochem. Biophys. 237: 124-134 (1985)

77. T. Hashimoto, T. Kuwabara, N. Usuda and T. Nagata. Purification of membrane polypeptides of rat liver peroxisomes. J. Biochem. (Tokyo) 100: 301-310 (1986)

78. Y. Fujiki, R.A. Rachubinski and P.B. Lazarow. Synthesis of a major integral membrane polypeptide of rat liver peroxisomes on free polysomes. Proc. Natl. Acad. Sci. 81: 7127-7131 (1984)

79. Y. Suzuki, T. Orii, M. Takaguchi, M. Mori, M. Hijikata and T. Hashimoto. Biosynthesis of membrane polypeptides of rat liver peroxisomes. J. Biochem. (Tokyo) 101: 491-496 (1987)

80. M.J. Santos, J.M. Ojeda, J. Garrido and F. Leighton. Peroxisomal organization in normal and cerebro-hepato-renal (Zellweger) syndrome fibroblasts. Proc. Natl. Acad. Sci. USA 82: 6556-6560 (1985)

81. P.B. Lazarow, Y. Fujiki, G.M. Small, P. Watkins and H.W. Moser. Presence of the peroxisomal 22 kDa integral membrane protein in the liver of a person lacking recognizable peroxisomes (Zellweger syndrome). Proc. Natl. Acad. Sci. USA 83: 9193-9196 (1986)

82. Y. Suzuki, N. Shimozawa, T. Orii, J. Aikawa, K. Tada, T. Kuwabara and T. Hashimoto. Biosynthesis of peroxisomal membrane proteins in infants with Zellweger syndrome. J. Inher. Metab. Dis. 10: 297-300 (1987)

83. E.A.C. Wiemer, S. Brul, W.W. Just, R. Van Driel, E. Brouwer-Kelder, M. Van Den Berg, P.J. Weijers, R.B.H. Schutgens, H. Van Den Bosch, A.W. Schram, R.J.A. Wanders and J.M. Tager. Presence of peroxiosmal membrane proteins in liver and fibroblasts from patients with the Zellweger syndrome and related disorders: evidence for the existence of peroxisomal ghosts. Submitted for publication

84. M.J. Santos, T. Imanaka, H. Shio, G.M. Small and P.B. Lazarow. Peroxisomal membrane ghosts in Zellweger syndrome-aberrant organelle assembly. Science 239: 1536-1538 (1988)

85. M.J. Santos, T. Imanaka, H. Shio and P.B. Lazarow. Peroxisomal integral membrane proteins in control and Zellweger fibroblasts. J. Biol. Chem. 263: 10502-10509 (1988)

MEMBRANE FLOW AND ENDOCYTOSIS

Dr. John Davey

Department of Biochemistry
University of Birmingham
Birmingham B15 2TT
Great Britain

SUMMARY

It is not my intention for this to be an exhaustive review of membrane transport or indeed of endocytosis. These topics have been dealt with extensively elsewhere and the reader is encouraged to browse through them at their leisure (Anderson and Kaplan, 1983; Goldstein et al., 1985; Silverstein et al., 1977; Steinman et al., 1983; Wileman et al., 1985). What I have tried to achieve is a representation of my personal views of membrane transport: highlighting those aspects of transport that are of particular interest to myself. Since my research interests are largely concerned with endocytosis it is from this pathway of membrane transport that I will draw the majority of examples and evidence. I will however include examples from other transport pathways where they are relevant or where they provide insights into transport events that are not observable on the endocytic pathway.

Such a short review cannot of course be exhaustive and I apologise in advance for any ommissions.

INTRODUCTION

One of the major characteristics of eukaryotic cells is their possession of intracellular membrane-bound compartments. These compartments are often represented as being static entities within the cell while in fact they are dynamic and there is a great deal of movement from one compartment to another. This movement is probably best illustrated by the pathways used to transport material into and out of the cell (see Figure 1).

There are essentially two main transport pathways; the first transports material into the cell and is called the endocytic pathway while the second transports material out of the cell and is called the exocytic or secretory pathway. What usually happens with the endocytic pathway is that extracellular material becomes trapped in vesicles at the plasma membrane and is transported to lysosmes where it is degraded. In the exocytic pathway material is synthesised at the endoplasmic reticulum and is transported in vesicles to the Golgi apparatus, through the Golgi and on to the plasma membrane. Neither pathway is however as simple as I have described and each main route through the cell has several possible alternative destinations. With the endocytic pathway, for example, not all of the internalised material is delivered to lysosomes, some is recycled back to the plasma membrane, some to other domains of the plasma membrane and some is transported to the Golgi complex. Similarly with the exocytic pathway not all the proteins are delivered to the plasma membrane and a major diversion of this route is the delivery of newly synthesised enzymes to the lysosomes.

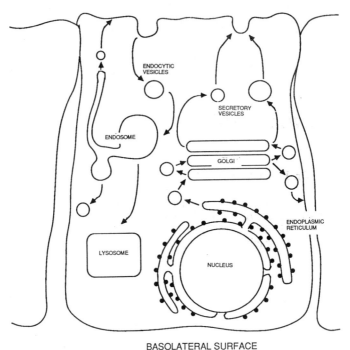

BASOLATERAL SURFACE

Figure 1

Summary of transport pathways.

One is left with a picture of vesicles forming at one cellular location, moving through the cytoplasm and fusing with their target membrane and this is what is commonly referred to as membrane traffic or membrane flow.

MEMBRANE TRANSPORT

When one dissects the many different transport events from the various pathways it becomes clear that although the components of each event are different the underlying mechanism for transport is very similar. This mechanism can be divided into a number of distinct steps and these are illustrated in Figure 2.

Figure 2 shows a schematic illustration of the vesicular transport of a molecule from one compartment to another. The compartment from which the molecule is transported is referred to as the donor and the compartment to which the molecule is transported is referred to as the acceptor. Any particular cellular compartment can simultaneously act as both a donor and an acceptor; it can accept molecules from an earlier compartment on a transport pathway and it can donate molecules to the next compartment on that pathway. A good example of this is the Golgi complex which accepts molecules from the endoplasmic reticulum and donates the same molecules to the plasma membrane.

Each transport event has to achieve two basic aims. Firstly it must transport the molecule that is to be transported but secondly it must retain the identity of the two compartments; the cell must avoid transporting molecules which confer identity on the donor compartment. Both of these aims are achieved by the scenario shown in Figure 2. The dark molecule represents the molecule to be transported from the donor compartment to the acceptor while the lighter molecules represent those molecules which confer identity on the donor compartment, that is the molecules which are not to be transported. The first step on the transport pathway is to sort

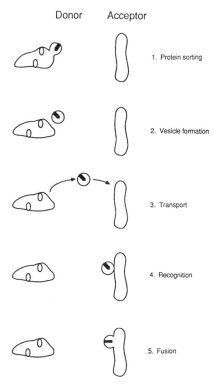

Donor   Acceptor

1. Protein sorting

2. Vesicle formation

3. Transport

4. Recognition

5. Fusion

Figure 2

Generalised transport event from a donor compartment to an acceptor compartment.

these two types of molecules. When this has been achieved the transport vesicle itself has to be formed. This vesicle is then transported through the cytoplasm to the acceptor compartment. To avoid the molecule being transported to the wrong compartment there must be some recognition event between the transport vesicle and the acceptor compartment. When the transport vesicle is satisfied that it is adjacent to the correct acceptor compartment then membrane fusion occurs and the molecule is delivered to the acceptor.

Following transport from the donor to the acceptor one could envisage several reasons why transport in the opposite direction, from acceptor to donor, might occur. Such a transport event might be required to replenish the membrane content of the donor compartment; if transport was continually from the donor to the acceptor then on a simplistic basis one would eventually run out of donor compartment. Transport from the acceptor to the donor might also be necessary to recapture molecules which have been wrongly transported to the acceptor. Such a transport event could be considered a checking mechanism. Alternatively transport from the acceptor to the donor might not occur. It is possible that in certain circumstances the donor compartment is replenished by a transport event from a compartment earlier in the pathway and if no molecules are incorrectly transported there would be no need for a checking mechanism. It would appear that examples of both instances exist in cells; transport from endosomes to lysosomes is probably unidirectional while recent work concerning the selective retention of GRP78 in the endoplasmic reticulum reveals bidirectional transport between the endoplasmic reticulum and the Golgi complex (Pelham, 1988).

What I would like to do for the remainder of this report is consider current ideas about each of the steps in Figure 2. Most of the examples I will use relate to endocytosis since this is the pathway in which I have a research interest but I will include other transport events where relevant.

ENDOCYTOSIS

At a very elementary level endocytosis can be divided into two types; fluid-phase endocytosis and receptor-mediated endocytosis. In fluid-phase endocytosis the molecules being internalised are dissolved in the extracellular medium and taken up by the cell as it internalises the medium. The rate of uptake of a particular molecule is therefore dependent upon its concentration in the medium. In receptor-mediated endocytosis however the molecule to be internalised is first concentrated for internalisation by binding to specific high-affinity receptors on the cell surface. Without this prior enrichment many of the molecules required by the cell would be internalised at a rate which would be too slow to allow cell growth or cell development. Although not yet unequivocally proven it does seem that both types of endocytosis share common routes of internalisation and that the vesicles used for receptor-mediated endocytosis are the same as those used for fluid-phase endocytosis. The best evidence for a common pathway is provided by experiments which follow the simultaneous internalisation of markers specific for both pathways. One such study monitored the internalisation of Semliki Forest virus (receptor-mediated marker) and ^{3}H-sucrose (fluid-phase marker) and showed that the decrease in the volume of sucrose endocytosed was equivalent to the volume of virus particles internalised during the same time (Helenius et al., 1980). Since it is likely that both pathways share common mechanisms I will limit future discussions to the receptor-mediated pathway simply because most of the work has been performed on this route.

The most common pathway of internalisation is receptor-mediated endocytosis in which the receptor is recycled back to the plasma membrane and the ligand is delivered to lysosomes for degradation and although there are many variations to this basic theme they are not relevant to our present discussion.

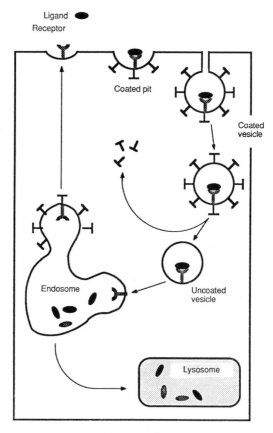

Figure 3

Schematic illustration of receptor-mediated endocytosis leading to degradation of ligand in lysosomes.

Figure 3 outlines the endocytic pathway and shows the major features encountered on the pathway. The ligand in the medium binds to specific high-affinity receptors on the cell surface and the ligand-receptor complex becomes concentrated in coated pits. These are specialised regions of the plasma membrane which are characterised by having a fuzzy coat in electron micrographs. The pit invaginates and pinches off to form a coated vesicle. The coat is removed and the uncoated vesicle fuses with the endosome. This compartment has a slightly acidic pH and this causes the ligand to be released from the receptor. The receptor is recycled back to the plasma membrane and the ligand is delivered to lysosomes where it is degraded. Some of the best visual evidence for this pathway comes from studies on the entry of viruses into cells during infection. Certain viruses exploit the endocytic pathway as their means of entering cells and electron microscopic observation of this process has generated many persuasive photographs (see, for example, Helenius et al., 1980; Matlin et al., 1981).

Figure 4 combines the endocytic pathway of Figure 3 with the general transport illustration of Figure 2 and my intention is to deal with each step in more detail. As for all good tours we shall have a guide and the guide I have chosen is the low-density lipoprotein (LDL) particle because its internalisation is arguably the best studied example of endocytosis.

Receptors, coated pits and coated vesicles

The LDL receptor is a trans-membrane protein which spans the membrane a single time. The majority of the receptor is extracellular and there is a short tail domain exposed on the cytoplasmic side of the membrane. The receptor binds the LDL particle and the receptor-LDL complex is internalised via a coated pit . These pits are so called because of the fuzzy coat-like

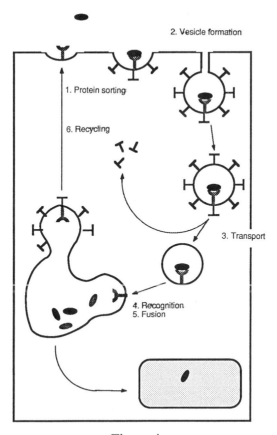

Figure 4

Dissection of the endocytic pathway into general transport events.

structure that is often visible on the cytoplasmic side of the mebrane following preparation of samples for electron microscopy. They are highly specialised structures which appear to play a pivitol role in the sorting and packaging of molecules during membrane transport. Although the pits account for only about 2% of the surface area of an average cell they appear to be the starting point for almost all endocytic entry. The reader should be aware of reports in which perturbation of the cytoplasm appears to inhibit the formation of coated pits without a complete inhibition of internalisation (Moya et al., 1985; Sandvig et al., 1987). While such work suggests that endocytosis via non-coated pits does or, at least under certain circumstances, can occur the cells are being studied under stressed conditions and the results may not reflect the situation in a normal cell. I will therefore limit my discussion to endocytosis through coated pits.

Coated pits are able to discriminate between those molecules which are to be internalised into the cell and those molecules which are to be left on the cell surface, in effect it is a filter which determines what goes into the cell and what does not (Bretscher, 1980). The receptor-ligand complexes for example become incorporated into coated pits while enzymes which function on the plasma membrane, such as aminopeptidase, do not become incorporated. An obvious and very interesting question is what determines whether a particular membrane protein at the cell surface will become incorporated into a coated pit and subsequently internalised into the cell? It is reasonable to suggest that the sorting mechanism is a two-way interaction, involving some feature of the pit and some feature of the membrane proteins themselves and it is these proteins which I would like to consider first.

The best examples of proteins which are internalised via pits are the receptors themselves but even if one limits a discussion to these examples they are found to have a number of different properties. Some receptors, such as the LDL receptor, are concentrated in coated pits even when they do not have their ligand bound to them. Others however, such as the EGF receptor, are more randomly distributed over the cell surface and concentrate within coated pits only when complexed with their ligand. In such cases it is likely that the binding of the ligand changes the conformation of the receptor such that the receptor can now be included into the pit. Whatever feature of the receptors is responsible for their incorporation into coated pits it is likely to be localised in the cytoplasmic tail region of the receptor since this is the only part which has access to the coat proteins. Consequently there had been the expectation that comparison of cytoplasmic sequences would reveal a sequence or structure which could be recognised by coated pits. The receptors do not appear to possess a common amino acid sequence specific for inclusion into coated pits similar to the signal sequences which determine the translocation of proteins across the endoplasmic reticulum and the incorporation of proteins into certain cytoplasmic organelles such as mitochondria and chloroplasts. The receptors so far studied have very different tail regions, not only varying in sequence but even with regard to whether they represent the carboxyl- or amino-terminus of the protein. In the absence of a consensus sequence many groups have begun to study the endocytosis of receptors possessing altered cytoplasmic sequences.

The importance of the tail domain in endocytosis of receptors is shown by the fact that removal of this region from the receptors for LDL, transferrin, epidermal growth factor or poly-Ig prevent their internalisation (Lehrman et al., 1985; Prywes et al., 1986; Mostov et al., 1986; Rothenberger et al., 1987). Possibly the best example of this approach has been carried out on the LDL receptor. The tail region of this receptor contains 50 amino acids and the groups of Michael Brown and Joseph Goldstein have studied over 25 different mutations that affect this region of the molecule (Brown et al., 1983; Lehrmann et al., 1985; Davis et al., 1986, 1987). These mutations are either naturally occurring or have been artificially generated by the mutagenesis of the cloned receptor gene which is then expressed in cells which lack an endogenous receptor. These changes indicate that the first 22 amino acids are important for clustering in coated pits and the tyrosine residue at position 807 is particularly important. This suggestive role for tyrosine residues in internalisation is strengthened by the findings that mutations which remove a single tyrosine residue from the cytoplasmic domain of either the vesicular stomatitis G glycoprotein (Michael Roth, unpublished results) or the the poly-Ig receptor prevent their efficient internalisation (Keith Mostov, unpublished results) while the insertion of a tyrosine residue into the tail region of the influenza virus haemagglutinin dramatically increases its entry into coated pits and subsequent internalisation into the cell (Lazarovits and Roth, 1988). Although other proteins known to be internalised through coated pits contain tyrosines in their cytoplamic domains, no consensus sequence around the tyrosines has been identified and it is more likely that the tyrosine residues form part of a signal of non-contiguous amino acids.

An alternative model is that the internalisation signal is created posttranslationally and thus would not be detected in the amino acid sequences of the receptors. An attractive candidate could be phosphorylation of the cytoplasmic tail. A number of receptors are known to be phosphorylated and this has been implicated in the internalisation of the transferrin receptor and the EGF receptor (May et al., 1985) although several other receptors, including the LDL receptor, are not phosphorylated. Whatever the signal turns out to be it must be recognised by the coat proteins and it is to them that we now turn.

Analysis of the coats reveals a remarkably simple composition; there is a major protein of 180kD, a group of proteins of 100kD, another group of 50kD and two proteins of about 30kD. The main component of the coat is the 180kD polypeptide (clathrin) which exists in its trimeric form (the triskelion) (Ungewickell and Branton, 1981). Tightly bound to the triskelia are three polypeptides of 30kD, usually referred to as the clathrin light chains (Ungewickell, 1983) and the purified clathrin polymerises spontaneuosly in the presence of calcium to form cage-like structures that resemble the coat structures seen in intact cells (Keen et al., 1979). In the absence of calcium, the assembly of clathrin triskelia into cages is dependent upon the presence of the 50kD and 100kD polypeptides, which are thus referred to as the assembly polypeptides (Pearse and Robinson, 1984). The assembly polypeptides can be subdivided by hydroxyapatite adsorption chromatography into HA-I and HA-II (Pearse and Robinson, 1984). Both groups contain several 100kD polypeptides and probably two smaller componets and immunofluorescence microscopy reveals that the HA-I complex is specifically associated with clathrin-coated membranes in the Golgi complex, whereas the HA-II complex appears to be restricted to coated pits on the plasma membrane (Robinson, 1987; Ahle et al.,1988).

The interaction between receptors and the coat proteins can be studied by the use of cell-free systems which mimic the intact cell. Such a system has been used to demonstrate a direct association between polypeptides of the HA-II group and a receptor protein (Pearse, 1985), the mannose-6-phosphate receptor being incorporated into reassembled cages with approximately one receptor molecule per molecule of 100 kD protein. Barbara Pearse has recently developed an assay system which is more amenable to study the interaction between receptors and assembly poypeptides (unpublished observations). Receptors are immobilised on solid supports and challenged with assembly polypeptides. The HA-II group of polypeptides, but not HA-I, bind to the cytoplasmic tail of the LDL receptor and this binding is competed for by the cytoplasmic regions of other receptors which enter plasma membrane coated pits. The mannose-6-phosphate receptor, which is found in coated pits at both the plasma membrane and the Golgi complex, binds both HA-I and HA-II polypeptides and, since there is no competition between the diferent groups, there appear to be two separate sites for their binding. Further work will more closely define this interaction.

If, as seems likely, receptors are positively included into coated pits then no mechanism, other than steric exclusion, is necessary to explain how non-internalised membrane proteins remain at the cell surface. A similar mechanism for inclusion and exclusion is seen in the budding of certain animal enveloped viruses from infected cells. The budding virus is able to discriminate between viral proteins and host proteins and this ability to sort the proteins appears to involve an interaction between the viral membrane proteins and the proteins in the viral core.

The best information on how the coated pits and vesicles are formed in the intact cell comes from deep-etch rotary shadowing studies (Heuser and Evans, 1980). The pit is believed to grow by the addition of new clathrin triskelia at its edge. These new triskelia are originally arranged as hexagons but, as the pit grows and invaginates, some of the hexagons are converted to pentagons such that the final coat in the vesicle has both hexagons and pentagons. This conversion from hexagons to pentagons seems to proceed through a seven-sided intermediate and the whole process is probably controlled by a specific enzyme. No candidates for such an enzyme have yet been identified. Another enzyme, again as yet not identified, is probably responsible for the final pinching off of the pit to form a vesicle. A system in which coated pit formation is achieved in semi-intact cells has recently been developed to further investigate this process (Graham Warren, unpublished results).

Uncoating of coated vesicles

Once the pit has sealed to form a coated vesicle the coat is rapidly removed. While uncoating is believed to be a prerequisite for fusion of the vesicle with later compartments along the endocytic pathway we have only a few insights into how it is achieved. Again, the majority of available information has been achieved using cell-free systems. Exposure of coated vesicles to low concentrations of urea (Ungewickell, 1983), alkaline pH (Woodward and Roth, 1978) or high concentrations of protonated amines (Keen et al. 1979) causes disassembly and release of coat polypeptides. It is however difficult to envisage how such changes in the cytoplasm could be limited to dissassembling coated vesicles without causing the dissassembly of coated pits still forming at the plasma membrane. An alternative possibility is that coat dissassembly is triggered by acidification of the vesicle. The vesicle membrane contains a proton-ATPase which causes an acidification of the interior and this lowering of pH could be transmitted, through transmembrane proteins, to the cytoplasmic side of the vesicle and induce uncoating. Since the pits are not sealed units they will be unable to induce the pH-dependent uncoating.

As an alternative to physical methods, uncoating may be enzymically controlled. An ATP-dependent enzyme which can remove triskelia from isolated coated vesicles has been purified and characterised (Braell et al., 1984; Schlossman et al., 1984) but the specificity of this enzyme is not understood.

Regardless of how the vesicles uncoat it is likely that the subunits, possibly as triskelions, would be recycled back to the plasma membrane to become incorporated into new pits.

Fusion with endosomes

Once uncoated, or at least partially uncoated, the vesicle is able to fuse with the next compartment on the pathway. This fusion occurs soon after internalisation and appears to occur while the vesicles are still at the periphery of the cell. The recipient compartment is generally reffered to as the early or peripheral endosome. Fusion between the two compartments is specific and it is reasonable to assume that this specificity is achieved through the involvement of specific proteins as illustrated in Figure 5.

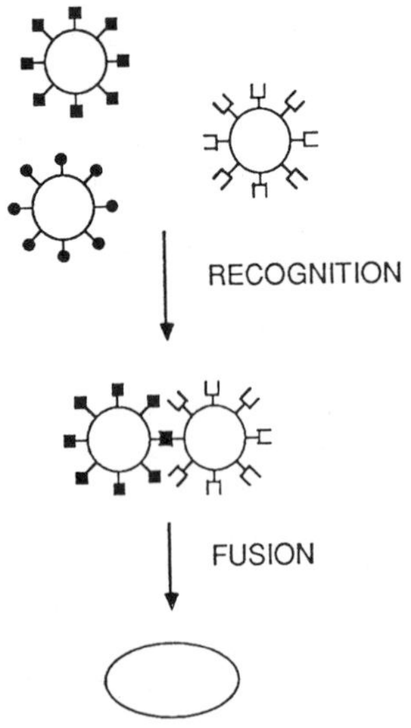

Figure 5

Schematic illustration of how the specificity of membrane fusion could be achieved.

On the right hand side of Figure 5 is the target compartment and on the left is a population of vesicles which in the cell would be those being used to transport material from one compartment to the next. The recognition is possibly achieved through the presence of receptor molecules on the cytoplasmic side of the vesicles and only vesicles displaying the correct molecules will bind to the target membrane. After recognition has occurred the two membranes fuse to a single entity. This fusion is likely to be mediated by proteins which are converted to potent fusogens following the recognition reaction.

Many groups, including ourselves, are currently attempting to characterise this fusion event at the molecular level and I would like to expand briefly on one of the methods used in my laboratory. Historically we have concentrated on using an *in vitro* approach in which the fusion event is faithfully recreated in a cell-free environment. It is only be removing the plasma membrane that one can gain direct access to the fusion event and apply classical biochemical techniques to the problem. The basic principle for such a system is shown in Figure 6.

Fusion is studied *in vitro* by an indirect assay. Conditions are arranged such that one component is placed in vesicle A and another component in vesicle B. When the two vesicles fuse, and only then, will the two components gain access to each other and interact to form a product. Since the formation of product is a consequence of fusion it can be used as a measure of the fusion. Some of the interactions one could imagine using are given in Figure 6. Several cell-free systems now exist for studying the fusion of uncoated vesicles with early endosomes (Braell, 1987; Davey et al., 1985; Davey, 1987; Diaz et al., 1987; Gruenberg and Howell, 1986; Woodman and Warren, 1988) and all have revealed the same characteristics. Fusion requires the hydrolysis of ATP and is dependent upon the presence of both soluble proteins and proteins on the cytoplasmic face of the membrane. The identity and function of these proteins are currently being investigated.

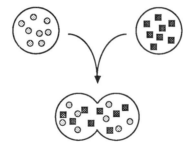

Possible interactions:
1. Enzyme - Substrate
2. Formation of complex,
   Antibody - Antigen
   Avidin - Biotin

Figure 6

Basic principle of cell-free assays to monitor membrane fusion.

Sorting in the endosome

Following fusion, the receptor-ligand complex is now located on the lumenal side of the endosomal membrane. Whilst lacking hydrolases, endosomes have an acidic interior of about pH 5.5 and this acidity is crucial for its role in the cell which appears to be the separation of the receptor from the ligand. In many cases the binding of the ligand to its receptor is sensitive to changes in pH and the ligand is released into the lumen of the endosome while the receptor

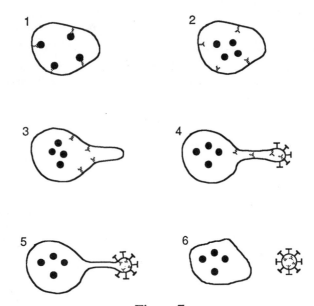

Figure 7
Schematic illustration of the uncoupling of receptor and ligand in the endosome.

remains attached to the membrane. The most evocative model which has been suggested for the subsequent separation of the receptor and the ligand has arisen from the immunocytochemical data of Hans Geuze and his colleagues (Geuze et al., 1983). This model is schematically illustrated in Figure 7.

The ligand is released from the receptor as a consequence of the acidic environment in the endosome and while the ligand remains in the main lumen of the endosome the receptor diffuses into a tubular extension of the membrane. Clustering of the receptors into the extension appears to involve a coated region of the membrane which is similar in function, but not in structure, to the coated pits on the plasma membrane. The coated region buds off to form a coated vesicle containing the receptor and this vesicle is then transported to the plasma membrane. Fusion of the two membranes will re-expose the receptors at the cell surface ready for binding another ligand molecule and further rounds of internalisation.

Transport to lysosomes

Once all of the receptors have been removed the endosome is transported to the perinuclear region of the cell. This transport appears to involve microtubules since it is inhibited by microtubule-disrupting drugs (Matteoni and Kreis, 1987). The exact identity of the perinuclear vesicles is unclear but they appear to be the intermediate between the early endosomes and lysosomes and have been referred to as late endosomes or multivesicular bodies. Cell-free systems suggest that while these vesicles are no longer capable of fusing with early endosomal vesicles (Jean Gruenberg, unpublished results) they are capable of transferring their contents to lysosomes (Barbara Mullock, unpublished results). The mechanism of transport from the late endosome to lysosomes is not yet known but two models have been suggested (Helenius et al., 1983). The first is the maturation model and proposes that the late endosome matures into a lysosome by fusion with a primary lysosome. The alternative is the vesicle-shuttle model and proposes that the late endosome is a permanent organelle, the contents being delivered to pre-existing lysosomes by transport vesicles. The recent development of a cell-free system for studying the transfer from endosome to lysosome should now allow identification and characterisation of the mechanism involved.

Once in the lysosome the ligand is degraded into its constituent parts which are then used by the cell.

FUTURE PROSPECTS

I believe we have now entered an exciting period of research into the endocytic pathway. Cell-free systems have been developed to study most elements of the pathway and these should reveal the identity and function of the many controlling proteins and factors. This biochemical approach to studying endocytosis is now being supplemented by genetic methods. Since endocytosis appears to play an important role in the life of most cells, the isolation of mutants defective in the pathway has been difficult and unrewarding. Recently however it has been shown that yeast cells possess a pathway that is very similar to if not identical with the endocytic pathway in higher eukaryotes (Riezman, 1985; Riezman et al., 1986; Davey, unpublished observations). The isolation of mutant cells defective in specific steps on the different pathways now provides us with a genetical method for identifying specific proteins involved in the transport events.

REFERENCES

Ahle, S. and Ungewickell, E. (1986) Purification and properties of a new clathrin assembly protein. EMBO J. 5, 3143-3149.

Ahle, S., Mann, A., Eichelsbacher, U. and Ungewickell, E. (1988) Structural relationships between clathrin assembly proteins from the Golgi and the plasma membrane. EMBO J. 7, 919-929.

Anderson, R.G.W. and Kaplan, J. (1983) Receptor-mediated endocytosis. Modern Cell Biol. 1, 1-52.

Braell, W.A. (1987) Fusion between endocytic vesicles in a cell-free system. Proc. Natl. Acad. Sci. USA 84, 1137-1141.

Braell, W.A., Schlossman, D.M., Schmid, S.L. and Rothman, J.E. (1984) Dissociation of clathrin coats coupled to the hydrolysis of ATP : Role of a uncoating ATPase. J. Cell Biol. 99, 734-741.

Brown, M.S., Anderson, R.G.W. and Goldstein, J.L. (1983) Recycling receptors: The round-trip itinerary of migrant membrane proteins. Cell 32, 663-667.

Davey, J. (1987) A cell-free analysis of the endocytic pathway. Bioscience Reports 7, 299-306.

Davey, J., Hurtley, S.M. and Warren, G. (1985) Reconstitution of an endocytic fusion event in a cell-free system. Cell 43, 643-652.

Davis, C.G., Lehrman, M.A., Russell, D.W., Anderson, R.G.W., Brown, M.S. and Goldstein, J.L. (1986) The JD mutation in familial hypercholesterolemia: Amino acid substitution in cytoplasmic domain impedes internalization of LDL receptors. Cell 45, 15-24.

Davis, C.G., van Driel, I.R., Russell, D.W., Brown, M.S. and Goldstein, J.L. (1987) The low density lipoprotein receptor. Identification of amino acids in cytoplasmic domain required for rapid endocytosis. J. Biol. Chem. 262, 4075-4082.

Diaz, R., Mayorga, L. and Stahl, P. (1987) In vitro fusion of endosomes following receptor-mediated endocytosis. J. Biol. Chem. 263, 6093-6100.

Geuze, H.J., Slot, J.M., Straus, J.A.M., Lodish, H.F. and Schwartz, A.L. (1983) Intracellular site of asialoglycoprotein receptor-ligand uncoupling: Double-label immunoelectron microscopy during receptor-mediated endocytosis. Cell 32, 277-287.

Goldstein, J.L., Brown, M.S., Anderson, R.G.W., Russell, D.W. and Schneider, W.J. (1985) Receptor-mediated endocytosis: Concepts emerging from the LDL receptor system. Ann. Rev. Cell Biol. 1, 1-39.

Gruenberg, J.E. and Howell, K.E. (1986) Reconstitution of vesicle fusions occurring in endocytosis with a cell-free system. EMBO J. 5, 3091-3101.

Helenius, A., Kartenbeck, J., Simons, K. and Fries, E. (1980) On the entry of Semliki Forest virus into BHK-21 cells. J. Cell Biol. 84, 404-420.

Helenius, A., Mellman, I., Wall, D. and Hubbard, A. (1983) Endosomes. Trends Biochem. Sci. 8, 245-250.

Heuser, J. and Evans, L. (1980) Three-dimensional visualisation of coated vesicle formation in fibroblasts. J. Cell Biol. 84, 560-583.

Iacopetta, B.J., Rothenberger, S. and Kühn, L.C. (1988) A role for the cytoplasmic domain in transferrin receptor sorting and coated pit formation during endocytosis. Cell 54, 485-489.

Keen, J.H., Willingham, M.C. and Pastan, I.H. (1979) Clathrin coated vesicles: Isolation, dissociation and factor dependent reassociation of clathrin baskets. Cell 16, 303-312.

Lazarovits, J. and Roth, M. (1988) A single amino acid change in the cytoplasmic domain allows the influenza virus hemagglutinin to be endocytosed through coated pits. Cell 53, 743-752.

Lehrman, M.A., Goldstein, J.L., Brown, M.S., Russel, D.W. and Schneider, W.J. (1985) Internalization-defective LDL receptors produced by genes with nonsense and frameshift mutations that truncate the cytoplasmic domain. Cell 41, 735-743.

Matteoni, R. and Kreis, T.E. (1987) Translocation and clustering of endosomes and lysosomes depends on microtubules. J. Cell Biol. 105, 1253-1265.

May, W.S., Sahyoun, N., Jacobs, S., Wolf, M. and Cuatrecasas, P. (1985) Mechanism of phorbol diester induced regulation of surface transferrin receptor involves the action of activated protein kinase C and an intact cytoskeleton. J. Biol. Chem. 260, 9419-9429.

Mostov, K.E., de Bruyn Kops, A. and Deicher, D.L. (1986) Deletion of the cytoplasmic domain of the polymeric immunoglobulin receptor prevents basolateral localization and endocytosis. Cell 47, 359-364.

Moya, M., Dautry-Varsat, A., Goud, B., Louzard, D. and Boquet, P. (1985) Inhibition of coated pit formation in Hep2 cells blocks the cytotoxicity of diptheria toxin but not that of ricin toxin. J. Cell Biol. 101, 548-559.

Pearse, B.M.F. (1985) Assembly of the mannose-6-phosphate receptor into reconstituted clathrin coats. EMBO J. 4, 2457-2460.

Pearse, B.M.F. and Robinson, M.S. (1984) Purification and properties of 100kd proteins from coated vesicles and their reconstitution with clathrin. EMBO J. 3, 1951-1957.

Pelham, H.R.B. (1988) Evidence that luminal ER proteins are sorted from secreted proteins in a post-ER compartment. EMBO J. 7, 913-918.

Prywes, R., Livneh, E., Ullrich, A. and Schlessinger, J. (1986) Mutations in the cytoplasmic domain of EGF receptor affect EGF binding and receptor internalization. EMBO J. 5, 2179-2190.

Riezman, H. (1985) Endocytosis in yeast: several of the yeast secretory mutants are defective in endocytosis. Cell 40, 1001-1009.

Riezman, H., Chvatchko, Y. and Dulic, V. (1986) Endocytosis in yeast. Trends Biochem. Sci. 11, 325-328.

Robinson, M.S. (1987) 100-kD coated vesicle proteins: Molecular heterogeneity and intracellular distribution studied with monoclonal antibodies. J. Cell Biol. 104, 887-895.

Rothenberger, S., Iacopetta, B.J. and Kühn, L.C. (1987) Endocytosis of the transferrin receptor requires the cytoplasmic domain but not its phosphorylation site. Cell 49, 423-431.

Sandvig, K., Olsnes, S., Petersen, O.W. and van Deurs, B. (1987) Acidification of the cytosol inhibits endocytosis from coated pits. J. Cell Biol. 105, 679-689.

Schlossman, D.M., Schmid, S.L., Braell, W.A. and Rothman, J.E. (1984) An enzyme that removes clathrin coats: Purification of an uncoating ATPase. J. Cell Biol. 99, 723-733.

Silverstein, S.C., Steinman, R.M. and Cohn, Z.A. (1977) Endocytosis. Ann. Rev. Biochem. 46, 669-722.

Steinman, R.M., Mellman, I.S., Muller, W.A. and Cohn, Z.A. (1983) Endocytosis and the recycling of plasma membrane. J. Cell Biol. 96, 1-27.

Ungewickell, E. (1983) Biochemical and immunological studies on clathrin light chains and their binding sites on clathrin triskelions. EMBO J. 2, 1401-1408.

Ungewickell, E. and Branton, D. (1981) Assembly units of clathrin coats. Nature 289, 420-422.

Wileman, T., Harding, C. and Stahl, P. (1985) Receptor-mediated endocytosis. Biochem. J. 232, 1-14.

Woodman, P.G. and Warren, G. (1988) Fusion between vesicles from the pathway of receptor-mediated endocytosis in a cell-free system. Eur. J. Biochem. 173, 101-108.

Woodward, M.P. and Roth, T.F. (1978) Coated vesicles: Characterization, selective dissassociation and reassembly. Proc. Natl. Acad. Sci. USA 75, 4394-4398.

TRANSPORT OF LIPIDS AND PROTEINS

DURING MEMBRANE FLOW IN EUKARYOTIC CELLS

Dick Hoekstra, Sinikka Eskelinen and Jan Willem Kok

Laboratory of Physiological Chemistry
University of Groningen, Bloemsingel 10
9712 KZ Groningen, The Netherlands

INTRODUCTION

Phospholipid molecules constitute the essential building blocks of biological membranes. They are arranged according to a bilayer structure while membrane proteins are embedded in or attached to this double layer of lipid molecules. Besides phospholipids, eukaryotic cell and organelle membranes may contain two other lipid classes: cholesterol and glycosphingolipids. As may be anticipated, proteins and the various lipid classes are not uniformly distributed among membranes, given the distinct functions of the intracellular organelles. Yet, *de novo* biosynthesis of both lipids and proteins occurs primarily at the endoplasmic reticulum. This implies that specific mechanisms must exist that determine the fate and distribution of these compounds among membranes. Obviously, this sets particular requirements to intracellular sorting and transport phenomena.

Apart from the question as to how newly synthesized lipids and proteins may reach their various destinations, another intriguing question is how each (intra)-cellular membrane maintains its characteristic distinction in molecular composition, given the fact that during the biological life of a cell there exists a continuous flow of membranes. Thus, membranes shuttle from the cell surface to interior compartments and back, between organelles and, especially in polarized cells, from one cell surface to another. It is apparent that also during these events specific sorting and other regulatory mechanisms must be operative to safeguard and maintain the identity of individual organelles and their membranes.

Much of our current knowledge on membrane trafficking has been derived from studies involving intracellular routing of proteins. In particular, the synthesis and assembly of viral proteins at the endoplasmic reticulum and their transfer to the cell surface via the Golgi complex is now fairly well understood. Also, studies of internalization and recycling of cell surface receptors during endocytosis have greatly contributed towards obtaining insight into the complex pathways of intracellular membrane traffic. Although knowledge of protein trafficking is rapidly advancing, very little attention has been paid thusfar to the (accompanying) intracellular movement of lipids. These and various other aspects concerning membrane traffic will be the focus of this paper. Before describing the pathways of membrane flow, we first discuss some general features of lipid composition and organization in mammalian cell membranes.

LIPID DISTRIBUTION IN CELLULAR MEMBRANES

In mammalian cells, the enzymes responsible for the biosynthesis of many phospholipids are associated with the endoplasmic reticulum (ER). Thus, phosphatidylcholine (PC), phosphatidylethanolamine (PE), phosphatidylinositol (PI) and phosphatidylserine (PS) are primarily synthesized at the ER (Vance and Vance, 1985; Hawthorne and Ansel, 1985; Kennedy, 1986). However, cardiolipin (CL) and phosphatidylglycerol (PG) are synthesized in the mitochondria (Kennedy, 1986). A particular phospholipid can also serve as a substrate in a so-called base-exchange reaction by which, for example, PS can be converted to PE or PC to sphingomyelin (SM) via ceramide. Sometimes these conversions can take place at the site of synthesis, i.e. the endoplasmic reticulum (PS to PE), in the plasma membrane (PC to SM) or in the mitochondria (PS to PE). Cholesterol is generally thought to be synthesized at the ER (Chesterton, 1968; Reinhardt et al., 1987). It has also been suggested, however, that the complete biosynthesis may not take place at the ER alone but rather, might be coupled to movement through a membrane pathway that eventually delivers cholesterol to the plasma membrane (Lange and Muraski, 1987, 1988). In a certain sense, the synthesis of glycosphingolipids would resemble such a biosynthetic pathway (Wiegandt, 1985). The backbone structure of these lipids, ceramide, is presumably synthesized at the ER. Through the action of glycosyltransferases, located in the Golgi complex, synthesis of the lipids is completed. Gangliosides, i.e. sialic acid-containing glycosphingolipids, can be further modified at the cell surface and converted into neutral glycosphingolipids (i.e. devoid of sialic acid) by plasma membrane-associated sialidase (Usuki et al., 1988).

Glycosphingolipids, the biological roles of which gradually emerge (Hakomori, 1981, 1983; Hoekstra and Düzgüneş, 1989, and references therein), are thought to be primarily located in the external leaflet of the plasma membrane. After intracellular synthesis, the gangliosides reach their destination at the cell surface within some 30 min (Miller-Podraza and Fishman, 1982). For the vectorial movement of newly synthesized cholesterol to the plasma membrane, half-times varying from 10 min to 1 h have been reported (DeGrella and Simoni, 1982; Lange and Matthies, 1984).

As already indicated by the preferential localization of glycosphingolipids, lipids are not uniformly distributed among membranes. This is also true for cholesterol and the phospholipids. On a molar basis, the amount of cholesterol relative to that of phospholipids is high (0.8) for plasma membranes, but substantially lower for the organelle membranes (0.1-0.24) (Dawidowicz, 1987). Parameters that determine this remarkable distribution have yet to be determined. With respect to the phospholipids, PC, PE, PI and SM are present in all intracellular membranes as well as in the plasma membrane. Quantitatively, differences may exist in the distribution of the lipids in the various membranes of the organelles and the plasma membrane. Both CL and PG are almost exclusively found in mitochondria where, as noted above, their synthesis takes place. Taken together, it appears, however, that as far as the heterogeneity of membrane lipid is concerned, it is not as much as the *absolute* difference in lipid species that determines the unique characteristics of (intra)cellular membranes. Rather, it is the *ratio* between the various lipid species, that primarily distinguishes the various (intra-)cellular membranes from each other with respect to lipid composition. In this regard, differences in protein composition are much more pronounced.

Within a given membrane, the lipids are furthermore asymmetrically distributed, i.e. the composition of the inner leaflet is different from that of the outer leaflet (Op den Kamp, 1979; Rothman and Lenard, 1981; Van Meer et al., 1987). Thus, PC and SM are predominantly located in the outer leaflet, while the aminophospholipids PE and PS are found in the inner leaflet (figure 1). Extensive studies of erythrocyte membranes led several investigators to suggest that the main cytoskeleton component, spectrin, ensured the presence of PE and PS in the inner leaflet (Haest et al., 1981; Williamson et al., 1982). More recently, evidence has accumulated that supports the involvement of specific proteins in maintaining lipid asymmetry, the so-called 'flippases' (Seigneuret and Deveaux, 1984; Connor and Schroit, 1987; Tilley et al., 1986; Martin and Pagano, 1987; Calvez et al., 1988). Thus, using spin-labeled, fluorescently tagged or photoactivatable PS or PE lipid analogs, it has been demonstrated that their insertion into the outer leaflet of the plasma membrane results in a rapid translocation to the inner leaflet. This movement appears to be inhibited by SH-reactive agents while the flop itself is ATP-dependent. A 31 kDa

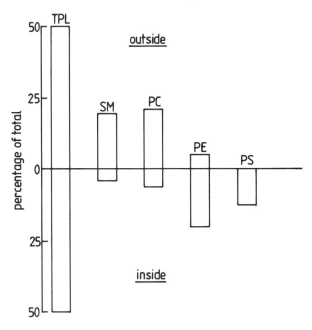

Figure 1. Schematic representation of the asymmetric distribution of phospholipids in human erythrocyte membranes.
TPL, total phospholipid (Data from Op den Kamp, 1979).

protein has been proposed to maintain the asymmetry of PS in erythrocyte membranes (Schroit et al., 1987; Connor and Schroit, 1988).

Once inserted into a membrane structure, phospholipid molecules show very little tendency to move spontaneously as individual monomers through the aqueous phase (Dawidowicz, 1987). How would these molecules then reach their destinations after synthesis? At least two other mechanisms are available. One possibility involves the transfer of lipids, mediated by specific exchange proteins, the so-called PLEP's (Wirtz, 1982; Dawidowicz, 1987a). The other possibility is a transport mechanism involving participation of membrane vesicles. Although a variety of PLEP's have been isolated, the evidence that would support a major role of this mechanism in intracellular bulk movement of lipids, is still lacking. Yet, as will be described below, it is rather premature to entirely exclude their involvement in lipid traffic as some experimental results can best be reconciled with lipid movement, mediated by specific proteins. On the other hand, the intracellular traffic of proteins, including their transfer from the site of synthesis (ER) to their sites of destination is largely mediated by *membrane* vesicles, implying that lipids are an inherent part of this transportation system. It is therefore that investigations aimed at clarifying lipid traffic in animal cells may benefit from studies focusing on protein traffic. The latter, in particular, have revealed that there exists an extensive flow of membranes (Farquhar, 1983) during processes such as endo- and exocytosis.

PATHWAYS OF INTRACELLULAR MEMBRANE TRAFFIC

Roughly, two major pathways of intracellular membrane flow in animal cells can be distinguished (figure 2). First, substantial amounts of membrane flow from the plasma membrane to intracellular organelles during endocytosis. Cultured cells may internalize approximately half their surface area per hour (Steinman et al., 1983). Second, in the opposite direction there is a flow of membrane vesicles that transfers newly synthesized proteins from the ER toward the cell surface. As becomes apparent from figure 2, there exist a number of intermediate compartments in these pathways.

These intermediate structures - the endosomes during entry and the Golgi complex during exit - play crucial roles in that they are capable of retrieving at least those proteins which are not supposed to be degraded in the lysosomes or those which are destined for membranes other than the plasma membrane per se. In other words,

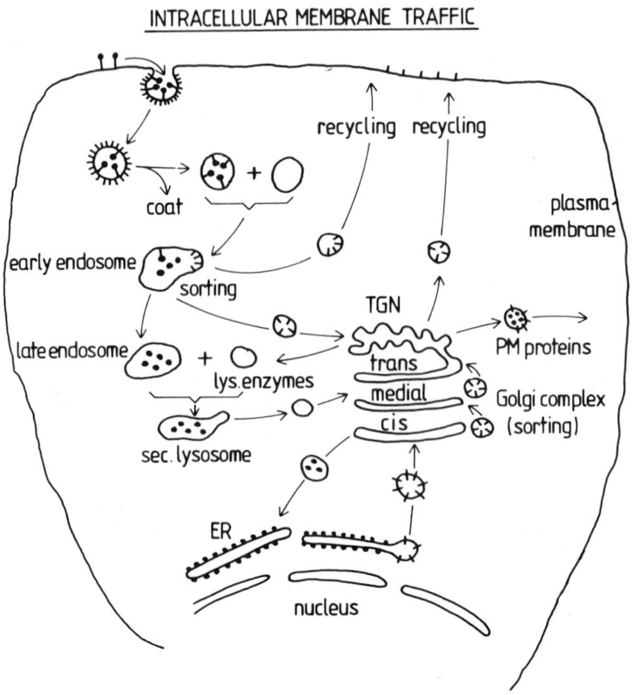

INTRACELLULAR MEMBRANE TRAFFIC

Figure 2. Membrane flow in animal cells. Substantial areas of the plasma membrane are internalized by endocytosis, particularly during receptor-mediated endocytosis (left part). In this process, extracellular macromolecules (●) bind to specific receptors (l). The ligand-receptor complexes (↑) migrate to coated pit regions which invaginate, resulting in formation of coated vesicles. Note that some receptors (f.e. LDL receptors) permanently reside in coated regions. Within ca. 30 secs, the coated vesicles loose their (clathrin) coat, giving rise to formation of uncoated vesicles. These vesicles may fuse with each other and, presumably, with other vesicles, eventually resulting in formation of early endosomes. In this compartment, a dissociation of ligand and receptor occurs (except for transferrin, see text), triggered by a gradual decrease of the pH. Subsequently, or perhaps concomitantly, the receptors are sorted into specific membrane regions of the early endosome. Small, receptor-enriched vesicles bud, which return to the cell surface: "recycling". The early endosomes develop into late endosomes, the membrane containers that eventually merge with primary lysosomes, causing degradation of the ligands in the secondary lysosomes. Some receptors (f.e. the transferrin receptor) may partly recycle via a membrane traffic pathway that leads from endosome (early?) to *trans* Golgi network (TGN) and subsequently to the plasma membrane. The TGN seems to act as a major sorting site in the cell. Proteins, synthesized at the endoplasmic reticulum reach the TGN after being processed through *cis*, *medial* and *trans* Golgi compartments, where a series of glycosylation reactions take

both pathways act by default and unless specifically 'tagged', the components 'caught' in the membrane structures participating in both routes, will arrive at their endpoints, i.e. the lysosomes (inbound) or the plasma membrane (outbound). In spite of the extensive trafficking of membranes, it is rather surprising that the distinct intracellular membranes, such as those of, for example, the lysosomes maintain their characteristic lipid and protein composition. Hence, specific mechanisms must be operative to prevent intermixing. It appears that endosomes play an important role in this respect.

Receptor-mediated endocytosis

To a considerable extent, studies of the cell-biological and molecular events during receptor-mediated endocytosis have benefitted from using specific 'markers' such as low density lipoprotein (LDL), transferrin (Tf) and epidermal growth factor (EGF). LDL acts as a carrier for serum cholesterol (Goldstein et al., 1985). The protein-lipid complex is internalized via endocytosis and cholesterol is subsequently released in the lysosomal compartment. Eventually, the cholesterol becomes accessible for membrane biogenesis, in addition to newly synthesized cholesterol.
The primary function of Tf, a metal-binding glycoprotein, is to act as a transporter for iron from the extra- to the intracellular milieu (Baker et al., 1987). The iron is released at an earlier stage in the endocytic pathway than cholesterol, i.e. at the endosomal level. EGF influences cell growth by causing an enhanced DNA synthesis and cell division (Carpenter and Cohen, 1979; Cohen, 1983; Schlessinger, 1986). Also this protein is internalized via the endocytic pathway and finally degraded in the lysosomes. All three proteins enter the cell by binding to a specific cell surface receptor, which is followed by internalization via receptor-mediated endocytosis. In general, receptor-mediated endocytosis involves the process in which extracellular ligands bind to specific cell surface receptors, followed by transport into the cell in clathrin-coated vesicles. The coated vesicles (Linden and Roth, 1983) originate from specialized regions of the cell surface: the coated pits, recognized as electron-dense areas by electronmicroscopy. In this area, the receptors are concentrated while other proteins appear to be excluded. At any given time, ca. 70% of the LDL receptors (Anderson et al., 1977), ca. 24-45% of the Tf receptors (Watts, 1985) and ca. 35% of the EGF receptors (Gordon et al., 1978) can be found in the coated pits. The signal that 'captures' the receptors in these regions is thought to be located in the cytoplasmic domains of the (transmembrane) receptors. Thus, single amino acid changes in these domains by site-specific mutagenesis suffice to severely inhibit internalization (Davis et al., 1986). Moreover, internalization occurs, irrespective of whether the receptor is occupied by its ligand (Watts, 1985), i.e. structural changes in the receptor, induced upon ligand binding, are not essential for triggering receptor-mediated endocytosis. The unique features of the cytoplasmic domain of a receptor

---

(**Figure 2. cont'd**) place. Transfer between the different compartments is mediated by membrane vesicles, that bud from one compartment and subsequently fuse with the next. Some ER specific proteins appear to be transferred to the *cis* compartment of Golgi prior to resuming permanent residence in the ER, a process that involves vesicular "backtransport" from *cis* Golgi to ER. From the TGN, membrane vesicles depart, containing proteins destined for secretion and plasma membrane (PM). Furthermore, vesicles pinch off that contain lysosomal hydrolases. These vesicles ("primary lysosomes") presumably fuse with late endosomes, or a more matured stage thereof, to form the "secondary" lysosomes, or a precursor thereof. After delivery, the receptors (Man 6-phosphate receptors) that sort the lysosomal enzymes in the TGN, return to the *trans* Golgi compartment. For further details and references, see text.

has been further emphasized by constructing hybrid proteins, consisting of the cytoplasmic domain of a receptor and the external domain of a non-related protein. It turns out that such proteins follow the internalization pattern observed for the parent protein that donated the cytoplasmic domain (Roth et al., 1986).

The coated vesicles formed consist of an inner shell, composed of 100 kDa and 50 kDa proteins (Pearse, 1982; Zaremba and Keen, 1983) and an outer shell, which primarily contains clathrin (Pearse, 1982). It has been suggested that the inner shell of proteins, but not clathrin, may interact directly with the cytoplasmic domain of the receptors (Pearse, 1987). Once formed, coated vesicles loose their coat within seconds, mediated by a cytosolic uncoating enzyme that couples ATP hydrolysis to removal of clathrin (Rothman and Schmid, 1986). The removal of clathrin is essential because in subsequent steps, several membrane fusion processes take place in which the presence of clathrin may act as a steric barrier (Davey et al., 1985). It is not clear to what extent uncoated vesicles subsequently fuse with each other and/or with other (intracellular) membrane vesicles. In spite of this uncertainty, the final result of vesicle fusion is the formation of a heterogeneous class of organelles: the endosomes (Tycko and Maxfield, 1982; Helenius et al., 1983; Geuze et al., 1983). At least two functionally different but kinetically related subpopulations of endosomes can be distinguished and it has been suggested, based on differences in protein composition, that these populations have distinct patterns of biogenesis (Schmid et al., 1988). Internalized ligand-receptor complexes rapidly (1-2 min) reach a compartment known as 'early endosomes'. It is thought that in this compartment the ligands dissociate from their receptors, which is triggered by a mild acidic pH generated after biogenesis of the endosomes. Subsequently, the receptors are 'sorted' and from the tubular extensions of the endosomes, (Geuze et al., 1983, 1987), small, receptor-enriched vesicles are pinched off, that return to the cell surface. By this *recycling* mechanism the receptors become available again for the next round of internalization. In the case of the Tf-Tf receptor complex, only iron, bound to apoTf, is released in the early endosomes. The apoTf remains associated with its receptor and recycles back to the cell surface. At the cell surface the apoTf dissociates from the Tf receptor and is released in the extracellular environment. Both Tf and the Tf receptor have also been detected in the Golgi complex (Hedman et al., 1987; Fishman and Fine, 1987), implying that there exists a membrane traffic route from the plasma membrane to the Golgi complex, presumably involving the early endosomes as an intermediate station (figure 3). That at least part of the Tf receptor complex can recycle via this route was furthermore demonstrated by a resialylation of Tf and the Tf receptor after endocytosis of their desialylated precursors (Regoeczi et al., 1982; Snider and Rogers, 1985). The enzyme (sialyltransferase), carrying out this reaction, is localized in the *trans* Golgi region (Roth et al., 1985). Indeed, more recently, it was shown that horse radish peroxidase-labeled Tf colocalizes after internalization with secretory albumin in vesicles located at the *trans* Golgi region (Stoorvogel et al., 1988). It was proposed that a merging of the 'recycling' endocytic pathway and the secretory pathway takes place in the *trans* Golgi network (TGN, figure 2). Although it is clear, therefore, that recycling of the Tf-Tf receptor complex can occur via at least two different routes (Hopkins, 1983), the biological rationale for recycling divergence remains to be resolved.

While the early endosomes are involved in receptor recycling, the *'late endosomes'* appear to be the compartment through which ligands and other biomaterials pass, destined for degradation in the lysosomes. As for the early endosomes, the late endosomes also contain an ATP-dependent proton pump, but late endosomes are more acidic than early endosomes. It is likely that the cytoplasmic delivery of nucleocapsids of certain enveloped viruses, such as Semliki Forest virus, vesicular stomatitis virus and influenza virus, is accomplished via this compartment (cf. Schmid et al., 1988). These viruses enter cells via receptor-mediated endocytosis and fuse with the endosomal membrane 'from within', which results in the expulsion of the viral genome into the cytoplasm (White et al., 1983; Hoekstra et al., 1988; Hoekstra, 1988). The fusion event is triggered by a mild acidic pH. The final step in the endocytic pathway involves the merging of, presumably, late endosomes with primary lysosomes, eventually resulting in degradation of the late endosomal contents.

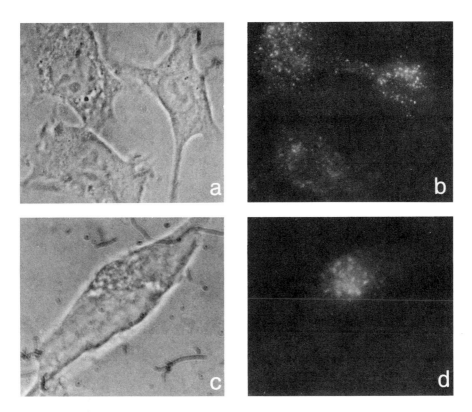

Figure 3. **Intracellular traffic of fluorescently-tagged transferrin upon receptor-mediated endocytosis in baby hamster kidney (BHK) cells.** Transferrin, labeled with fluorescein-isothiocyanate, was incubated with the cells at 2°C. After removal of unbound Tf, the cells were warmed to 37°C. 4 min after triggering endocytosis, fluorescence is seen in small vesicles, that are distributed throughout the cell (b; a, phase contrast micrograph). A tendency of the vesicles to concentrate at Golgi regions becomes apparent. After 10 min, only the Golgi area is labeled (d; c, phase contrast micrograph).

The primary lysosomes originate from the TGN, the intracellular site where sorting and packaging of proteins occurs, destined for various intracellular sites and the plasma membrane. The TGN appears to be an intracellular compartment of major importance as it is thought to regulate the delivery of newly synthesized proteins to their appropriate sites.

The biosynthetic pathway

After their synthesis, proteins destined for Golgi complex, lysosomes and plasma membrane as well as those which are secreted into the extracellular environment, leave the ER, packaged in membrane vesicles. As argued by Rothman and colleagues (Rothman, 1987; Pfeffer and Rothman, 1987), the traffic routes of proteins can be distinguished in a signal-mediated route and a signal-independent or default pathway. Thus, unless a protein contains a specific signal after translocation in the ER, it will move with the bulk of membrane and fluid, derived from the ER lumen, to the plasma membrane where it is either inserted (in case of a membrane protein) or excreted when it concerns an aqueous soluble protein. Indeed, it has been claimed (Rindler and Traber, 1988) that secretory proteins need not possess specific sorting signals for transport along the basolateral secretory pathway in intestinal cells, i.e. these proteins are transported via a default pathway.

Membrane vesicles, pinched off from the ER, first reach the *cis* Golgi compartment (figure 2) before the proteins sequentially pass through the *medial* and *trans* cisternae. Transfer between the cisternae is again accomplished by membrane vesicles that pinch off and fuse with the compartment next in line. This has been elegantly demonstrated by studying the traffic of a viral protein, G, of vesicular stomatitis virus (VSV) (Balch et al., 1984). The design of the experiment involved the ability to measure the transfer of the protein between successive Golgi compartments in vitro. Each compartment has a different set of carbohydrate-modifying enzymes. The high mannose carbohydrate structure of glycoproteins arriving in the *cis* compartment, is trimmed to $Man_5GlcNAc_2$ (Man, mannose; GlcNAc, N-acetylglucosamine) by mannosidase I (Kornfeld and Kornfeld, 1985). In the *medial* compartment, GlcNAc is added by a GlcNAc transferase while in the *trans* compartment galactose residues are added to the Asn-linked oligosaccharides of the transported glycoproteins. In a mutant Chinese hamster ovary (CHO) cell line (clone 15B; Gottlieb et al., 1975; Tabas and Kornfeld, 1978), GlcNAc transferase is missing. Consequently, when a Golgi fraction is isolated from the mutant, infected with VSV, the G protein will not become labeled with radioactive GlcNAc, present in the incubation medium. However, when a Golgi fraction from uninfected wild-type cells was included in the incubation mixture, radioactive GlcNAc was added to the G protein, implying that a transfer between the 'mutant' *cis* compartment and the wild-type *medial* compartment has taken place. Hence, in the cell-free system the sequence of transfer specificity is maintained. In subsequent work evidence has accumulated, showing that the routing between the various Golgi compartments involves the budding and fusion of transport vesicles. This transport phenomenon is energy dependent and requires the presence of cytosol and proteins located on the cytoplasmic surface of the Golgi membrane.

Viral membrane proteins have contributed substantially as model for biosynthesis and transport from the ER to the plasma membrane. The rationale for the application of these proteins as a tool relies on the ability of the virus to eliminate the synthesis of proteins other than the viral proteins. In the case of VSV, the G protein is the only membrane protein that is synthesized upon infection with this virus. The distinct steps in synthesis, assembly and transport of another viral protein (HA, hemagglutinin from influenza virus) has also been studied in detail (Copeland et al., 1986, 1988). The results of this work are summarized in table 1, which includes the kinetics of the processing.

Trafficking of proteins through the Golgi complex serves, in essence, two purposes. First, in this stack, the carbohydrate structure is further modified by a series of glycosylation reactions (Kornfeld and Kornfeld, 1985; Hoekstra and Düzgüneş, 1989). Second, it is in this organelle that proteins, destined for lysosomes, plasma membrane and secretory vesicles, including the "storage" vesicles that release their contents by an extracellular stimulus, are sorted. More specifically, it is thought that the trans Golgi network represents the Golgi region were sorting and packaging takes place (Griffiths and Simons, 1986; Pfeffer and Rothman, 1987). The signal specificity that triggers the sorting of lysosomal hydrolases is the best characterized system thusfar and involves recognition of a mannose 6-phosphate (Man 6-P) residue by a Man 6-P receptor (Von Figura and Hasilik, 1986). Presumably, lysosomal hydrolases acquire a Man 6-P label in the *cis* Golgi (Pohlmann et al., 1982; Goldberg and Kornfeld, 1983). After passing through *medial* and *trans* compartments, for further oligosaccharide processing, the Man 6-P-tagged enzymes encounter the Man 6-P receptors, that presumably reside in the TGN (Duncan and Kornfeld, 1988). Conflicting reports have been published, however, on the exact location of these receptors (Brown and Farquhar, 1984; Geuze et al., 1985; Griffiths et al., 1988). There appear to be two distinct classes of Man 6-P receptors, which differ in molecular weight (215 kDa and 46 kDa, Hoflack and Kornfeld, 1985; Von Figura and Hasilik, 1986). In addition, one of them (46-kDa) requires divalent cations for optimal binding of the lysosomal enzymes. It is not known whether the different receptors, both containing multiple carbohydrate binding sites, also display distinct functions.

Table 1. **Synthesis, assembly and transport to the cell surface of a viral membrane glycoprotein**

Time	Molecular and/or biological events.
<1 min	-Synthesis of HA_0 monomers at ER ribosomes -Co-translational insertion into ER membrane, cleavage of signal peptide and coreglycosylation
~3 min	-Formation of HA_0 trimers, presumably in the ER (half time is 7-10 min).
7-15 min	-Trimming of high-mannose carbohydrate to $Man_5GlcNAc_2$ in *cis* compartment of the Golgi complex -Addition of GlcNAc in the *medial* compartment; transfer to the *trans* Golgi compartment.
20-40 min	-Appearance of HA_0 trimers at the cell surface.

HA_0 is a membrane glycoprotein, present in the envelope of influenza virus. By proteolytic cleavage HA_0 is converted into HA, the active form of the virus protein that mediates binding and fusion of influenza virus (White et al., 1983; Wiley and Skehel, 1987). The quaternary structure of the protein is a trimer. Its synthesis, assembly and trafficking was monitored by chasing virus-infected cells (CV-a, monkey kidney cells; CHO, Chinese hamster ovary cells) after pulse labeling with ^{35}S-methionine. Immunological and density gradient techniques were used to follow protein processing. For details, see Copeland et al., 1986, 1988.

At any rate, after attachment according to a lectin-like interaction (cf. Hoekstra and Düzgüneş, 1989), the enzyme-receptor complex segregates somehow from the bulk phase followed by pinching off of transport vesicles. These vesicles, which can be regarded as primary lysosomes, merge with, in all likehood, the *late* endosomes which eventually results in their conversion to secondary lysosomes. Binding of the Man 6-P tagged proteins to the receptors is pH-sensitive, i.e. at pH 5.5 or below dissociation occurs (Von Figura and Hasilik, 1986), analogous to ligand-receptor dissociation, occurring in the early endosomes during receptor-mediated endocytosis (see above). Moreover, also in this case, the Man 6-P receptors, after delivery of the ligands, recycle back to the *trans* Golgi region for another round of transport (Von Figura and Hasilik, 1986; Braulke et al., 1987; Duncan and Kornfeld, 1988). The cycling pathway of the Man 6-P-receptors is constitutive, i.e. trafficking is independent of ligand binding.

Recently, Pfeffer and Rothman (1987) have proposed and discussed the idea that within ER and Golgi distinct phases might exist that regulate selective *vs* non-selective membrane flow. Proteins destined for transfer from ER and Golgi would enter a so-called mobile phase, i.e. a region or domain from which membrane vesicles can bud. Immobile phases would retain those proteins which are perminant residents of either ER or Golgi. As long as a protein would reside in the mobile phase it will be transported with the bulk membrane flow. Thus a protein would need a signal

(Munro and Pelham, 1987) to be retained in the immobile phase while transiting proteins would remain in or diffuse into the mobile phase by default. Interestingly, at least some of the resident proteins of ER appear to be selectively retrieved from a post-ER compartment (presumably Golgi) before they resume their permanent location (Pelham et al., 1988). According to the afore-mentioned proposal this would imply that the proteins would initially enter the mobile phase in the ER.

In the foregoing paragraphs we have given a bird's-eye view of the abundance of membrane traffic pathways in a eukaryotic cell. Many details have yet to be resolved, especially questions that are dealing with the regulation of the trafficking of the various membrane vesicles, continuously pinching off and fusing with distinct target membranes. Although beyond the scope of this paper, some brief comments on intracellular membrane fusion seem appropriate.

INTRACELLULAR MEMBRANE FUSION

Details of the molecular mechanisms of membrane fusion are gradually emerging (Düzgüneş, 1985; Wilschut and Hoekstra, 1986; Blumenthal, 1987; Doms and Helenius, 1988; Hoekstra et al., 1988; Hoekstra and Wilschut, 1988). As for the biosynthetic pathway of protein traffic, viruses are widely used as a model system for studying the mechanisms of fusion of biological membranes. It is well-established that membrane fusion of a variety of enveloped viruses is mediated by viral transmembrane glycoproteins (White et al., 1983; Hoekstra, 1988). The virions bind to cell surface receptors and, depending on the virus family, fusion takes place at either neutral pH (i.e. at the plasma membrane; Sendai virus) or at mild acidic pH (i.e. in the late endosomal compartment from "within"; influenza virus, VSV). Presumably, the mechanism of fusion involves the penetration of an exposed hydrophobic amino acid region of the viral fusion protein into the target membrane (Novick and Hoekstra, 1988), which is then followed by the merging process of both membranes. The subsequent deposition of the viral nucleocapsid into the cytoplasm is necessary for initiation of the replication of viral proteins. A priori, it would appear very likely that intracellular membrane fusion events, which, obviously, must be very carefully regulated and controlled, proceed according to a very similar type of mechanism, involving specific recognition of the target membrane and specific proteins that trigger the fusion reaction. Indeed, various studies of the fusion stages during endocytosis and in the secretory pathway, carried out in cell-free systems, have revealed that both cytosolic and membrane-associated proteins are involved (Davey et al., 1985; Balch and Rothman, 1985; Gruenberg and Howell, 1986; Balch et al., 1986; Braell, 1987; Diaz et al., 1988). In addition to proteins, ATP appears to be necessary. Specific fusion proteins, as found for enveloped viruses, have not yet been identified. This also holds for factors that regulate membrane traffic. In this respect, based on a number of studies (Melançon et al., 1987; Segev et al., 1988; Schmitt et al., 1988), Bourne (1988) has developed the hypothesis that GTP-binding proteins could specify the direction traveled by vesicles between appropriate cellular compartments. The essence of the model (figure 4) is, that GTP/GDP-induced conformational switches may mediate vectorial transport of secretory vesicles. Clearly, solid evidence will be necessary to prove or disprove the model. Yet, it may appear quite instructive in setting the principles of how intracellular membrane traffic *could* be regulated.

INTRACELLULAR FLOW OF LIPIDS

The fact that intracellular translocation of proteins, occurring during endocytosis or in the biosynthetic secretory pathway, is mediated by membrane vesicles, evidently implies that lipids accompany this transport process. Given the continuous and substantial flow of lipids, forming the bilayer structure of these vesicles, this raises immediate questions as to how the various organelle membranes as well as the plasma membrane maintain their characteristic lipid composition. Hence, as for proteins, one would thus anticipate that specific mechanisms are operational that properly sort the lipids and determine their fate during membrane flow. Yet, although knowledge on protein trafficking in the cell is rapidly advancing, virtually nothing is known about the flow of lipids, let alone their sorting and recycling. However, developments in this field are gradually emerging, as reflected by a number of recent reviews (Pagano and Sleight, 1985; Sleight, 1987; Dawidowicz, 1987; Simons and Van Meer, 1988).

To reveal the flow of lipids in cells, the fate of radiolabeled and fluorescently tagged lipid analogs (figure 5) is usually monitored. Particularly the latter class of molecules has become a popular tool to study lipid flow. Such derivatized lipids can

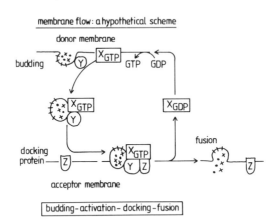

Figure 4. Hypothetical mechanism for regulation of membrane traffic.
The upper part of the figure represents the donor compartment from which a vesicle is about to bud (donor membrane) while the lower part is the compartment to which the vesicle should be directed (acceptor membrane). A specific membrane protein, Y, of the budding vesicle is recognized by a specific GTP binding protein, X_{GTP}. Note that budding itself is not triggered by X_{GTP} binding. In the biosynthetic pathway this phenomenon may require ATP (Pfeffer and Rothman, 1987). X_{GTP} binding to Y will switch on ("activation") the latter. The activated complex is recognized by a specific docking protein Z, located on the membrane of the acceptor compartment. Following GTP hydrolysis, the binding protein dissociates and recycles back to the acceptor membrane for another round of vesicle trafficking. The next step at the acceptor compartment will be fusion as a result of which Y and Z randomize by lateral diffusion. The fusion process need not neccessarily be instantaneous (see for example Wattenberg et al., 1986) nor need it be mediated by the proteins (YZ complex) involved in targeting. (For details, cf. Bourne, 1988).

be readily inserted into the plasma membranes of cultured cells and allow one to monitor their fate in a direct manner by fluorescence microscopy. This approach has the additional advantage that randomization during homogenization of cells for subcellular fractionation can be excluded. On the other hand, the lipids have been "probed", implying that their structure is no longer identical to that of the parent lipid. Yet, as argued by Pagano and Sleight (1985), in many respects their behavior appears to be similar or identical to isotopically labeled lipids. Thus, in conjunction with biochemical and biophysical techniques, as well as colocalization studies with well-characterized protein receptors or fluorescently tagged antibodies, such fluorescent lipid analogs could be helpful tools in obtaining insight in the selectivity of lipid traffic.

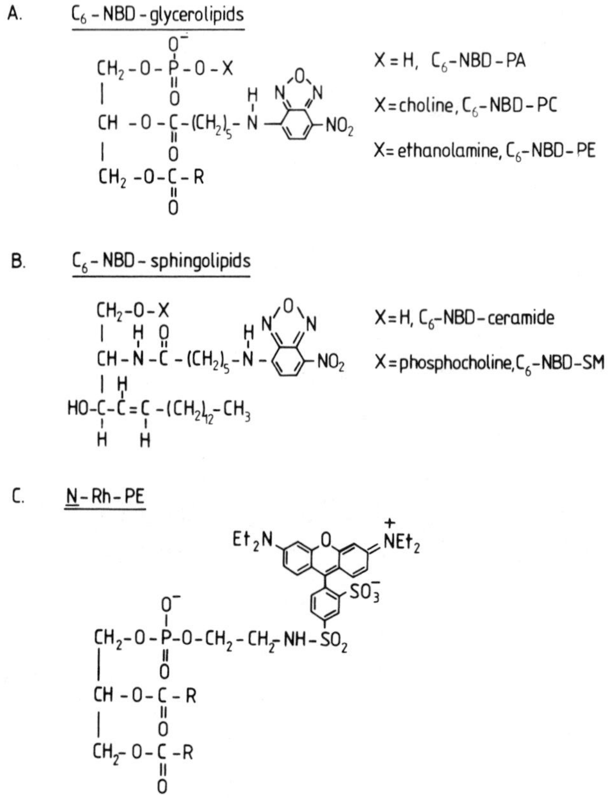

A. C_6 – NBD – glycerolipids

$$CH_2-O-\overset{\overset{O^-}{|}}{\underset{\underset{O}{||}}{P}}-O-X$$

$$CH-O-\underset{\underset{O}{||}}{C}-(CH_2)_5-\overset{H}{N}-\langle NBD \rangle-NO_2$$

$$CH_2-O-\underset{\underset{O}{||}}{C}-R$$

X = H, C_6–NBD–PA

X = choline, C_6–NBD–PC

X = ethanolamine, C_6–NBD–PE

B. C_6 – NBD – sphingolipids

$$CH_2-O-X$$
$$CH-\overset{H}{N}-\underset{\underset{O}{||}}{C}-(CH_2)_5-\overset{H}{N}-\langle NBD \rangle-NO_2$$
$$HO-\underset{\underset{H}{|}}{C}-\underset{\underset{H}{|}}{C}=C-(CH_2)_{12}-CH_3$$

X = H, C_6–NBD–ceramide

X = phosphocholine, C_6–NBD–SM

C. N – Rh – PE

$$Et_2N\cdots O \cdots \overset{+}{N}Et_2$$
$$SO_3^-$$

$$CH_2-O-\overset{\overset{O^-}{|}}{\underset{\underset{O}{||}}{P}}-O-CH_2-CH_2-NH-SO_2$$

$$CH-O-\underset{\underset{O}{||}}{C}-R$$

$$CH_2-O-\underset{\underset{O}{||}}{C}-R$$

Figure 5.   Structures of fluorescent lipid analogs.
A. Phospholipids, acyl-chain labeled with N-4-nitrobenzo-2-oxa-1,3-
    diazole-aminocaproic acid (NBD; C_6-NBD-phospholipids).
B. NBD-labeled ceramide and glycolipids.
C. Polar head-labeled derivatives of PE. Only rhodamine-labeled PE is shown
    (N-Rh-PE). The head group can also be labeled with NBD.

It is clear that the flow of newly synthesized phospholipids will start from the site of synthesis, i.e. the endoplasmic reticulum. It appears equally apparent that this flow can at least partly be accomplished by membrane vesicle-mediated transfer, together with proteins (see above). Whether additional transfer mechanisms are involved, is less clear. Spontaneous transfer of phospholipids, as monomers through the aqueous phase, is extremely unlikely, since it occurs at extremely slow rates (hours to days; Roseman and Thompson, 1980; McLean and Phillips, 1981) while, in addition, such a mechanism would not provide the required specificity of lipid flow, as reflected by the difference in phospholipid composition of the intracellular membranes. That lipid movement, mediated by PLEPs, cannot be ignored can be concluded from observations on the kinetics of lipid traffic from ER to the cell surface upon *de novo* synthesis. In hamster fibroblasts (V79) newly synthesized PE begins to appear at the cell surface after a lag time of only 2.5 min (Sleight and Pagano, 1983). This was seen when using radiolabeled oleic acid or ethanolamine as precursors. With labeled glycerol, detectable amounts were found after about 20 min, a time interval more closely resembling that of protein traffic (see table 1). However, chemical agents known to inhibit secretion did not block PE appearance, implying that the lipid was not cotransported (in membrane vesicles) with membrane glycoproteins. Similarly, in CHO cells, newly synthesized PC rapidly moves to the cell surface with a halftime of approximately 2 min at 25°C (Kaplan and Simoni, 1985). On the other hand, transfer of newly synthesized PE in *Dictiostelium discoideum* is blocked by colchicine and vesicles have been isolated from these cells that might be involved in lipid transport (De Silva and Siu, 1981).

It is clear that a number of questions on the fate and processing of newly synthesized PC and PE remain to be answered. Given the rapid kinetics, the transfer obviously differs from that of secretory proteins. A question that remains to be answered is whether the observed movement represents a net transfer or an exchange process. In addition, one might wonder whether vesicles still could be involved that by-pass the Golgi system. Alternatively, if these rapid translocation processes are mediated by PLEPs, how then would targeting be accounted for?

After insertion the lipids participate in the dynamics of the membranes, involving the continuous flow and shuttling events that occur during the biological life of a cell. The next issue thus focuses on the question as to how the membranes maintain their distinct lipid composition and where the lipids flow during these shuttling processes. There exist several possibilities to unravel the specificity of lipid traffic and to monitor the flow of membranes in animal cells. A very promising approach relies on the use of fluorescent lipid analogs (figure 5) as outlined above. The fluorescent probe, usually N-4-nitrobenzo-2-oxa-1,3 diazole amino caproic acid (NBD), is attached to the glycerol (or ceramide, see below) backbone at the sn-2 position via a C_6-carbon spacer (C_6-NBD-lipid). The properties of the analog differ from the parent structure in that the C_6-NBD-lipids can rapidly transfer between membranes through the aqueous phase (Struck and Pagano, 1981; Nichols and Pagano, 1981; Hoekstra, 1982). This property can be exploited, however, as it allows one to insert these derivates (using liposomes as donor membranes) in the outer leaflet of the plasma membrane from which they can be removed again by a "backexchange" in the presence of (non-labeled) acceptor membranes (Struck and Pagano, 1981) or by washing the cells with a bovine serum albumin solution. This makes it possible to quantify the amount of lipid analog that becomes internalized and/or reappears at the cell surface at appropriate experimental conditions. When labeling is carried out at 2°C, both C_6-NBD-PC and C_6-NBD-PE remain in the outer leaflet of the cell surface, as seen for hamster fibroblasts (Struck and Pagano, 1981; see figure 6). Upon warming the cells to 37°C, part of the labeled lipids is internalized and, moreover, the intracellular destination of the derivatives is different. Approximately 20% of the C_6-NBD-PC becomes internalized, which is inhibited in the presence of metabolic inhibitors, suggesting participation of an endocytic mechanism. The

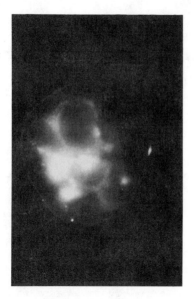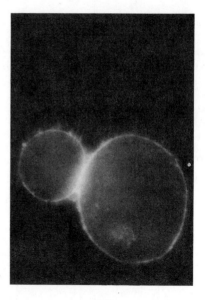

Figure 6.   Redistribution and internalization of C_6-NBD-PC during virus-induced cell fusion.
C_6-NBD-PC was inserted into the outer leaflet of Chinese hamster fibroblasts (V79) at 2°C. Subsequently, Sendai virus was added and after 5 min, cell fusion was triggered by an incubation at 37°C for 10 min (left). At least five nuclei are discernable in the fused cell. Alternatively, the C_6-NBD-PC labeled cells were fused in the presence of poly(ethylene) glycol (Right). Note that in this case, internalization of the fluorescent lipid analog is minimal, i.e. most fluorescence is still plasma membrane-associated, very similar as observed for non-treated cells. For details see Hoekstra, 1983.

fluorescent lipid accumulates intracellularly in the peri-nuclear region, colocalizing with the Golgi complex as revealed by specific antibodies, directed against this organelle (Sleight and Pagano, 1984). The transfer of C_6-NBD-PC was mediated by small vesicles in which plasma membrane proteins, labeled with a fluorescent lectin, comigrated. However, the final destination of the lipid and the proteins differed, indicating that the internalization pathways diverged at a stage yet to be identified. Eventually, C_6-NBD-PC is lost from the cells (half time ca. 90 min). The loss of fluorescence could be accounted for by degradation at the cell surface which results in the release of the NBD-fatty acid in the medium. The labeled fatty acid cannot be reutilized for *de novo* synthesis of lipids thus excluding ambiguities due to deacylation-reacylation reactions. How Golgi-localized lipid is removed from the cells is not clear. Since the cellular-associated lipid only consists of C_6-NBD-PC, i.e. intracellular degradation could not be detected, it is possible that the analog recycled back to the cell surface followed by degradation at the cell surface.

In this context it is interesting to note that internalization of C_6-NBD-PC can be *induced* upon virus fusion (Hoekstra, 1983). When hamster fibroblasts (V79) containing C_6-NBD-PC in their outer leaflets after incubation at 2°C, were fused at 37°C in the presence of Sendai virus, approximately 40-50% of the probe accumulated in peri-nuclear regions, possibly reflecting the Golgi area (figure 6). This process could not be blocked in the presence of metabolic inhibitors, suggesting that an endocytic mechanism did not target delivery to the Golgi area. Rather, it was proposed that the virus triggered a flip-flop of the lipid which could imply that once in the inner leaflet, the PC-derivate is rapidly transferred to the Golgi area by an, as yet, unknown mechanism. It would thus appear that in spite of the different mechanisms involved (endocytosis *vs* virus-induced flip-flop) the final destination for C_6-NBD-PC remains the same.

In addition to labeling the Golgi region, C_6-NBD-PE is also found in mitochondria and the nuclear envelope when warming the cells from $2^{\circ}C$ to $37^{\circ}C$ (Sleight and Pagano, 1985). Interestingly, when the cells were pretreated with metabolic inhibitors, both mitochondria and nuclear envelope were stained, but *not* the Golgi region. Presumably, this phenomenon can be related to the presence of an aminophospholipid translocase activity in the membrane, establishing and maintaining phospholipid asymmetry (see above). Thus, this activity will translocate PE from the outer leaflet to the inner leaflet. When present in the inner leaflet, the probe may move as a monomer to (unlabeled) acceptor membranes within the cell. However, since only mitochondrial and nuclear membranes are labeled one would anticipate that transfer occurs via a mechanism more specific than spontaneous transfer. Whatever this mechanism could be, a vesicle-mediated transfer mechanism in this case could be excluded as the movement of vesicles was not observed (Sleight and Pagano, 1985).

In contrast to the NBD-derivatives of PC and PE, C_6-NBD-PA *enters* the cell at $2^{\circ}C$. It appears, however, that this lipid is converted (80-90%) to C_6-NBD-diacylglycerol, occurring at the plasma membrane which, in turn, is followed by internalization (Pagano et al., 1983). At $37^{\circ}C$, the diacylglycerol is further converted into C_6-NBD-triacylglycerol and C_6-NBD-PC. The former lipid exclusively associates with lipid droplets while the latter stains ER, mitochondrial and nuclear envelope membranes.

The work discussed thusfar clearly demonstrates that upon internalization, the fluorescent lipids do not randomize over internal membranes. C_6-NBD-PC bears some striking similarity to the movement of glycoprotein receptors in that it passes, after endocytosis, the Golgi complex (see above). Whether the lipid is actually recycled, analogous to glycoprotein receptors has not yet been firmly established. With respect to "flip-flop" properties, the PE-derivative undergoes a very similar fate as that seen for spin-labeled PE derivatives. The membrane ensures its flip to the inner leaflet and, assuming that flippases catalyze this reaction, the NBD-derivative of PE is recognized by this translocator. The observations of the behavior of C_6-NBD-PA in particular would suggest that the cell recognizes the different classes of lipids and "sorts" them to different intracellular locations. As yet, it is not evident that each of these steps is mediated by vesicular transport. Rather, proteins may be involved as well, as suggested in the case of intracellular movement of C_6-NBD-PE (Sleight and Pagano, 1985).

In contrast to acyl chain-labeled lipid derivatives, head group-labeled phospholipid analogs, such as N-Rh-PE and N-NBD-PE (figure 5), are *non-exchangeable* lipid probes (Struck and Pagano, 1980; Struck et al., 1981; Hoekstra, 1982, 1982a; Nichols and Pagano, 1983). Thus, the probes do not move as monomers through the aqueous phase, nor do they transfer during aggregation of labeled and non-labeled membranes. These properties have been exploited by using N-NBD-PE and N-Rh-PE in an assay to monitor membrane fusion, based on the principle of resonance energy transfer (Struck et al., 1981; Hoekstra, 1982a). These derivatives can be inserted into membranes, however, when the probes are added to the cells in an ethanolic phase (unpublished observation, cf. Hoekstra et al., 1984). At $2^{\circ}C$, the probe remains in the plasma membrane of BHK cells and its diffusion rate constant is virtually identical to that of C_6-NBD-PC, indicating a proper insertion in the membrane. Upon raising the temperature to $37^{\circ}C$ the probe is rapidly cleared from the cell surface and, interestingly, accumulates in the *lysosomal* compartment (figure 7; Kok, J.W., Ter Beest, M., Scherphof, G. and Hoekstra, D., in preparation). Metabolic inhibitors prevent internalization but colocalization with the Tf/Tf receptor complex was never observed (Eskelinen, Kok, J.W. and Hoekstra, D., in preparation). Yet, the transfer from the surface was mediated by vesicles but the exact mechanism, presumably endocytic in nature, remains to be established. Although fluorescence microscopy revealed a random distribution at the cell surface, fluorometric measurements demonstrated that N-Rh-PE was present at the cell surface (at $2^{\circ}C$) as small aggregates. A comparison with C_6-NBD-PC in these cells showed that while

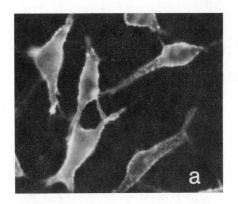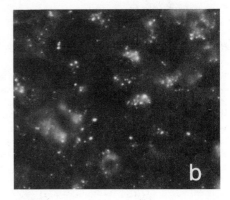

Figure 7. Intracellular fate of N̲-Rh-PE after insertion into plasma membranes of BHK cells.
N̲-Rh-PE was inserted into BHK plasma membranes at 2°C, using an ethanol injection procedure (a; cf. Hoekstra et al., 1984). After raising the temperature to 37°C, the fluorescent lipid analog is internalized and localized in the lysosomes (b), as evidenced by Percoll gradient analysis and by colocalization with fluorescent markers that concentrate in the lysosomes (unpublished observations). Internalization is inhibited by metabolic inhibitors. Detectable amounts of fluorescence in the lysosomes can be seen after approximately 10-15 min at 37°C.

essentially all NBD-lipid is still at the cell surface, significant amounts of N-Rh-PE are already located in the lysosomal compartment. Hence, we assume that the aggregated nature of N-Rh-PE elicites a signal which causes its rapid (endocytic) elimination from the cell surface. Given the artificial nature of this PE derivative, no conclusions can be drawn as to the traffic of natural lipids at physiological conditions. Yet, the probe triggers a physiological *response* and that, by itself, merits further investigation in defining and characterizing the membrane traffic pathways in animal cells.

Cholesterol

Cholesterol can transfer spontaneously between membranes, i.e., without direct intermembrane interactions (Backer and Dawidowicz, 1981). In addition, there exists evidence that such a transfer can also be mediated by sterol carrier proteins (Chanderbhan et al., 1982; McLean and Phillips, 1981; Noland et al., 1980; Tanaka et al., 1984; Vahouny et al., 1984). It would appear, however, that newly synthesized cholesterol is predominantly delivered to the plasma membrane by a vesicle-mediated transport mechanism (DeGrella and Simono, 1982; Lange and Matthies, 1984; Kaplan and Simoni, 1985a). In contrast to the rapid movement of PE or PC (see above), the time course of transfer of cholesterol between its site of synthesis (ER) and the plasma membrane is considerable. Half times of 10 min up to at least 1 h have been reported (DeGrella and Simoni, 1982; Kaplan and Simoni, 1985a; Mills et al., 1984; Lange and Matthies. 1984). Energy poisons and low temperature ($<15^{\circ}$C) block the transport process. However, reagents that perturb the cytoskeletal meshwork (colchicine, cytochalasin B, etc.) or monensin do not arrest transfer. As will be described below, monensin does inhibit the traffic of ceramide-containing lipids from the Golgi to the plasma membrane. This could imply that the Golgi complex might be bypassed in the cholesterol traffic pathway. These observations also emphasize that vesicular traffic may not always be regulated by elements of the cytoskeletal network (cf. Dawidowicz, 1987). Furthermore, it has been shown recently (Lange and Muraski, 1988) that cholesterol biosynthesis is topographically heterogeneous.

When cholesterol biosynthesis in cultured human fibroblasts was interrupted, followed by isopycnic sucrose gradient analysis of homogenates, various membrane fractions could be isolated that contained (labeled) cholesterol and its precursors in different ratios (Lange and Muraski, 1988). At least three membrane fractions, characteristic for the presence of different amounts of precursors and end product, were identified. Transfer of the intermediates between successive compartments was disrupted in homogenized cells when the temperature was kept below 10°C. Based on enzyme activity profiles, the authors suggested that the precursors specifically tested, were not located in the Golgi complex but rather at a separate intracellular locus (Lange and Muraski, 1988). This would be consistent with the absence of inhibition of cholesterol transport upon monensin treatment (see above). Overall, the synthesis of cholesterol does not seem to take place exclusively at the ER, but rather, appears to proceed via intermediates that move between discrete membranes until eventually the plasma membrane is reached. Given the fact that the Golgi complex is intimately involved in protein traffic, it seems reasonable to conclude that at least part of the biosynthetic cholesterol pathway is independent of that of membrane glycoproteins.

Plasma membranes maintain a high level of cholesterol, relative to the levels of cholesterol found in organelle membranes. When radiolabeled cholesterol is inserted in the plasma membrane of human fibroblasts, there is essentially no flow of cholesterol from the cell surface to intracellular membranes (Lange and Matthies, 1984; Wattenberg and Silbert, 1983). The mechanism by which the high ratio of cholesterol to other lipids in plasma membranes is maintained is obscure. The constitutive, endocytic pathway appears to internalize cholesterol, however, as rat liver endocytic vesicles have been isolated with a cholesterol/phospholipid ratio similar to that of the plasma membrane (Helmy et al., 1986). This would imply that cholesterol is not excluded from being internalized. It is not known whether the extent of internalization may be cell-type dependent and further work will be needed to clarify this apparent controversy.

Glycosphingolipids

Glycosphingolipids are the class of lipids which is generally believed to be localized predominantly on the outer leaflet of cellular plasma membranes. They can be distinguished into two types. One type contains, in addition to the neutral carbohydrate portion, sialic acid (gangliosides) while the other lacks the acidic carbohydrate (neutral glycosphingolipids). For both types, the ceramide backbone structure (figure 5) functions as the precursor for their biosynthesis. Glycolipids have been related to many biological functions, including intercellular communication, the regulation of cell growth and differentiation, the immune response and the pathogenesis of cancer (Hakomeri, 1981, 1983; Nudelman et al., 1983; Wiegandt, 1985). In contrast to the synthesis of other lipids, the synthesis of glycolipids requires a major involvement of the Golgi complex. After formation of the ceramide backbone structure, which presumably occurs in the ER, subsequent glycosylation reactions take place in the Golgi complex (Wiegandt, 1985), analogous to carbohydrate trimming and modification of glycoproteins (see above). The common precursor for all the gangliosides of the ganglio series is lactosylceramide. Addition of a sialic acid, occurring in the *trans* Golgi compartment leads to formation of GM_3, a ganglioside that causes a negative modulation of growth of various cell lines. GM_3 is transferred to the cell surface, presumably following the common pathway for the biosynthetic delivery of membrane glycoproteins (Miller-Podraza and Fishman, 1983). GM_3, once in the plasma membrane, inhibits cell growth by interfering with EGF receptor phosphorylation (Bremer et al., 1986; see above). This inhibition can be relieved when GM_3 is converted to lactosylceramide, which can be mediated by a plasma membrane-associated sialidase activity. Based on pulse chase experiments, using radiolabeled N-acetylmannosamine and serine to study the turnover of endogenous GM_3, it was proposed (Usuki et al., 1988) that GM_3-dependent regulation of cell growth could be accomplished by a mechanism involving degradation of plasma membrane GM_3 to lactosylceramide. The latter would then return to the Golgi complex where GM_3 is synthesized again by addition of sialic acid, followed by its return to the cell surface. Whether this type of 'recycling' involves the endocytic pathway has not yet been determined. If so, the lactosylceramide has to be sorted during the inbound traffic in order to avoid degradation in the lysosomal compartment. In this context, exogeneous insertion of various radiolabeled gangliosides in the plasma membrane of cultured human fibroblasts has been shown to result predominantly in degradation in the lysosomal compartment (Sonderfeld et al., 1985; Ghidoni et al., 1986). This would imply that during or shortly after internalization 'sorting' of the different glycolipids may occur.

Degradation products of the endocytic pathway can be reutilized, however, for biosynthesis, which presumably takes place in the Golgi complex. Interestingly, in liver, gangliosides are only resynthesized from 'recycled' glucosylceramide, and not from galactosylceramide, while apparently ceramide can also serve as the precursor for synthesis of the phospholipid sphingomyelin (Tettamanti et al., 1988). The latter has also been shown to occur after internalization of C_6-NBD-ceramide (figure 5), inserted into plasma membranes of hamster fibroblasts. In contrast to C_6-NBD-PC, the ceramide is rapidly internalized in the cells at 2^oC, presumably mediated by a flip-flop mechanism, and colocalizes with mitochondria (Lipsky and Pagano, 1983). Upon raising the temperature to 37^oC, fluorescence associates with the Golgi complex (after ca. 10 min) and subsequently a recycling occurs to the cell surface (ca. 20-30 min). Analysis of the cell-surface associated fluorescent products revealed that equal amounts of NBD-sphingomyelin and NBD-glucosylceramide had formed. It was found that the synthesis of both lipids took place in the Golgi and that transport to the cell surface was blocked by monensin. The kinetics of transfer of both lipids is very similar to those of secretory and plasma membrane proteins and, in conjunction with the effect of monensin, would support a vesicular transport mechanism (Lipsky and Pagano, 1985).

These observations were further extended at the level of lipid 'sorting' in epithelial MDCK cells. The surface membrane of these cells is polarized, consisting of two domains: the apical and the basolateral membrane. The two plasma membrane domains have, in a quantitative sense, different lipid compositions, in that the apical domain is particularly enriched in glycolipids (Simons and Van Meer, 1988). Analogous to the experiment described in the previous paragraph, NBD-ceramide was converted into NBD-sphingomyelin and NBD-glucosylceramide at 37^oC, after insertion and subsequent internalization at 2^oC. Interestingly, the latter lipid was preferentially delivered to the apical membrane, while, as anticipated, NBD-sphingomyelin was delivered in about equal amounts to the apical and basolateral membranes (Van Meer et al., 1987). The evidence supports a transport from the Golgi to the plasma membrane by vesicular carriers (Simons and Virta, 1987).

It was proposed that sorting of the NBD-lipids, destined for apical and basolateral membranes, occurs in the trans Golgi network (Simons and Van Meer, 1988). The type of sorting could be more or less analogous to that proposed for glycoprotein sorting in that trafficking to the basolateral membrane proceeds according to a default pathway, while that directed toward the apical membrane could be signal-mediated (Pfeffer and Rothman, 1987; see above). Whether proteins ('sorting proteins') are involved as signals in sorting, giving rise to formation of microdomains that resemble the properties of the membranes of destination, remains to be determined. It is worthwhile to note in this respect that by using photoreactive glycolipids it has been shown that a distinct glycolipid may display a particular affinity for a specific (set of) protein(s) (Sonnino et al., 1988) after insertion in mammalian cell membranes. This could suggest that a glycolipid may indeed be located in well-defined regions of a membrane, as triggered by distinct proteins. It would appear therefore that further application of photoactivatable glycosphingolipids may prove to be a highly useful tool in analyzing questions dealing with sorting mechanisms of these lipids.

In summary, it would appear that the overall picture of glycolipid transfer is more comprehensible than that observed for phospholipids, in spite of the numerous questions that remain to be answered. Glycolipids move via the Golgi network to the cell surface, mediated by vesicular transport, and return intracellularly via endocytosis. As discussed, sorting probably occurs at the level of the endosome, directing at least some glycolipids (lactosylceramide) toward the Golgi where they can be reutilized for biosynthesis. Others are degraded in the lysosomes, for reasons to be determined, but also in this case the degradation products can be reutilized for biosynthesis. How these compounds reach the Golgi, where resynthesis takes place, remains to be elucidated. At any rate, in many respects the trafficking of glycosphingolipids seems to resemble the routes of glycoprotein traffic. It would thus appear that further advances in both 'fields' will be mutually beneficial.

ACKNOWLEDGMENTS

Part of the authors' work cited in this paper was carried out under the auspices of the Netherlands Foundation for Chemical Research (SON) with financial support from the Netherlands Organization for Scientific Research (NWO).

One of us (S.E.) was supported by a post-doctoral fellowship from the Natural Science Research Council of the Academy of Finland. The expert secretarial assistance of Mrs Lineke Klap and Mrs Rinske Kuperus is very much appreciated.

REFERENCES

Anderson, R.G.W., Goldstein, J.L. and Brown, M.S., 1977, A mutation that impairs the ability of lipoprotein receptors to localize in coated pits on the cell surface of human fibroblasts, Nature 270:695-699.

Backer, J.M. and Dawidowicz, E.A., 1981, Mechanism of cholesterol exchange between phospholipid vesicles, Biochemistry 20:3805-3810.

Baker, E.N., Rumball, S.V., and Anderson, B.F., 1987, Transferrins: insights into structure and function from studies on lactoferrin, Trends Biochem. Sci. 12:350-353.

Balch, W.E., Dunphy, W.G., Braell, W.A. and Rothman, J.E., 1984, Reconstitution of the transport of protein between successive compartments of the Golgi measured by the coupled incorporation of N-acetylglucosamine, Cell 39:405-416.

Balch, W.E. and Rothman, J.E., 1985, Characterization of protein transport between successive compartments of the Golgi apparatus; asymmetric properties of donor and acceptor activities in a cell-free system, Arch. Biochem. Biophys. 240:413-425.

Balch, W.E., Elliot, M.M. and Keller, D.S., 1986, ATP-coupled transport of vesicular stomatitis virus G protein between the endoplasmic reticulum and the Golgi, J. Biol. Chem. 261:14681-14689.

Blumenthal, R., 1987, Membrane fusion, Curr. Top. Membr. Transp. 29:203-254.

Bourne, H.R., 1988, Do GTPases direct membrane traffic in secretion? Cell 53:669-671.

Braell, W.A., 1987, Fusion between endocytic vesicles in a cell-free system, Proc. Natl. Acad. Sci. USA 84:1137-1141.

Braulke, T., Gartung, C., Hasilik, A. and Von Figura, K., 1987, Is movement of mannose 6-phosphate specific receptor triggered by binding of lysosomal enzymes? J. Cell Biol. 104:1735-1742.

Bremer, E.G., Schlessinger, J. and Hakomori, S.-I., 1986, Ganglioside-mediated modulation of cell growth, J. Biol. Chem. 261:2434-2440.

Brown, W.J. and Farquhar, M.G., 1984, The mannose 6-phosphate receptor for lysosomal enzymes is concentrated in cis-Golgi cisternae. Cell 36:295-307.

Calvez, J.-Y., Zachowski, A., Herrmann, A., Morrot, G. and Deveaux, P., 1988, Asymmetric distribution of phospholipids in spectrin-poor erythrocyte vesicles, Biochemistry 27:5666-5670.

Carpenter, G. and Cohen, S., 1979, Epidermal growth factor, Ann. Rev. Biochem. 48:193-216.

Chanderbhan, R., Noland, B.J., Scallen, T.J. and Vahouny, G.V., 1982, Sterol carrier protein$_2$: Delivery of cholesterol from adrenal lipid droplets to mitochondria for pregnenolone synthesis, J. Biol. Chem. 257:8928-8934.

Chesterton, C.J., 1968, Distribution of cholesterol precursors and other lipids among rat liver intracellular structures, J. Biol. Chem. 243:1147-1151.

Cohen, S., 1983, The epidermal growth factor (EGF), Cancer 51:1787-1791.

Connor, J. and Schroit, A.J., 1987, Determination of lipid asymmetry in human red cells by resonance energy transfer, Biochemistry 26:5099-5105.

Connor, J. and Schroit, A.J., 1988, Transbilayer movement of phosphatidylserine in erythrocytes: inhibition of transport and preferential labeling of a 31.000-Dalton protein by sulfhydryl reactive reagents, Biochemistry 27:848-851.

Copeland, C.S., Doms, R.W., Bolzau, E.M., Webster, R.G. and Helenius, A., 1986, Assembly of influenza hemagglutinin trimers and its role in intracellular transport, J. Cell Biol. 103:1179-1191.

Copeland, C.S., Zimmer, K.-P., Wagner, K.R., Healey, G.A., Mellman, I. and Helenius, A., 1988, Folding, trimerization, and transport are sequential events in the biogenesis of influenza virus hemagglutinin, Cell 53:197-209.

Davey, J., Huntley, S.M. and Warren, G., 1985, Reconstitution of an endocytic fusion event in a cell-free system, Cell 43:643-652.

Davis, C.G., Lehrman, M.A., Russell, D.W., Anderson, R.G.W., Brown, M.S. and Goldstein, J.L., 1986, The J.D. mutation in familial hypercholesterolemia: amino acid substitution in cytoplasmic domain impedes internalization of LDL receptors, Cell 45:15-24.

Dawidowicz, E.A., 1987, Dynamics of membrane lipid metabolism and turnover, Ann. Rev. Biochem. 56:43-61.

Dawidowicz, E.A., 1987a, Lipid exchange: transmembrane movement, spontaneous movement and protein-mediated transfer of lipids and cholesterol, Curr. Top. Membr. Transp. 29:175-202.

DeGrella, R.F. and Simoni, R.D., 1982, Intracellular transport of cholesterol to the plasma membrane, J. Biol. Chem. 257:14256-14262.

De Silva, N.S. and Siu, C.-H., 1981, Vesicle-mediated transfer of phospholipids to plasma membrane during cell aggregation of Dictyostelium discoideum, J. Biol. Chem. 256:5845-5850.

Diaz, R., Mayorga, L. and Stahl, P., 1988, In vitro fusion of endosomes following receptor mediated endocytosis, J. Biol. Chem. 263:6093-6100.

Doms, R.W. and Helenius, A., 1988, Properties of a viral fusion protein, in: Molecular mechanisms of membrane fusion, Ohki, S., Doyle, D., Flanagan, T.D., Hui, S.W. and Mayhew, E., eds., pp. 385-398, Plenum Press, New York.

Duncan, J.R. and Kornfeld, S., 1988, Intracellular movement of two mannose 6-phosphate receptors: return to the Golgi apparatus, J. Cell Biol. 106:617-628.

Düzgüneş, N., 1985, Membrane fusion, Subcell. Biochemistry 11:195-286.

Farquhar, M.G., 1983, Multiple pathways of exocytosis, endocytosis and membrane recycling: validation of a Golgi route, Fed. Proc. 42:2407-2413.

Fishman, J.B. and Fine, R.E., 1987, A trans Golgi derived exocytotic, coated vesicle can contain both newly synthesized cholinesterase and internalized transferrin, Cell 48:157-164.

Geuze, H.J., Slot, J.W., Strous, G.J., Lodish, H.F. and Schwartz, A.L., 1983, Intracellular site of asialoglycoprotein receptor-ligand uncoupling: double label immuno-electron microscopy during receptor-mediated endocytosis, Cell 32:277-287.

Geuze, H.J., Slot, J.W., Strous, G.J., Hasilik, A. and Von Figura, K., 1985, Possible pathways for lysosomal enzyme delivery, J. Biol. Biol. 101:2253-2262.

Geuze, H.J., Slot, J.W. and Schwartz, A.L., 1987, Membranes of sorting organelles display lateral heterogeneity in receptor distribution, J. Cell Biol. 104:1715-1723.

Ghidoni, R., Trinchera, M., Venerando, B., Fiorilli, A., Sonnino, S. and Tettamanti, A., 1986, Incorporation and metabolism of exogenous GM_1 ganglioside in rat liver, Biochem. J. 237:147-155.

Goldberg, D.E. and Kornfeld, S., 1983, Evidence for extensive subcellular organization of asparagine-linked oligosaccharide processing and lysosomal enzyme phosphorylation, J. Biol. Chem. 258:3159-3165.

Goldstein, J.L., Brown, M.S., Anderson, R.G.W., Russell, D.W. and Schneider, W.J., 1985, Receptor-mediated endocytosis, Ann. Rev. Cell Biol. 1:1-39.

Gordon, P., Carpentier, J.-L., Cohen, S. and Orci, L., 1978, Epidermal growth factor: morphological demonstration of binding, internalization and lysosomal association in human fibroblasts, Proc. Natl. Acad. Sci. USA 75:5025-5029.

Gottlieb, C., Baenziger, J. and Kornfeld, S., 1975, Deficient uridine diphosphate-N-acetylglucosamine glycoprotein N-acetylglucosaminyl-transferase activity in a clone of Chinese hamster ovary cells with altered surface glycoproteins, J. Biol. Chem.

250:3303-3309.

Griffiths, G. and Simons, K., 1986, The trans-Golgi network: sorting at the exit site of the Golgi complex, Science 234:438-443.

Griffiths, G., Hoflack, B., Simons, K., Mellman, I. and Kornfeld, S., 1988, The mannose 6-phosphate receptor and the biogenesis of lysosomes, Cell 52:329-341.

Gruenberg, J.E. and Howell, K.E., 1986, Reconstitution of vesicle fusions occurring in endocytosis with a cell-free system, EMBO J. 5:3091-3101.

Haest, C.W.M., Plasa, G. and Deuticke, B., 1981, Selective removal of lipids from the outer membrane layer of human erythrocytes without hemolysis, Biochim. Biophys. Acta 649:701-708.

Hakomori, S.-I., 1981, Glycosphingolipids in cellular interaction, differentiation, and oncogenesis, Ann. Rev. Biochem. 50:733-764.

Hakomori, S.-I. and Kannagi, R., 1983, Glycosphingolipids as tumor-associated and differentiation markers, J. Natl. Cancer Inst. 71:231-251.

Hawthorne, J.N. and Ansell, G.B., 1985, eds. Phospholipids, New Comprehensive Biochemistry, Vol. 14, Elsevier Biomedical Press, Amsterdam.

Hedman, K., Goldenthal, K.L., Rutherford, A.V., Pastan, I. and Willingham, M.C., 1987, Comparison of the intracellular pathways of transferrin recycling and vesicular stomatitus virus membrane glycoprotein exocytosis by ultra-structural double-label cytochemistry, J. Histochem. Cytochem. 35:233-243.

Helenius, A., Mellman, I., Wall, D. and Hubbard, A., 1983, Endosomes, Trends Biochem. Sci. 8:245-250.

Helmy, S., Porter-Jordan, K., Dawidowicz, E.A., Pilch, P., Schwartz, A.L. and Fine, R.E., 1986, Separation of endocytic from exocytic coated vesicles using a novel cholinesterase mediated density shift technique, Cell 44:497-506.

Hoekstra, D., 1982, Fluorescence method for measuring the kinetics of Ca^{2+}-induced phase separations in phosphatidylserine-containing lipid vesicles, Biochemistry 21:1055-1061.

Hoekstra, D., 1982a, Role of lipid phase separations and membrane hydration in phospholipid vesicle fusion, Biochemistry 21:2833-2840.

Hoekstra, D., 1983, Topographical distribution of a membrane-inserted fluorescent phospholipid analogue during cell fusion, Exp. Cell Res. 144:482-488.

Hoekstra, D., De Boer, T., Klappe, K. and Wilschut, J., 1984, Fluorescence method for measuring the kinetics of fusion between biological membranes, Biochemistry 23:5675-5681.

Hoekstra, D., Klappe, K., Stegmann, T. and Nir, S., 1988, Parameters affecting the fusion of viruses with artificial and biological membranes, in: Molecular mechanisms of membrane fusion, Ohki, S., Doyle, D., Flanagan, T.D., Hui, S.W. and Mayhew, E., eds., pp. 399-412, Plenum Press, New York.

Hoekstra, D. and Wilschut, J., 1988, Membrane fusion of artificial and biological membranes. Role of local membrane dehydration, in: Water transport in biological membranes, Benga, G., ed. pp. 143-176, CRC Press Inc. Boca Raton.

Hoekstra, D., 1988, Glycolipids, glycoproteins and membrane fusion, Indian J. Biochem. Biophys. 25:76-84.

Hoekstra, D. and Düzgüneş, N., 1989, Lectin-carbohydrate interactions in model and biological membrane systems, Subcell. Biochem. 14:229-278.

Hoflack, B. and Kornfeld, S., 1985, Lysosomal enzyme binding to mouse P338D1, macrophage membranes lacking the 215kDa mannose 6-phosphate receptor: evidence for the existence of a second mannose 6-phosphate receptor, Proc. Natl. Acad. Sci. USA 82:4428-4432.

Hopkins, C.R., 1983, Intracellular routing of transferrin and transferrin receptors in epidermoid carcinoma A431 cells, Cell 35:321-330.

Kaplan, M.R. and Simoni, R.D., 1985, Intracellular transport of phosphatidylcholine to the plasma membrane, J. Cell Biol. 101:441-445.

Kaplan, M.R. and Simoni, R.D., 1985a, Transport of cholesterol from the endoplasmic reticulum to the plasma membrane, J. Cell Biol. 101:446-453.

Kennedy, E.P., 1986, The biosynthesis of phospholipids, in: Lipids and Membranes: Past, present and future, Op den Kamp, J.A.F., Roelofsen, B. and Wirtz, K.W.A., eds., pp. 171-206, Elsevier, Amsterdam.

Kornfeld, R. and Kornfeld, S., 1985, Assembly of asparagin-linked oligosaccharides, Ann. Rev. Biochem. 54:631-664.

Lange, Y. and Matthies, H.J.G., 1984, Transfer of cholesterol from its site of synthesis to the plasma membrane, J. Biol. Chem. 259:14624-14630.

Lange, Y. and Muraski, M.F., 1987, Cholesterol is not synthesized in membranes bearing 3-hydroxy-3-methylglutaryl coenzyme A reductase, J. Biol. Chem. 262:4433-4436.

Lange, Y. and Muraski, M.F., 1988, Topographic heterogeneity in cholesterol biosynthesis, J. Biol. Chem. 263:9366-9373.

Linden, C.D. and Roth, T.F., 1983, The structure of coated vesicles, in: Receptor-mediated endocytosis, series B, Vol. 15, Cuatrecasas, P. and Roth, T.F., eds., pp. 21-44, Chapman and Hall, London.

Lipsky, N.G. and Pagano, R.E., 1983, Sphingolipid metabolism in cultured fibroblasts: microscopic and biochemical studies employing a fluorescent ceramide analogue, Proc. Natl. Acad. Sci. USA 80:2608-2612.

Lipsky, N.G. and Pagano, R.E., 1985, Intracellular translocation of fluorescent sphingolipids in cultured fibroblasts: Endogenously synthesized sphingomyelin and glucocerebroside analogues pass through the Golgi apparatus en route to the plasma membrane, J. Cell Biol. 100:27-34.

Martin, O.C. and Pagano, R.E., 1987, Transbilayer movement of fluorescent analogs of phosphatidylserine and phosphatidylethanolamine at the plasma membrane of cultured cells, J. Biol. Chem. 262:5890-5898.

McLean, L.R. and Phillips, M.C., 1981, Mechanism of cholesterol and phosphatidylcholine exchange or transfer between unilamellar vesicles, Biochemistry 20:2893-2900.

Melançon, P., Glick, B.S., Malhotra, V., Weidman, P.J., Serafini, T., Gleason, M.L., Orci, L. and Rothman, J.E., 1987, Involvement of GTP-binding 'G' proteins in transport through the Golgi stack, Cell 51:1053-1062.

Miller-Podraza, H. and Fishman, P., 1982, Translocation of newly synthesized gangliosides to the cell surface, Biochemistry 21:3265-3270.

Mills, J.T., Furlong, S.T. and Dawidowicz, E.A., 1984, Plasma membrane biogenesis in eukaryotic cells: Translocation of newly synthesized lipid, Proc. Natl. Acad. Sci. USA 81:1385-1388.

Munro, S. and Pelham, H.R.B., 1987, A C-terminal signal prevents secretion of luminal ER proteins, Cell 48:899-907.

Nichols, W.J. and Pagano, R.E., 1981, Kinetics of soluble lipid monomer diffusion between vesicles, Biochemistry 20:2783-2789.

Nichols, J.W. and Pagano, R.E., 1983, Resonance energy transfer assay of protein-mediated lipid transfer between vesicles, J. Biol. Chem. 258:5368-5371.

Noland, B.J., Arebalo, R.E., Hansbury, E. and Scallen, T.J., 1980, Purification and properties of sterol carrier protein$_2$, J. Biol. Chem. 255:4282-4289.

Novick, S.L. and Hoekstra, D., 1988, Membrane penetration of Sendai virus glycoproteins during the early stages of fusion with liposomes as determined by hydrophobic photoaffinity labeling, Proc. Natl. Acad. Sci. USA 85:7433-7437.

Nudelman, E., Kannagi, R., Hakomori, S.-I., Parsons, M., Lipinski, M., Wiels, J., Fellous, M. and Tursz, T. 1983, A glycolipid antigen associated with Burkitt lymphoma defined by a monoclonal antibody, Science 220:509-511.

Op den Kamp, J.A.F., 1979, Lipid asymmetry in membranes, Ann. Rev. Biochem. 48: 47-71.

Pagano, R.E., Longmuir, K.J. and Martin, O.C., 1983, Intracellular translocation and metabolism of a fluorescent phosphatidic acid analogue in cultured fibroblasts, J. Biol. Chem. 258:2034-2040.

Pagano, R.E. and Sleight, R.G., 1985, Defining lipid transport pathways in animal cells, Science 229:1051-1057.

Pearse, B.M.F., 1982, Coated vesicles from human placenta carry ferritin, transferrin and immunoglobulin G, Proc. Natl. Acad. Sci. USA 79:451-455.

Pearse, B.M.F., 1987, Clathrin and coated vesicles, EMBO J. 6:2507-2512.

Pelham, H.R.B., Hardwick, K.G. and Lewis, M.J., 1988, Sorting of soluble ER proteins in yeast, EMBO J. 7:1757-1762.

Pfeffer, S.R. and Rothman, J.E., 1987, Biosynthetic protein transport and sorting by the endoplasmic reticulum and Golgi, Ann. Rev. Biochem. 56:829-852.

Pohlmann, R., Waheed, A., Hasilik, A. and Von Figura, K., 1982, Synthesis of phosphorylated recognition marker in lysosomal enzymes is located in the cis part of Golgi apparatus, J. Biol. Chem. 257:5323-5325.

Regoeczi, E., Chindemi, P.A., Debanne, M.T. and Charlwood, P.A., 1982, Partial resialylation of human asialotransferrin type 3 in the rat, Proc. Natl. Acad. Sci. USA 79:2226-2230.

Reinhart, M.P., Billheimer, J.T., Faust, J.R. and Gaylor, J.L., 1987, Subcellular localization of the enzymes of cholesterol biosynthesis and metabolism in rat liver, J. Biol. Chem. 262:9649-9655.

Rindler, M.J. and Traber, M.G., 1988, A specific sorting signal is not required for the polarized secretion of newly synthesized proteins from cultured intestinal epithelial cells, J. Cell Biol. 107:471-479.

Roseman, M.A. and Thompson, T.E., 1980, Mechanism of the spontaneous transfer of phospholipids between bilayers, Biochemistry 19:439-444.

Roth, J., Taatjes, D.J., Lucocq, J.M., Weinstein, M. and Paulson, J.C., 1985, Demonstration of an extensive trans-tubular network continuous with the Golgi apparatus stack that may function in glycosylation, Cell 43:287-295.

Roth, M.G., Gundersen, D., Patil, N. and Rodriguez-Boulan, E., 1986, The large external domain is sufficient for the correct sorting of secreted or chimeric influenza virus hemagglutinins in polarized monkey kidney cells, J. Cell Biol. 104:769-782.

Rothman, J.E. and Lenard, J., 1977, Membrane asymmetry: The nature of membrane asymmetry provides clues to the puzzle of how membranes are assembled, Science 195:743-753.

Rothman, J.E. and Schmid, S.L., 1986, Enzymatic recycling of clathrin from coated vesicles, Cell 46:5-9.

Rothman, J.E., 1987, Protein sorting by selective retention in the endoplasmic reticulum and Golgi stack, Cell 50:521-522.

Schlessinger, J., 1986, Allosteric regulation of the epidermal growth factor receptor kinase, J. Cell Biol. 103:2067-2072.

Schmid, S.L., Fuchs, R., Male, P. and Mellman, I., 1988, Two distinct subpopulations of endosomes involved in membrane recycling and transport to lysosomes, Cell 52:73-83.

Schmitt, H.D., Puzicha, M. and Gallwitz, D., 1988, Study of a temperature-sensitive mutant of the ras-related YPT1 gene product in yeast suggests a role in the regulation of intracellular calcium, Cell 53:635-647.

Schroit, A.J., Madsen, J. and Ruoho, A.E., 1987, Radioiodinated, photoactivatable phosphatidylcholine and phosphatidylserine: Transfer properties and differential photoreactive interaction with human erythrocyte membrane proteins, Biochemistry 26:1812-1819.

Segev, N., Mulholland, J. and Botstein, D., 1988, The yeast GTP-binding YPT1 protein and a mammalian counterpart are associated with the secretion machinery, Cell 52:915-924.

Seigneuret, M. and Deveaux, P.F., 1984, ATP-dependent asymmetric distribution of spin-labeled phospholipids in erythrocyte membrane: relation to shape change, Proc. Natl. Acad. Sci. USA 81:3751-3755.

Simons, K. and Virta, H., 1987, Perforated MDCK cells support intracellular transport, EMBO J. 6:2241-2247.

Simons, K. and Van Meer, G., 1988, Lipid sorting in epithelial cells, Biochemistry 27:6197-6202.

Sleight, R.G. and Pagano, R.E., 1983, Rapid appearance of newly synthesized phosphatidylethanolamine at the cell surface, J. Biol. Chem. 258:9050-9058.

Sleight, R.G. and Pagano, R.E., 1984, Transport of a fluorescent phosphatidylcholine analog from the plasma membrane to the Golgi apparatus, J. Cell Biol. 99:742-751.

Sleight, R.G. and Pagano, R.E., 1985, Transbilayer movement of a fluorescent phosphatidylethanolamine analogue across the plasma membrane of cultured mammalian cells, J. Biol. Chem. 260:1146-1154.

Sleight, R.G., 1987, Intracellular lipid transport in eukaryotes, Ann. Rev. Physiol.

49:193-208.

Snider, M.D. and Rogers, O.C., 1985, Intracellular movement of cell surface receptors after endocytosis: resialylation of asialo-transferrin receptor in human erythroleukemia cells, J. Cell Biol. 100:826-834.

Sonderfeld, S., Conzelmann, E., Schwarzmann, G., Burg, J., Hinrichs, U. and Sandhoff, K., 1985, Incorporation and metabolism of ganglioside GM_2 in skin fibroblasts from normal and GM_2 gangliosidosis subjects, Eur. J. Biochem. 149:247-255.

Soninno, S., Acquotti, D., Chigorno, V., Pitto, M., Montecucco, C., Schiavo, G., Kirschner, G. and Tettamanti, G., 1988, Recent advances in the chemistry and technology of gangliosides, Indian J. Biochem. Biophys. 25:144-149.

Steinman, R.M., Mellman, I.S., Muller, W.A. and Cohn, Z.A., 1983, Endocytosis and recycling of plasma membrane, J. Cell Biol. 96:1-27.

Stoorvogel, W., Geuze, H.J., Griffith, J.M. and Strous, G.J., 1988, The pathways of endocytosed transferrin and secretory protein are connected in the trans-Golgi reticulum, J. Cell Biol. 106:1821-1829.

Struck, D.K. and Pagano, R.E., 1980, Insertion of fluorescent phospholipids into the plasma membrane of a mammalian cell, J. Biol. Chem. 255:5405-5410.

Struck, D.K., Hoekstra, D. and Pagano, R.E., 1981, Use of resonance energy transfer to monitor membrane fusion, Biochemistry 20:4093-4099.

Tabas, I. and Kornfeld, S., 1978, The synthesis of complex-type oligosaccharides, J. Biol. Chem. 253:7779-7786.

Tanaka, T., Billheimer, J.T. and Strauss, J.F., 1984, Luteinized rat ovaries contain a sterol carrier protein, Endocrinology 114:533-540.

Tettamanti, G., Ghidoni, R. and Trinchera, M., 1988, Recent advances in ganglioside metabolism, Indian J. Biochem. Biophys. 25:106-111.

Tilley, L., Cribier, S., Roelofsen, B., Op den Kamp, J.A.F. and Van Deenen, L.L.M., 1986, ATP-dependent translocation of amino phospholipids across the human erythrocyte membrane, FEBS Lett. 194:21-27.

Tycko, B. and Maxfield, F.R., 1982, Rapid acidification of endocytic vesicles containing α_2-macroglobulin, Cell 28:643-651.

Usuki, S., Lyu, S.-C. and Sweeley, C.C., 1988, Sialidase activities of cultured human fibroblasts and the metabolism of GM_3 ganglioside, J. Biol. Chem. 263:6847-6853.

Vahouny, G.V., Dennis, P., Chanderbhan, R., Fiskum, G., Noland, B.J. and Scallen, T.J., 1984, Sterol carrier protein$_2$ (SCP$_2$)-mediated transfer of cholesterol to mitochondria inner membranes, Biochem. Biophys. Res. Commun. 122:509-515.

Vance, D.E. and Vance, J.E., 1985, eds. Biochemistry of lipids and membranes, Cummings, Menlo Park, CA.

Van Meer, G., Stelzer, E.H.K., Wijnaendts-Van Rosandt, R.W. and Simons, K., 1987, Sorting of sphingolipids in epithelial (Madin-Darby-canine kidney) cells, J. Cell Biol. 105:1623-1635.

Von Figura, K. and Hasilik, A., 1986, Lysosomal enzymes and their receptors, Ann. Rev. Biochem. 55:167-193.

Wattenberg, B.W. and Silbert, D.F., 1983, Sterol partitioning among intracellular membranes, J. Biol. Chem. 258:2284-2289.

Wattenberg, B.W., Balch, W.E. and Rothman, J.E., 1986, A novel prefusion complex formed during protein transport between Golgi cisternae in a cell-free system, J. Biol. Chem. 261:2202-2207.

Watts, C., 1985, Rapid endocytosis of the transferrin receptor in the absence of bound transferrin, J. Cell Biol. 100:633-637.

White, J., Kielian, M. and Helenius, A., 1983, Membrane fusion proteins of enveloped animal viruses, Quart. Rev. Biophys. 16:151-195.

Wiegandt, H., 1985, ed., Glycolipids, New Comprehensive Biochemistry, Vol. 10, Elsevier, Amsterdam.

Wiley, D.C. and Skehel, J.J., 1987, The structure and function of the hemagglutinin membrane glycoprotein of influenza virus, Ann. Rev. Biochem. 56:365-394.

Williamson, P., Bateman, J., Kozarsky, K., Mattocks, K., Hermanowics, N., Choe, H.-R. and Schlegel, R.A., 1982, Involvement of spectrin in the maintenance of phase-state asymmetry in the erythrocyte membrane, Cell 30:725-733.

Wilschut, J. and Hoekstra, D., 1986, Membrane fusion: Lipid vesicles as a model system, Chem. Phys. Lip. 40:145-166.

Wirtz, K.W.A., 1982, Phospholipid transfer proteins, <u>in</u>: Lipid-protein interactions, Jost, P.C. and Griffith, O.H., eds., pp. 151-231, Wiley, New York.

Zaremba, S. and Keen, J.H.., 1983, Assembly polypeptides from coated vesicles mediate reassembly of unique clathrin coats, J. Cell Biol. 97:1339-1347.

CONTROL ANALYSIS: PRINCIPLES AND APPLICATION TO EXPERIMENTAL

PRACTICE

Albert K. Groen[1] and Joseph M. Tager[2]

[1]Department of Gastroenterology and Hepatology and
[2]E.C. Slater Institute for Biochemical Research
University of Amsterdam, Academic Medical Centre
Meibergdreef 15, 1105 AZ Amsterdam
The Netherlands

INTRODUCTION

Cellular metabolism is brought about by a complex network of diverse metabolic pathways each of which is composed of a series of enzyme-catalyzed reactions. Many of the enzymes have been well characterized; their kinetic properties are known and it has been established that certain cellular metabolites regulate their activity in vitro. It is however, often not known if these metabolites also control enzyme activity under in vivo conditions. This is due to the fact that it is very difficult to study control of enzyme activity under in vivo conditions. It is usually impossible to vary the concentrations of substrates, products and allosteric effectors independently under intracellular conditions so that the study of the influence of each individual effector is hampered.

During the last two decades several theories have been developed that enable one to approach this problem. The most important are listed in Table 1.

Table 1. Important control theories.

Name of theory	Originator(s)	Year	Ref.
Control Analysis	Kacser and Burns	1973	1
	Heinrich and Rapoport	1974	2
Metabolic Control Theory	Crabtree and Newsholme	1987	3
Biochemical Systems Theory	Savageau	1969	4

Kacser and Burns (1) and Heinrich and Rapaport (2) developed Control Analysis in 1973-1974. The most recent version of the Metabolic Control Theory of Newsholme and Crabtree appeared in 1987 (3). Savageau (4) first published the principles of Biochemical Systems Theory in 1969. All three theories contain elements which appear to be based on principles published earlier by Higgins (5). Application of the theories to experimental practice has been relatively rare.

Thus far only Control Analysis has been applied by authors other than the originators of the theory. This is probably due to the fact that it is only with Control Analysis that theoretical concepts can readily be translated into actual experiments.

In this paper we shall review some of the work we have done in the past few years on the experimental application of Control Analysis (6-8).

DEFINITION OF CONTROL PARAMETERS

A fundamental characteristic of Control Analysis is the distinction between system (global) parameters and local parameters. The system parameters have been called control coefficients and are defined in general terms as

$$C_P^Y = \frac{P}{Y} \cdot \left(\frac{dY}{dP}\right)_{ss} \tag{1}$$

For instance, the control by an enzyme E_i on a system parameter such as the steady state flux J is given by the fractional increase in flux J induced by a fractional increase in the activity of enzyme E_i. There are two boundary conditions.

Firstly, the system should be in a steady state; and secondly, the activity of the enzyme must be varied independently of other parameters. For instance, if there is enzyme-enzyme interaction in the pathway it is not possible to vary the activity of one enzyme without influencing also the activity of a second enzyme. When the two boundary conditions are fulfilled, there is a summation theorem which states that the sum of the flux control coefficients is equal to unity.

$$\Sigma C_{E_i}^J = 1 \tag{2}$$

Another important control parameter is the concentration control coefficient. For this coefficient, too, there is a summation theorem.

$$\Sigma C_{E_i}^{S_j} = 0 \tag{3}$$

Where S_j is the concentration of an intermediate. The local control coefficients have been called elasticity coefficients and are defined under more restricted boundary conditions. The generalized definition is as follows.

$$\varepsilon_P = \frac{P}{Z} \cdot \left(\frac{\delta Z}{\delta P}\right) \tag{4}$$

Here P refers to parameter P and Z is local property Z.

The elasticity coefficients most often used in practice are those of enzymes towards their substrate(s) and product(s). The boundary conditions for enzyme elasticity coefficients are strict. Only one effector of enzyme activity may vary, whereas all other effectors, thus other substrates, products, allosteric effectors and environmental factors such as pH and ionic strength, must be kept constant. The elasticity coefficients are linked to control coefficients via connectivity theorems. For the flux control coefficients the flux control connectivity theorem states (1) :

$$\sum_{i=1}^{n} C^{J}_{E_i} \cdot \varepsilon^{vi}_{S_j} = 0 \tag{5}$$

For concentration control coefficients the connectivity theorem is (9):

$$\sum_{i=1} C^{S_k}_{E_i} \cdot \varepsilon^{v_i}_{S_j} = -\delta^{k}_{j} \tag{6}$$

$\delta^{k}_{j}=1$ if j=k and 0 if J≠k

In a linear pathway all flux and concentration control coefficients can be calculated directly from the connectivity relations and the summation theorems (1). Since the elasticity coefficients are properties of the isolated components it is clear that in a linear pathway the control coefficients are generated from the elasticity coefficients. In a branched pathway the situation is more complex since at the branch points three or more enzymes share common intermediate. This reduces the number of connectivity relations and extra equations are required to be able to calculate all control coefficients. The extra equations are generated by the branching theorems which give the relations between the control coefficients and the fluxes through the branches. They are defined as follows For the flux control coefficients (10)

$$J_2 \cdot \sum_{\text{branch 1}} C^{J}_{E_i} - J_1 \cdot \sum_{\text{branch 2}} C^{J}_{E_j} = 0 \tag{7}$$

and for the concentration control coefficients (10,11)

$$J_2 \cdot \sum_{\text{branch 1}} C^{S_k}_{E_i} - J_1 \cdot \sum_{\text{branch 2}} C^{S_k}_{E_j} = 0 \tag{8}$$

Since in branched pathways there is more than one flux, a new type of control coefficient can be formulated for such a pathway. This is the flux ratio control coefficient which gives the control exerted by an enzyme on the ratio of two pathway fluxes. It has been defined by Westerhoff and Kell (11) as follows

$$C^{J_r}_{E_i} = \frac{E_i}{J_r} \cdot \left(\frac{dJ_r}{dE_i}\right)_{ss} \tag{9}$$

where J_r refers to the ratio of two fluxes in the pathway.

An essential first step in application of Control Analysis is the delineation of the metabolic pathway under consideration. The boundary condition for determination of control coefficients is that the system should be in a steady state. The translation to experimental practice is to keep the concentrations of the initial substrate and the end product constant. Of course, if pathway flux is not sensitive to the end product or is saturated with the initial substrate the boundary condition is also fulfilled. It is not always possible to predict whether the second boundary condition (no enzyme-enzyme interaction) can be fulfilled. Enzyme-enzyme interactions have been reported to take place in a number of metabolic pathways (13). However, in most of these cases conclusive evidence is lacking. It has proved to be extremely difficult to actually demonstrate the effects of enzyme-enzyme interaction under physiological conditions. When enzyme-enzyme interactions occur the summation theorems are no longer valid and a modified summation theorem has to be used (H. Kacser, personal communication).

Direct determination of control coefficients of enzymes

The simplest way to determine control coefficients of enzymes is to modulate enzyme activity by using a specific inhibitor. The most convenient to use are irreversible inhibitors because in that case the concentration of active enzyme is proportional to the concentration of the inhibitor. If non-competitive, mixed type or competitive inhibitors are used the inhibition kinetic constants have to be known. For all the different types of inhibitors the appropriate equations to be used for the calculation of the control coefficients have been derived and are given in ref. 6. An example of how to determine the flux control coefficient of an enzyme experimentally using an inhibitor is given in the contribution by Meijer (14). When instead of the flux through an enzyme the concentration of intermediary metabolites is measured the concentration control coefficients can be calculated. Note that an enzyme in a linear pathway has only one flux control coefficient but as many concentration control coefficients as there are intermediary metabolites.

The concentration control coefficients on the initial substrate and end product of the pathway are, of course, zero because these concentrations are kept constant. In principle the inhibitor method can only be used if the inhibitor is specific for one enzyme. However, Rigoulet et al. (15) have shown recently that less specific inhibitors can be utilized, too, under some conditions.

Another attractive way to determine control coefficients directly is to vary enzyme activity by genetic manipulation. This method has been employed by Flint et al. (16) and Walsh and Koshland (17) and will probably be the method of choice in the future. Determination of control coefficients by calculation from elasticity coefficients.

As discussed above, control coefficients can be determined indirectly by calculation from elasticity coefficients and the flux distribution in a metabolic pathway. This method is in general more laborious than the direct methods described in the preceding section but it has an important advantage. It not only gives the values for the control coefficients but also provides insight into the mechanisms underlying control by enzymes.

Elasticity coefficients can either be calculated from enzyme kinetic data or be determined experimentally. For enzymes of which the numerator of the rate equation is of the form VF((ABC...-PQR../Keq)) the elasticity coefficient for the substrates is given by:

$$\varepsilon_S^v = \frac{1}{1-\Gamma/K_{eq}} - \frac{V_f}{V'_F} \qquad (10)$$

where vf refers to the rate through the enzyme in the forward direction and V'F to the maximal velocity in the forward direction at a saturating concentration of S but with the other reactants held at the steady state concentrations prevailing during pathway flux. The elasticity coefficient for the products is similarly given by:

$$\varepsilon_P^v = \frac{\Gamma/K_{eq}}{1-\Gamma/K_{eq}} - \frac{V_r}{V'_R} \qquad (11)$$

When the enzyme under consideration operates close to thermo-dynamic equilibrium calculation of the elasticity coefficients is very simple. In that case the first term in eqs. 10 and 11 is very high whereas the second term will be low and can in most cases be neglected. Thus in the case of near-equilibrium reactions elasticity coefficients can simply be calculated from the Γ/K_{eq} ratio. This also holds when two or more near-equilibrium reactions are combined (10). For non-equilibrium enzymes the Γ/K_{eq} is very low; consequently the first term is 1 in equation 10 and 0 in equation 11. Then the second term determines the value of the elasticity coefficient. This term, which, of course, varies between -1 and 0, can either be calculated or determined experimentally. The elasticity coefficients for allosteric enzymes or other enzymes with complex kinetics have to determined experimentally in almost all cases. A detailed example of how the elasticity coefficient of a complex enzyme such as carbamoylphosphate synthetase can be measured experimentally is given in reference (18).

Substrate elasticity coefficients for non-equilibrium enzymes can often be estimated quite simply when it is possible to determine the steady-state pathway flux as a function of the concentration of the substrate. Since under physiological conditions many non-equilibrium enzymes are not inhibited appreciably by their product the elasticity coefficient for the substrate can be calculated directly from a plot of ln(v) as a function of ln(S).

An example is given in Fig 1. Here the rate of glucose formation is plotted as a function of the intracellular glucose-6-phosphate concentration. The relationship is a proportional one; clearly glucose-6-phosphatase is not saturated under these conditions. Hence the elasticity coefficient is equal to 1. By determining or calculating all the elasticity coefficients of the enzymes in a metabolic pathway the control coefficients of these enzymes can be calculated.

We have carried out such a procedure for the complex pathway of gluconeogenesis from lactate. This procedure has

also been applied to simpler pathways such as citrulline synthesis by isolated mitochondria (18) or amino acid metabolism
(19,20). In experimental practice it is not always possible to
measure the elasticity coefficients of all individual enzymes.
Fortunately, some of the methods to determine elasticity coefficients are also applicable to combined sets of reactions.
The flux control coefficient of a combined reaction is equal
to the sum of the flux control coefficients of the individual
enzymes (10). We will give an example of how control of gluconeogenesis from lactate can be studied by dividing the pathway
into four segments that all play an important role in control-

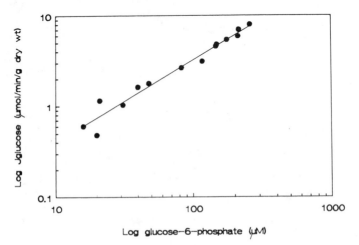

Fig. 1. Relationship between the rate of glucose formation and
the concentration of glucose-6-phosphate. Rat liver
cells (200 mg dry wt.) were perifused with different
concentrations of lactate and pyruvate in the presence
of 0.1 mM oleate and 0.1 µM glucagon. In each steady
state of glucose formation a sample of the cell suspension was taken for the determination of intracellular glucose-6-phosphate.

ling flux. The "sub" pathways which we have constructed are
given in Fig. 2.
      The first segment is formed by the reactions between
pyruvate and phosphoenolpyruvate. Pyruvate kinase constitutes
the second segment in the pathway. We have combined the near-
equilibrium enzymes between phosphoenolpyruvate and glycer-
aldehyde-3-phosphate in segment 3. Segment 4 is the remainder
of the pathway: the reactions between glyceraldehyde-3-phos-
phate and glucose. The elasticity coefficients of these combined reactions can be determined experimentally in a
straightforward way.

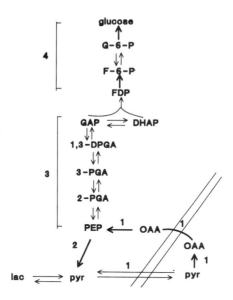

Fig. 2 Simplified scheme of the gluconeogenic pathway.

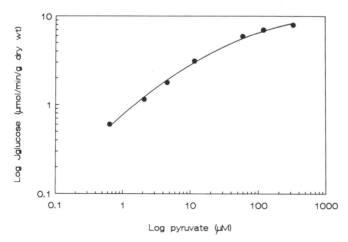

Fig. 3 Relationship between the extracellular pyruvate concentration and gluconeogenic flux. Rat-liver cells (280-300mg dry weight) from starved rats were perifused with different concentrations of lactate and pyruvate in the presence of 0.1 mM oleate and 0.1 μM glucagon.

Fig. 3 depicts the rate of gluconeogenesis from lactate in rat
liver cells as a function of the extracellular pyruvate concentration. The cells were incubated in the presence of 0.1
μM glucagon; under these conditions pyruvate kinase is inhibited and gluconeogenesis is a linear pathway. Since both phosphoenolpyruvate carboxykinase and pyruvate carboxylase are
almost irreversible under intracellular conditions and are not
product inhibited, the combined reactions between pyruvate and
phosphoenolpyruvate are not influenced by the concentration of
phosphoenolpyruvate. Therefore, the elasticity coefficient of
the combined reaction to extracellular pyruvate can be determined from Fig. 2. The value varies from 1 at low pyruvate
concentrations to 0.05 at the highest concentration, indicating that the combined reaction is then almost saturated with
substrate. The elasticity coefficient of the combined reactions towards their product phosphoenolpyruvate can not be
determined in such a simple way. We will assume that it is
equal to the elasticity coefficient of pyruvate carboxylase;
this elasticity coefficient can be calculated from eqs. 10 and
11. Later in this section we will evaluate the influence of
this assumption on the results.

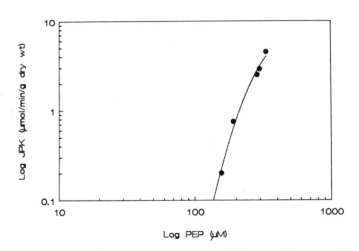

Fig. 4 Relationship between pyruvate kinase flux and the concentration of phosphoenolpyruvate. Rat-liver cells
(280-300mg dry weight) from starved rats were perifused
with different concentrations of lactate and pyruvate
in the presence of 0.1 mM oleate. Pyruvate kinase flux
was determined as described in ref. 21.

Fig. 4 depicts the flux through pyruvate kinase as a function
of the concentration of phosphoenolpyruvate. Pyruvate was
kept constant in the experiment of Fig. 4, hence the elasticity coefficient of pyruvate kinase to phosphoenolpyruvate can
be determined directly from the curve in Fig. 4.
     The elasticity coefficient of the combined near- equili-
brium reactions between phosphoenolpyruvate and glyceralde-
hyde-3-phosphate, can be calculated simply from the combined
Γ/Keq ratio using equations 8 and 9. Fig. 5 shows gluconeo-

genic flux as a function of the concentration of dihydroxyacetonephosphate, the substrate for the combined reactions between glyceraldehyde-3-phosphate and glucose. The glucose concentration was very low in the experiments of Fig. 5 and did
not influence the flux. Assuming that the triosephosphateiso-
merase reaction is in equilibrium the elasticity coefficient
of the combined reaction to glyceraldehyde-3-phosphate can
directly be calculated from Fig. 5.

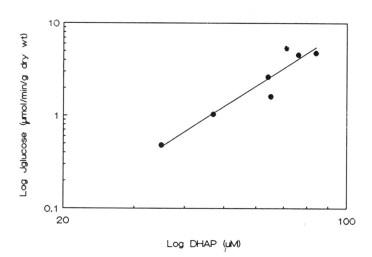

Fig. 5. Gluconeogenic flux as a function of the intracellular
dihydroxyacetone-phosphate concentration. For experimental details see legend to Fig. 4.

The four segments of the gluconeogenic pathway are connected via the following connectivity relations.

$$C_1^{J_g} \varepsilon_{PEP}^1 + C_2^{J_g} \varepsilon_{PEP}^2 + C_3^{J_g} \varepsilon_{PEP}^3 = 0$$

$$C_3^{J_g} \varepsilon_{GAP}^3 + C_4^{J_g} \varepsilon_{GAP}^4 = 0$$

The summation theorem states

$$C_1^{J_g} + C_2^{J_g} + C_3^{J_g} + C_4^{J_g} = 1$$

The branching theorem provides an additional equation

$$J_2 C_1^{J_g} - J_1 C_2^{J_g} = 0$$

The resulting four equations with four unknowns can be solved
for the flux control coefficients.

The procedure for the calculation of the concentration control coefficients is analogous to that discussed above. For the concentration control coefficients, too, connectivity, branching and summation theorems exist. In fact, it turns out that the concentration control coefficients and flux control coefficients can be calculated simultaneously when the equations are grouped in matrix form. For the simplified gluconeogenic pathway in Fig. 2. the full matrix equation is.

$$
\begin{bmatrix}
C_1^{J_4} & C_1^{PEP} & C_1^{GAP} & C_1^{J_2/J_4} \\
C_2^{J_4} & C_2^{PEP} & C_2^{GAP} & C_2^{J_2/J_4} \\
C_3^{J_4} & C_3^{PEP} & C_3^{GAP} & C_3^{J_2/J_4} \\
C_4^{J_4} & C_4^{PEP} & C_4^{GAP} & C_4^{J_2/J_4}
\end{bmatrix}
=
\begin{bmatrix}
1 & 1 & 1 & 1 \\
\varepsilon_{PEP}^1 & \varepsilon_{PEP}^2 & \varepsilon_{PEP}^3 & 0 \\
0 & 0 & \varepsilon_{GAP}^3 & \varepsilon_{GAP}^4 \\
J_2/J_1 & 1 & 0 & 0
\end{bmatrix}^{-1}
\cdot
\begin{bmatrix}
1 & 0 & 0 & 0 \\
0 & 1 & 0 & 0 \\
0 & 0 & 1 & 0 \\
0 & 0 & 0 & 1
\end{bmatrix}
$$

Inversion of the matrix containing the elasticity coefficients yields the "control" matrix containing the values for all the flux control coefficients and concentration control coefficients. The last column of this matrix gives the values for the flux ratio control coefficients.

We have determined values for the elasticity coefficients at different concentrations of lactate and pyruvate. In addition, the flux through the different segments of the pathway was determined under the same conditions. This enabled us to calculate all the control coefficients. For instance at saturating concentrations of lactate and pyruvate, the matrix of control coefficients is as follows

	C^{J_4}	C^{PEP}	C^{GAP}	C^{J_2/J_4}
Segment 1	0.59	0.54	0.49	1.31
Segment 2	-0.17	-0.16	-0.15	0.61
Segment 3	0.31	-0.2	0.26	-1.0
Segment 4	0.27	-0.18	-0.61	-0.91

Under these conditions control on flux is distributed among all segments of the pathway. Pyruvate kinase, of course, exerts, negative control. Surprisingly, the near-equilibrium enzymes between phosphoenolpyruvate and glyceraldehyde-3-phosphate also exert control. Their control is of the same order of magnitude as the control by the enzymes between glyceraldehyde-3-phosphate and glucose although this combined reaction is very much displaced from equilibrium. Clearly the extent to which an enzyme is displaced from equilibrium has little predictive value with respect to the amount of control it exerts. The control exerted by the near-equilibrium enzymes is a consequence of their concentration control coefficient on the concentration of phosphoenolpyruvate. The value of the concentration control coefficient of the reactions between glyceraldehyde-3-phosphate and glucose is very similar and therefore the flux control coefficient of these reactions is also similar to the flux control coefficient of the near-equilibrium reactions.

We have also calculated the values for the control coefficients at lower gluconeogenic fluxes, i.e. at lower concentrations of lactate and pyruvate. In Fig. 6 the values for the flux control coefficients of the different segments of the gluconeogenic pathway are plotted as a function of the flux. It is clear that at lower fluxes almost all control shifts to the first segment of the pathway. This is due to pyruvate kinase. The enzyme shows allosteric kinetics with respect to its substrate phosphoenolpyruvate. At low fluxes the concentration of phosphoenolpyruvate is low and thus the elasticity coefficient of pyruvate kinase to phosphoenolpyruvate is low. This not only leads to a decrease in flux through the enzyme but also lessens the role the enzyme plays in control of gluconeogenesis. In the absence of pyruvate kinase flux, gluconeogenesis is a simple linear pathway. In such a pathway almost inevitably the first non-equilibrium enzyme will control the flux. In the gluconeogenic pathway this role is played by pyruvate carboxylase.

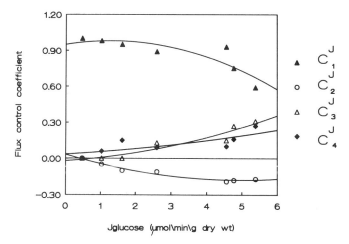

Fig. 6 The flux control coefficients of the four segments of the gluconeogenic pathway as a function of gluconeogenic flux. For experimental details see the legend to Fig. 4. The flux control coefficients were calculated as described in the text.

Only if there were significant product control of pyruvate carboxylase by oxaloacetate could a predominant role of the enzyme be prevented. Since the concentration of oxaloacetate is kept low under intracellular conditions an important role of pyruvate carboxylase is certain. Fig 7a and 7b depict the concentration control coefficients on phosphoenolpyruvate and glyceraldehyde-3-phosphate as a function of gluconeogenic flux. Because the enzymes between phosphoenolpyruvate and glyceraldehyde-3-phosphate are in near-equilibrium the concentration control coefficients of the different pathway segments on the two intermediates are similar. The first pathway segment has a relatively high concentration control coefficient at all fluxes. Pyruvate kinase has a low concentration control coefficient on both intermediates because of the very high elasticity coefficient to phosphoenolpyruvate. This very high elasticity coefficient leads to a buffering of the concentrations of phosphoenolpyruvate and glyceraldehyde-3-phosphate. Fig 7a and 7b show the concentration control coefficients of segment 4 on phosphoenolpyruvate. At low gluconeogenic flux the elasticity coefficient of pyruvate kinase to phosphoenolpyruvate is low. Then the concentration control coefficient of segment 4 on phosphoenolpyruvate is relatively high. At increasing flux, hence at an increasing elasticity coefficient of pyruvate kinase to phosphoenolpyruvate, the concentration control coefficient of segment 4 on phosphoenolpyruvate decreases, showing the buffering effect of pyruvate kinase on the concentration control coefficient of phosphoenolpyruvate.

The effect of experimental error.

As yet, there exists no direct method to estimate the influence of experimental error in the values of the elasticity coefficients on the values of the control coefficients. The influence of experimental error can be assessed indirectly by calculating the effect of a change in the value of an elasticity coefficient on the control distribution. In the calculation given above we used a value for the elasticity coefficient of segment 1 to phosphoenolpyruvate of -0.04. The accuracy of this value is questionable. We have therefore evaluated the importance of this elasticity coefficient on the control distribution. It turns out that a fivefold change in its value changes the values of the control coefficients by only 10%. Clearly, this elasticity coefficient does not play an important role in the control distribution.

CONCLUDING REMARKS

In this paper we have shown that insight into the control structure of a metabolic pathway can be obtained in a relatively simple way by application of Control Analysis. By dividing a metabolic pathway into segments very simple control equations are obtained that can be solved easily provided that the combined elasticity coefficient of the lumped enzymes can be determined. As we have shown this is easy for near-equilibrium segments in the pathway. In general also the substrate elasticity coefficient for a non-equilibrium segment can be determined easily. However, the elasticity coefficient for the product of a non-equilibrium segment is more difficult to determine. These elasticity coefficients have to be determined experimentally or must be calculated from enzyme kinetic data.

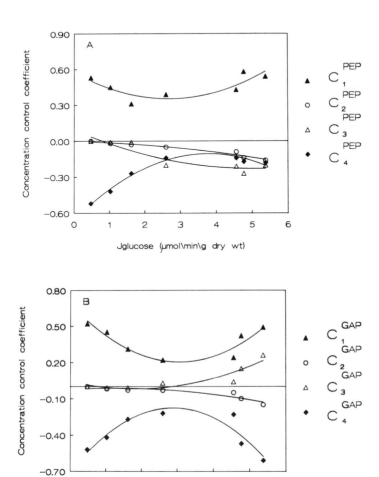

Fig. 7. Relationship between gluconeogenic flux and the concentration control coefficients of four gluconeogenic
pathway segments on the concentration of phosphoenolpyruvate (A) and dihydroxyacetonephosphate (B). For
experimental details see the legend to Fig. 4. The
concentration control coefficients were calculated as
described in the text.

Fortunately, in many cases the influence of these elasticity coefficients on the control distribution is not very large.

REFERENCES

1. H. Kacser and J.A. Burns. The control of flux. Symp. Soc. Exp. Biol. 27: 65-104 (1973).
2. R. Heinrich and T.A. Rapoport. A linear steady state treatment of enzymatic chains. Eur. J. Biochem. 42: 89-95 (1974).
3. B. Crabtree and E.A. Newsholme. The derivation and interpretation of control coefficients. Biochem. J. 247, 113-120 (1987).
4. M. A. Savageau. Concepts relating the behaviour of biochemical systems to their underlying molecular properties. J. Theoret. Biol 25, 365-369 (1969).
5. J. Higgins. Analysis of sequential reactions. In "Control of Energy Metabolism" (B. Chance, R. Estabrook, and J. R. Williamson, eds.), pp 13-46 Academic Press, New York (1963).
6. A.K. Groen, R. van der Meer, H.V. Westerhoff, R.J.A. Wanders, T.P.M. Akerboom, and J.M. Tager. Control of metabolic fluxes in : Metabolic Compartmentation (Sies, H., ed.) pp. 9-37 Acad. Press N.Y. (1982).
7. A.K. Groen, R.J.A. Wanders, H.V. Westerhoff, R. Van der Meer, and J.M. Tager. Quantification of the contribution of various steps to the control of mitochondrial respiration. J. Biol. Chem. 257, 2754- 2757 (1982).
8. A.K. Groen. Quantification of control in studies on intermediary metabolism, Ph. D. Thesis. University of Amsterdam (1984).
9. H.V. Westerhoff and Y-D Chen. How do enzyme activities control metabolite concentrations. Eur. J. Biochem. 142, 425-430 (1984).
10. D.A. Fell and H.M. Sauro. Metabolic control and its analysis. Eur. J. Biochem. 148 555-561 (1985).
11. H.V. Westerhoff and D.B. Kell. Matrix method for determining steps most rate-limiting to flux in biotechnological processes. Biotechnol. Bioeng. 30, 1101-1107 (1987).
12. H.M. Sauro, J.R. Small and D.A. Fell. Metabolic control and its analysis. Eur. J. Biochem. 165, 215-221 (1987).
13. P.A. Srere. Why are enzymes so big? Trends Biochem. Sci. 9, 387-390 (1984)
14. A.J. Meijer. Enzymic reactions in ureogenesis: analysis of the control of citrulline synthesis in isolated rat-liver mitochondria. this book.
15. M. Rigoulet, N. Averet, J-P. Mazat, B. Guerin, and F. Cohadon. Redistribution of the flux control coefficients in mitochondrial oxidative phosphorylations in the course of brain edema.Biochem. Biophys. Acta 932, 116-123 (1988).
16. H.J. Flint, R.W. Tateson, J.B. Barthelmess, D.J. Porteous, W.D. Donachie, and H. Kacser. Control of the flux in the arginine pathway of Neurospora Crassa. Biochem. J. 200, 231-246 (1981).
17. K. Walsh and D.E. Koshland. Characterization of rate-controlling steps in vivo by use of an adjustable expression vector. Proc. Natl. Acad. Sci. USA 82, 3577-3582 (1985).

18. R.J. A Wanders, C.W.T. van Roermund and A.J. Meijer. Analysis of the control of citrulline synthesis in isolated rat-liver mitochondria Eur. Biochem. J. 142, 247-254 (1984).

19. M. Salter, R.G. Knowles and C.I. Pogson. Quantification of the importance of individual steps in the control of aromatic amino acid metabolism. Biochem. J. 234, 635-647 (1986).

20. D.A. Fell and K. Snell. Control analysis of mammalian serine metabolism. Biochem J. 256, 97-101 (1988).

21. A.K. Groen, R.C. Vervoorn, R. Van der Meer and J.M. Tager. Control of gluconeogenesis in rat liver cells. J. Biol. Chem. 258, 14346-14353 (1983).

22. A.K. Groen, C.W.T. Roermund, R.C. Vervoorn and J.M. Tager. Control of gluconeogenesis in rat liver cells. Flux control coefficients of the enzymes in the gluconeogenic pathway in the absence and presence of glucagon. Biochem. J. 237, 379-389 (1986).

CELLULAR ORGANIZATION OF NITROGEN METABOLISM

Dieter Häussinger

Medizinische Universitätsklinik
Hugstetterstrasse 55
D-7800 Freiburg
West Germany

INTRODUCTION

Nitrogen metabolism in mammals is organized at the interorgan, the intraorgan-intercellular, the cellular and the subcellular level. This article focusses mainly on the intercellular and intracellular organization of nitrogen metabolism in the liver and reviews its major structural-functional aspects and implications. However, one should not overlook that also other organs, such as brain and kidney, exhibit a similarly sophisticated organization and regulation of nitrogen metabolism (for reviews see (1,2)). The importance of hepatic nitrogen metabolism is underlined by at least two aspects: (i) following the blood flow the liver is the first organ to process the amino acid and ammonia load derived from intestinal absorption and metabolism (3), (ii) the formation of urea is a liver-specific process with an important role not only for maintenance of ammonia, but also of bicarbonate homeostasis.

The early work of Hans Krebs identified the liver not only as having a particular role in urea synthesis, but also in glutamine metabolism, thereby already emphasizing the different properties of hepatic and renal glutaminases, respectively (4). Much effort has been devoted since then to the understanding of glutamine metabolism in different tissues and its relation to ammonia and acid-base homeostasis (for reviews see (1,2)). Balance studies across the whole liver gave either conflicting results or demonstrated no net glutamine turnover at all, favouring a view in which the liver did not play a major role in glutamine metabolism, although the activities of glutamine metabolizing enzymes were found to be high. Later, a fundamental conceptional change in the field of hepatic glutamine metabolism was derived from the understanding of the unique regulatory properties of hepatic glutaminase (product activation by ammonia), the discovery of hepatocyte heterogeneities in nitrogen metabolism with metabolic interactions between differently localized subacinar hepatocyte populations and the role of an intercellular cycling of glutamine in the liver acinus for the maintenance of whole body ammonia and bicarbonate homeostasis.

The liver acinus is the functional unit of the liver (5). In the acinus, periportal (at the sinusoidal inflow) and perivenous (at the sinusoidal outflow) hepatocytes differ in their enzyme equipment and metabolic function. This functional hepatocyte heterogeneity holds for a variety of metabolic pathways (for reviews see (6), but is most markedly developed with respect to ammonia and glutamine metabolism (Fig. 1). The enzymes of the urea cycle and glutaminase are present in periportal hepatocytes, whereas glutamine synthetase is found only in perivenous hepatocytes. This was shown in the intact rat liver by comparing metabolic fluxes in antegrade and retrograde perfusions (7), in experiments on zonal cell damage (8), by immunohistochemistry (9-11), more recently with hepatocyte preparations enriched in periportal and perivenous cells (12) and by use of the dual-digitonin-perfusion technique (13). The borderline between the periportal urea-synthesizing compartment and the perivenous glutamine-synthesizing compartment is rather strict: glutamine synthetase is exclusively found in a small hepatocyte population (about 6-7% of all hepatocytes of an acinus) surrounding the terminal hepatic venules at the outflow of the sinusoidal bed (9) and these cells are virtually free of urea cycle enzymes (10). Most amino acids being delivered via the portal vein are predominantly taken up and processed in the much larger periportal compartment which contains in addition to urea cycle enzymes also high activities of gluconeogenic enzymes (14). One exception is the uptake of vascular glutamate, which occurs predominantly by the small perivenous hepatocyte population containing glutamine synthetase (15-17). This was underlined in recent studies on the

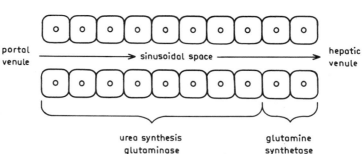

FIG. 1. The liver acinus

Blood flow is from the portal to the hepatic venule; portal blood gets first into contact with periportal hepatocytes containing urea cycle enzymes and glutaminase before perivenous cells capable of glutamine synthesis are reached. This latter cell population ("perivenous scavenger cells") is located at the outflow of the sinusoidal bed and comprises of only about 6-7% of all hepatocytes.

incorporation of added ^{14}C-glutamate into newly synthesized glutamine in perfused rat liver. Extrapolation to maximal rates of hepatic glutamine synthesis showed that about 90% of the labeled glutamate being taken up by the liver was released as labeled glutamine. This demonstrates that the small perivenous cell population containing glutamine synthetase accounts for at least 90% of total hepatic glutamate uptake (Stoll & Häussinger, manuscript in preparation). From the size of this cell population, i.e. about 6-7% of all acinar hepatocytes, a more than 100-fold higher glutamate uptake activity is calculated for perivenous hepatocytes compared to periportal ones, demonstrating also marked hepatocyte heterogeneities regarding plasma membrane transport systems. The joint localization of glutamate uptake and glutamine synthetase in perivenous hepatocytes on the one hand and of the uptake of other amino acids, ureogenesis and gluconeogenesis in periportal hepatocytes on the other hand may serve as examples for the more general rule that functionally linked processes exhibit a common subacinar localization. There are also some further remarkable features of the small perivenous hepatocyte population containing glutamine synthetase: (i) in addition to glutamate, another precursor for glutamine synthesis, i.e. vascular oxoglutarate is practically exclusively taken up by these perivenous cells (18), underlining again the specialization of these cells for glutamine synthesis. (ii) Furthermore, there is some evidence that this perivenous compartment plays an important role for the inactivation of signal molecules (19,20). (iii) The carbonic anhydrase isoenzymes II and III as well as the cytochrome P_{450} PB_2 isoenzyme exhibit an acinar distribution pattern similar to that of glutamine synthetase (21,22). On the other hand mitochondrial carbonic anhydrase V with its important role in providing bicarbonate for carbamoylphosphate synthesis (23,24) is functionally localized in periportal urea-synthesizing hepatocytes, but has not yet been localized histochemically. Whereas glutaminase and glutamine synthetase are heterogeneously distributed in the liver acinus, glutamine transaminases are present in the periportal as well as in the perivenous area of the acinus (25), although slight activity gradients between the respective subacinar zones may exist.

Several functional implications arise from the reciprocal acinar distribution of the two ammonia detoxication systems, urea and glutamine synthesis. The sinusoidal blood stream gets first into contact with cells capable of urea synthesis before the glutamine-synthesizing compartment is reached. In the intact perfused rat liver, the affinity of urea synthesis for ammonium ions is lower than that of glutamine synthesis with $K_{0.5}$ values of about 0.6 and 0.1 mM, respectively; i.e. similar to the differences in the $K_m(NH_4^+)$ values for isolated carbamoylphosphate synthetase and glutamine synthetase. This may explain why detoxication of a physiological portal ammonia load in the isolated perfused rat liver occurs by only about two-thirds through urea synthesis and by about one-third through glutamine synthesis, although the urea cycle capacity is by far not exceeded under these conditions. From a structural-functional point of view urea and glutamine synthesis represent in the intact liver acinus the sequence of a periportal low-affinity, but high-capacity system (urea synthesis) and a perivenous high-affinity system (glutamine synthesis) for ammonia detoxication (7,8) (Fig. 2). The role of perivenous glutamine synthetase is that of a scavenger for the ammonia which escaped periportal urea synthesis, before the sinusoidal blood enters the systemic circulation (7). A considerable fraction of the ammonia generated during the breakdown of amino acids inside the periportal hepatocytes is released into the sinusoidal space, despite the high urea cycle enzyme activity in this compartment. These ammonium ions, however, are transported with the blood stream to perivenous hepatocytes, where they are taken up and utilized for glutamine synthesis

(7). This demonstrates not only metabolic interactions between different subacinar hepatocyte populations, but also underlines the comparatively low affinity of periportal urea synthesis. The importance of perivenous glutamine synthesis as a scavenger for ammonia is evidenced by the finding that an efficient hepatic ammonia extraction at physiologically low ammonia concentrations requires an intact glutamine synthetase activity (7): after destruction of the perivenous area by CCl_4 treatment, glutamine synthesis

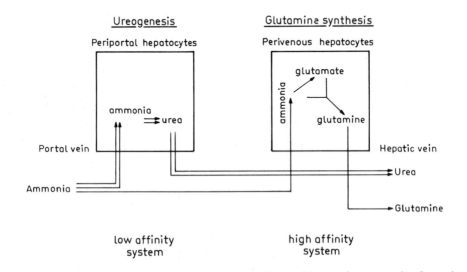

FIG. 2. Structural and functional organization of hepatic ammonia detoxication.
Urea and glutamine synthesis are anatomically switched behind each other and represent with respect to ammonia detoxication the sequence of a periportal "low affinity, but high capacity system" (ureogenesis) and a perivenous "high affinity system" (glutamine synthesis). At a constant portal ammonia load flux through periportal urea synthesis determines the amount of ammonia reaching the perivenous compartment. From ref. (2).

is virtually abolished and hyperammonemia ensues, although simultaneously periportal urea synthesis is not affected (8). An impairment of perivenous glutamine synthesis seems also to be one important pathogenetic factor for the development of hyperammonemia in cirrhotic humans (26). In addition, the perivenous compartment was shown not only to scavenge ammonia before the sinusoidal blood enters the systemic circulation, but has probably also a role in scavenging (i.e. "inactivating") a variety of signal molecules such as extracellular nucleotides and eicosanoids (19,20). This led to the "perivenous scavenger cell hypothesis" (20), with theoretically important implications for the pathophysiology of liver disease.

The structural and functional organization of the ammonia-detoxicating pathways in the liver acinus also implies, that with a constant ammonia supply via the portal vein, flux through the urea cycle in the whole peri-portal compartment will determine the amount of NH_4^+ reaching perivenous hepatocytes, thereby setting the substrate (ammonia) supply for perivenous glutamine synthetase. Accordingly, all factors regulating urea cycle flux indirectly exert control on the more downstream located glutamine syn-thetase. This is of special importance in metabolic acidosis, when peri-portal urea synthesis is switched off in order to spare bicarbonate (for regulatory and mechanistic details see below) and ammonia is increasingly delivered to the perivenous hepatocytes. Thereby ammonia homeostasis is maintained by an increase in glutamine synthesis (2,27). Thus, the structural and functional acinar organization of the ammonia-detoxicating pathways including the scavenger role of perivenous glutamine synthetase allows modulation of urea synthesis without the threat of hyperammonemia and provides a means for switching hepatic ammonia detoxication rapidly from urea synthesis to net glutamine synthesis and vice versa.

HEPATIC GLUTAMINASE: PROPERTIES, CONTROL AND AMPLIFIER FUNCTION

Phosphate-dependent glutaminase (in the following referred to as gluta-minase) is the major glutamine-degrading enzyme in liver. It has a joint localization together with carbamoylphosphate synthetase I inside the mito-chondria of the periportal hepatocytes. The enzyme is both, immunologically and kinetically different from kidney glutaminase (for reviews see (28)). Liver glutaminase is not inhibited by its product glutamate (4), but requi-res its other product ammonia as an essential and obligatory activator. In the absence of ammonia, the enzyme is virtually inactive (29). In the intact liver, half-maximal and maximal activation of glutaminase is obser-ved with ammonia concentrations of 0.2 and 0.6 mM, respectively, indicating that the portal ammonia concentration within the physiological range is an important regulator of hepatic glutamine breakdown (30,31). Regarding the gut-liver axis, intestinal ammonia production was soon recognized as an important signal for glutamine degradation in the liver (socalled "interor-gan feed-forward activation" (30)). Flux through hepatic glutaminase is very sensitive to small pH changes with inhibition at low pH (27,29,31-33), due to a diminished ammonia activation of the enzyme (29,32) and an inhibi-tion of glutamine transport across the plasma and mitochondrial membrane resulting in a decreased mitochondrial steady state concentration of gluta-mine (33). The pH-sensitivity of hepatic glutaminase markedly differs from the kidney enzyme: whereas kidney glutaminase is activated in acidosis, the liver enzyme is inhibited and vice versa in alkalosis.

Flux through hepatic glutaminase is an important factor regulating urea cycle flux (7) because of its structural and functional link to carbamoyl-phosphate synthesis. This occurs at the level of ammonia provision inside the mitochondria for carbamoylphosphate synthetase (27), the rate-control-ling step of the urea cycle (34) with its K_m(ammonia) far above the physio-logical ammonia concentrations (35). The unique regulatory properties of glutaminase (i.e. activation by its product ammonia and pH sensitivity) enable this enzyme to function as a pH-and hormone-modulated ammonia ampli-fying system (2,27) inside the mitochondria (Fig. 3). This amplification allows the urea cycle to operate efficiently even at physiologically low portal ammonia concentrations, what would otherwise be impossible in view of the high K_M(ammonia) of carbamoylphosphate synthetase: ammonia is generated from glutamine inside the mitochondria in parallel to the ammonia delivered via the portal vein (7). Accordingly, a comparatively high urea cycle flux, despite the low ammonia affinity of carbamoylphosphate, can be maintained. In perfused rat liver and with physiological portal ammonia and glutamine concentrations, up to 20-30% of the urea produced is due to intramitochondrial ammonia formation by periportal glutamine breakdown. In addition, some evidence for a direct channelling of glutamine-amide derived nitrogen into carbamoylphosphate synthetase has been presented (36). The outstanding role of glutamine as a substrate for ureogenesis is also underlined by the finding that glutamine-derived nitrogen, in contrast to portally delivered ammonia, is incorporated into urea without requirement of a mitochondrial carbonic anhydrase (37). A further coordination of delivery of glutamine-derived ammonia into carbamoylphosphate synthetase is obtained by the fact that N-acetylglutamate stimulates glutaminase as well as carbamoylphosphate synthetase (38).

Flux through mitochondrial glutaminase in liver is also under the control of glutamine transport across the plasma and mitochondrial membrane. Specific glutamine transport systems are involved (for review see (39)). In the plasma membrane two transport systems have been described: the Na^+-dependent transport system "N" (40), which transports also histidine and asparagine, and a Na^+-independent system, which is probably involved in facilitated diffusion of glutamine from hepatocytes (41). In studies on the initial rate of glutamine uptake into either whole cells (40,41) or mitochondria (43), glutamine influx and efflux rates far above the known rates of metabolic glutamine turnover were found. At first glance, these findings seemed to contradict a role of glutamine transport across biological membranes as an important site of control of glutamine metabolism (25,33,42,44). However, the activities of the glutamine transport systems in the plasma and mitochondrial membranes build up glutamine concentration gradients, which determine the flux through glutamine-metabolizing enzymes in the liver by setting the steady state glutamine concentrations in the respective subcellular compartments (33,42). With a physiological extracellular glutamine concentration of 0.6 mM and a physiological extracellular pH = 7.4, in rat liver in vivo and in the perfused rat liver the cytosolic and mitochondrial glutamine concentrations are about 6 and 20 mM (42). Accordingly, even in presence of a physiological extracellular glutamine concentration, the mitochondrial glutaminase operates at glutamine concentrations near the K_m(glutamine) of the enzyme. When the extracellular pH increases from 7.3 to 7.7 the mitochondrial glutamine concentration raises from 15 to 50 mM (33). Such pH-dependent fluctuations of the mitochondrial glutamine concentrations, despite constancy of the extracellular concentration, are critical for flux through mitochondrial glutaminase with its K_m(glutamine) of 22-28 mM. This also largely explains why flux through glutaminase is increased 4-5-fold upon raising the extracellular pH from 7.3 to 7.7. Thus, the control of glutamine metabolism by transport is exerted by the setting of the subcellular glutamine concentrations, rather than by the absolute velocity of initial glutamine import into the cells. However, also the latter can become controlling for glutamine metabolism in experimental systems: stimulation of glutamine transamination by unphysiologically high concentrations of ketomethionine decreased flux through mitochondrial glutaminase by about 80%, because glutamine taken up by the liver was already consumed in the cytosolic compartment (25).

THE INTERCELLULAR GLUTAMINE CYCLE

In the intact liver acinus, periportal glutaminase and perivenous glutamine synthetase are simultaneously active (7,30,31), resulting in an inter-

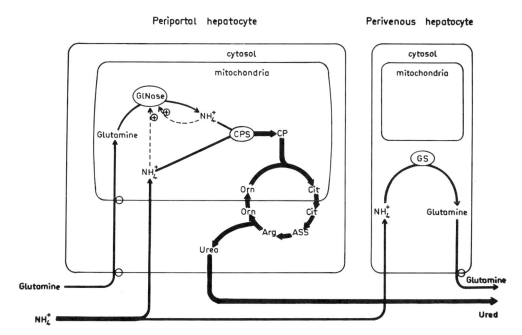

FIG. 3. Functional hepatocyte heterogeneity in glutamine and ammonia metabolism: the intercellular glutamine cycle

Periportal glutamine breakdown by ammonia-activated glutaminase
(GlNase) increases flux through the urea cycle by amplifying the
mitochondrial ammonia supply for carbamoylphosphate synthetase
(CPS). Periportally consumed glutamine is resynthesized perivenous-
ly from the ammonia which escaped periportal urea synthesis ("low
affinity system for ammonia detoxication"). With a well-balanced
acid-base status flux through glutaminase matches glutamine synthetase flux (intercellular glutamine cycling); there is no net turnover of glutamine, however, portal ammonia is completely converted
into urea despite the low affinity of CPS for ammonia. Perivenous
glutamine synthetase (GS) acts as a scavenger for ammonia at the
sinusoidal outflow and allows modulation of urea cycle flux without
threat of hyperammonemia. Periportal glutaminase acts as a pH- and
hormone-modulated ammonia amplifier thereby adjusting urea cycle
flux (besides other mechanisms, which were omitted for clearness
from the scheme) according the requirements of the hormonal and
acid-base status.

cellular glutamine cycle (7) with periportal glutamine breakdown and perivenous resynthesis of glutamine. Intercellular glutamine cycling also
occurs in vivo (16,45) and the flux through both enzymes underlies a complex regulation, resulting in either a net glutamine uptake or release by
the liver or no net glutamine turnover at all. This complex and simultaneous regulation of flux through both enzymes by hormones, pH, ammonia and
glutamine concentrations (31) explains the conflicting findings on the role
of the liver in glutamine metabolism in a variety of earlier studies. With
a physiological pH = 7.4 in the extracellular space, intercellular glutamine cycling allows the complete conversion of portal ammonia into urea
((7,27), for review see (40)) and the net glutamine balance across the
liver is close to zero: periportal glutamine consumption for ammonia amplification in the periportal compartment is matched by the simultaneous
resynthesis of glutamine in the perivenous compartment from the ammonia
which escaped the periportal low affinity system urea synthesis. These
relationships are schematically depicted in fig. 3. In line with an important role of intercellular glutamine cycling for effective ammonia detoxication via the urea cycle is also the finding that flux through the intercellular glutamine cycle increases under conditions known to stimulate urea
synthesis, such as glucagon treatment or an increase of the portal ammonia
load.

Intercellular glutamine cycling also implies opposite net movements of
glutamine across the plasma membrane of periportal (import) and perivenous
(export) hepatocytes, respectively. It is likely that glutamine uptake into
periportal cells is brought about by the Na^+ dependent system N, whereas
glutamine export from perivenous cells probably occurs by facilitated
diffusion, although some studies suggested also a role of Na^+ in the perivenous export process (42).

HEPATIC NITROGEN METABOLISM AND SYSTEMIC PH HOMEOSTASIS

General considerations

Mechanisms for the maintenance of the extracellular pH close to 7.4 must
keep the extracellular HCO_3^-/CO_2 ratio constant, as predicted by the Hen-
derson-Hasselbalch equation. Because CO_2 and HCO_3^- are continuously formed
during the complete oxidation of ingested foodstuffs, there is a need to
eliminate these metabolites from the organism at the same rate as they are
generated. The complete oxidation of carbohydrates and fat yields CO_2 which
is excreted by the lungs. Hydrolysis of proteins yields bipolar amino
acids, whose complete oxidation produces HCO_3^- and NH_4^+ in almost stoichiometric amounts (due to the chemical structure of an "average" amino acid).
In man, ingestion and complete oxidation of a 100g protein diet produces
about 1 mol HCO_3^- and NH_4^+ per day, each. The major pathway for the removal
of metabolically generated HCO_3^- is urea synthesis (46), which consumes
HCO_3^- and NH_4^+ in the same stoichiometry as they are generated during
protein breakdown:

$$2 \; HCO_3^- \; + \; 2 \; NH_4^+ \longrightarrow urea \; + \; CO_2 \; + \; 3 \; H_2O$$

This important role of urea synthesis in eliminating bicarbonate has
been overlooked in the past and the role of this liver-specific process was
identified exclusively with the detoxication of ammonia. In chemical terms,
urea synthesis may be viewed as an irreversible, energy-driven neutralization of the strong base HCO_3^- by the weak acid NH_4^+, yielding neutral and
excretable end-products. It should be noted that the need to neutralize
HCO_3^- by NH_4^+ is common to all land-living animals. Nature has developed
urea synthesis to solve this problem in mammals, whereas the pathway of

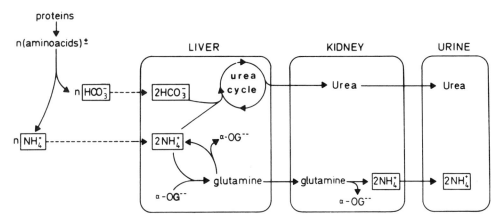

FIG. 4.   Interactions between liver and kidney in the maintenance of
ammonia and bicarbonate homeostasis

NH_4^+ and HCO_3^- are generated during protein breakdown in almost
stoichiometric amounts. Whereas urea synthesis consumes HCO_3^-,
HCO_3^- is spared when hepatic urea synthesis is switched off and
NH_4^+ is excreted as such into urine ("renal ammoniagenesis"), with
glutamine serving as non-toxic transport form of NH_4^+. When NH_4^+ is
excreted as such into urine, there is no net production or consumption of 2-oxoglutarate in the organism. Numbers in circles refer to
major points of flux control by the acid-base status. In metabolic
acidosis, flux through the urea cycle (reaction 1) and hepatic
glutaminase (reaction 2) is decreased, whereas flux through hepatic
glutamine synthetase (reaction 3) and renal glutaminase (reaction
4) is increased. This results in NH_4^+ disposal without concomitant
HCO_3^- removal from the organism. The liver becomes an important
organ in systemic pH regulation by adjusting ammonia flux into
either urea (irreversible HCO_3^- elimination) or glutamine according
the acid-base status and the concerted action between liver and
kidney allows maintenance of NH_4^+ and HCO_3^- homeostasis. From ref.
(55).

purine synthesis is used in birds, some insects and reptils . For the
evolutionary aspects of this topic the reader is referred to (46,47).

Whereas urea synthesis irreversibly consumes 1 mol HCO_3^- per mol NH_4^+
being disposed of, NH_4^+ can also be eliminated as such (without concomitant
HCO_3^- consumption) from the organism by excretion into the urine. This
process involves socalled renal ammoniagenesis (for reviews see (2)), i.e.
the breakdown of glutamine in the kidney with subsequent urinary excretion
of the NH_4^+ being produced. Under these conditions HCO_3^- arising during
protein breakdown is preserved in the organism. A concerted action between
liver and kidney including sophisticated regulatory mechanisms in both
organs allows the adjustment of NH_4^+ flux into either urea or into the
urine, providing the basis for an independent control of ammonia and bicarbonate homeostasis in the organism. These interorgan relationships in
maintaining bicarbonate and ammonia homeostasis are schematically depicted
in fig. 4. At a constant rate of protein breakdown (i.e. of NH_4^+ and HCO_3^-
formation) flux through the urea cycle determines the rate of irreversible
disposal of HCO_3^- from the body; ureogenesis is regulated by the acid-base
status in the extracellular fluid thereby creating a feed-back control
between the extracellular HCO_3^- concentration and the HCO_3^- consumption
parallelling urea synthesis. When bicarbonate has to be spared, for example
in metabolic acidosis, urea synthesis is switched off, resulting in an
increased ammonia delivery to perivenous hepatocytes containing glutamine
synthetase. Non-toxic ammonia levels are maintained by an increased glutamine formation in the perivenous "scavenger cells" of the liver acinus.
Glutamine serves then as a non-toxic transport form of ammonia to the
kidney, where it is liberated again and excreted into the urine. Thus, when
urea synthesis is switched off, the kidney overtakes the task of final
elimation of waste ammonia and glutamine may be seen as a non-toxic transport form of ammonia between the tissues. The rapid switching of hepatic
ammonia detoxication from urea to net glutamine synthesis in metabolic
acidosis has been observed by a variety of investigators (27,31,48-50).

Factors controlling urea cycle flux according the needs of the acid-base
status

Several mechanisms are listed below which participate in adjusting urea
cycle flux in the periportal acinar compartment according the requirements
of systemic bicarbonate homeostasis and contribute to a rapid switching of
hepatic ammonia detoxication from urea to net glutamine synthesis in acidosis. Acid-base control of urea cycle flux is largely exerted at the level
of the delivery of the mitochondrial substrates (ammonia and bicarbonate)
for carbamoylphosphate synthetase and the activity of this enzyme (51).

1. In acidosis flux through periportal glutaminase is markedly decreased
(27,29,31,33) due to a lowering of the mitochondrial steady state glutamine
concentration (33) and a decreased ammonia activation of the enzyme
(28,29,32) at low pH. Accordingly, this switching off of the mitochondrial
ammonia amplifier glutaminase results in a decrease of urea cycle flux
(fig. 3) (27,52). The diminished periportal glutamine consumption contribu-
tes in addition to an increased perivenous glutamine synthesis to net
glutamine release by the liver in acidosis (27).

2. Not only the uptake of glutamine, but also the uptake of other amino
acids by the liver is decreased in acidosis (53).

3. A decrease in extracellular pH increases the degree of protonation of
the NH_3/NH_4^+ system. This could decrease urea cycle flux (50), because NH_3
and not NH_4^+ is the true substrate for carbamoylphosphate synthetase (35).
However, it is unclear whether such an effect plays a major role at physiologically low ammonia concentrations, because compensation in vivo may
occur by involving an increased number of periportal hepatocytes in urea

synthesis (51).

4. A decreased rate of N-acetylglutamate formation at low pH may decrease the activity of carbamoylphosphate synthetase (54).

5. HCO_3^- provision for carbamoylphosphate synthesis inside the mitochondria decreases by inhibition of mitochondrial carbonic anhydrase flux in metabolic acidosis (23) and/or a low CO_2 tension in metabolic acidosis following respiratory compensation. A role of mitochondrial carbonic anhydrase V (24) in providing bicarbonate for biosynthetic processes inside the mitochondria is explained by the fact that the mitochondrial membrane is impermeable for HCO_3^- (not for CO_2) and the spontaneous, uncatalyzed CO_2 hydratation inside the mitochondria is not fast enough to meet the HCO_3^- requirements of urea synthesis. Accordingly, inhibitors of carbonic anhydrase, such as acetazolamide, markedly inhibit urea synthesis.

These mechanisms create a feed-back regulation between the extracellular HCO_3^- concentration, the pH in the extracellular space and the flux through the urea cycle (equivalent to irreversible HCO_3^- disposal), which may help to maintain the extracellular pH close to 7.4 and the extracellular bicarbonate concentration near 25 mM. The structural and functional organization of nitrogen metabolism in the liver acinus is one prerequisite for permitting a modulation of urea cycle flux according the needs of the acid-base status without the threat of hyperammonemia, because an efficient ammonia detoxication is garantueed by the perivenous high affinity system glutamine synthetase (fig. 3). Thus, intercellular glutamine cycling without net hepatic glutamine turnover represents only the special situation of a well balanced acid-base situation. This modern view on the regulation of pH homeostasis (fig. 4; for further details see (2)), being based on complex interorgan and intraorgan-intercellular interactions in nitrogen metabolism, represents a conceptional movement from a two-organ-understanding (lungs and kidneys) to a three-or-more organ-understanding of acid-base balance, including an important role of the liver.

ACKNOWLEDGEMENTS

Our own studies reported herein were supported by Deutsche Forschungsgemeinschaft, Sonderforschungsbereich 154, and the Heisenbergprogramm.

REFERENCES

1 Häussinger, D. & Sies, H., eds. (1984) "Glutamine Metabolism in Mammalian Tissues", Springer Verlag, Heidelberg.
2 Häussinger, D., ed. (1988) "pH Homeostasis - Mechanisms and Control", Academic Press, London.
3 Windmueller, H.G. & Spaeth, A.E. (1974) Uptake and metabolism of plasma glutamine by the small intestine. J. Biol. Chem. 249, 5070-5079
4 Krebs, H.A. (1935) Metabolism of amino acids. The synthesis of glutamine from glutamic acid and the enzymatic hydrolysis of glutamine in animal tissues. Biochem. J. 29, 1951-1959
5 Rappaport, A.M. (1976) The microcirculatory acinar concept of normal and pathological hepatic structure. Beitr. Pathol. 157, 215-243
6 Thurman, R.G., Kauffman, F.C. & Jungermann, K. (1986) Regulation of hepatic metabolism. Plenum Press New York

7   Häussinger, D. (1983) Hepatocyte heterogeneity in glutamine and ammonia metabolism and the role of an intercellular glutamine cycle during ureogenesis in perfused rat liver. Eur. J. Biochem. 133, 269-275

8   Häussinger, D. & Gerok, W. (1984) Hepatocyte heterogeneity in ammonia metabolism: impairment of glutamine synthesis in CCl_4 induced liver cell necrosis with no effect on urea synthesis. Chem. Biol. Interact. 48, 191-194

9   Gebhardt, R. & Mecke, D. (1983) Heterogeneous distribution of glutamine synthetase among rat liver parenchymal cells in situ and in primary culture. Embo J. 2, 567-570

10 Gaasbeek-Janzen, J.W., Lamers, W.H., Moorman, A.F.M., DeGraaf, J., Los, A. & Charles, R. (1984) Immunohistochemical localization of carbamoyl-phosphate synthetase (ammonia) in adult rat liver; evidence for a heterogeneous distribution. J. Histochem. Cytochem. 32, 557-564

11 Saheki, T., Yagi, Y., Sase, M., Nakano, K. & Sato, E. (1983) Immuno-histochemical localization of argininosuccinate synthetase in the liver of control and citrullinemic patients. Biomed. Res. 4, 235-238

12 Pösö, A.R., Penttilä, K.E., Suolinna, E.M. & Lindros, K.O. (1986) Urea synthesis in freshly isolated and in cultured periportal and peri-venous hepatocytes. Biochem. J. 239, 263-267

13 Quistorff, B. & Grunnet, N. (1987) Dual-digitonin-pulse perfusion. Biochem. J. 243, 87-95

14 Jungermann, K. (1986) Functional heterogeneity of periportal and peri-venous hepatocytes. Enzyme 35, 161-180

15 Häussinger, D. & Gerok, W. (1983) Hepatocyte heterogeneity in glutamate uptake by isolated perfused rat liver. Eur. J. Biochem. 136, 421-425.

16 Cooper, A.J., Nieves, E., Filc-DeRicco, S. & Gelbard, A.S. (1988) Short-term metabolic fate of ^{13}N-ammonia, ^{13}N-alanine, ^{13}N-glutamate and amide-^{13}N-glutamine in normal rat liver in vivo. In: Advances in ammonia metabolism and hepatic encephalopathy (Soeters, P.B., Wilson, J.H.P., Meijer, A.J. & Holm, E., eds.) pp. 11-25, Elsevier Science Publ. Amsterdam.

17 Taylor, P.M. & Rennie, M.J. (1987) Perivenous localization of Na-dependent glutamate transport in perfused rat liver. Febs Lett. 221, 370-374

18 Stoll, B., Stehle, T. & Häussinger, D. (1988) Hepatocyte heterogeneity in oxoglutarate metabolism: preferential uptake in perivenous glutami-ne-synthesizing cells. Eur. J. Biochem. submitted

19 Häussinger, D., Stehle, T., Gerok, W., Tran-Thi, T.-A. & Decker, K. (1987) Hepatocyte heterogeneity in response to extracellular ATP. Eur. J. Biochem. 169, 645-650

20 Häussinger, D. & Stehle, T. (1988) Hepatocyte heterogeneity in response to eicosanoids. The perivenous scavenger cell hypothesis. Eur. J. Biochem. 175, 395-403

21 Carter, N., Jeffery, S., Legg, R., Wistrand, P. & Lönnerholm, G. (1987) Expression of hepatic carbonic anhydrase isoenzymes in vitro and in vivo. Biochem. Soc. Trans. 15, 667-668

22 Wolf, C.R., Moll, E., Friedberg, T., Oesch, F., Buchmann, A., Kuhlmann, W.D. & Kunz, H.W. (1984) Characterization, localization and regulation of a novel phenobarbital-inducible form of cytochrom P_{450}, compared with three further P_{450}-isoenzymes, NADPH P_{450}-reductase, glutathione transferases and microsomal epoxide hydrolase. Carcinogenesis 5, 993-1001.

23 Häussinger, D. & Gerok, W. (1985) Hepatic urea synthesis and pH regula-tion. Role of CO_2, $HCO_3{}^-$, pH and the activity of carbonic anhydrase. Eur. J. Biochem. 152, 381-386

24 Dodgson, S.J., Forster, R.E., Schwed, D.A. & Storey, B.T. (1983) Con-tribution of matrix carbonic anhydrase to citrulline synthesis in

isolated guinea pig mitochondria. J. Biol. Chem. 258, 7696-7701

25 Häussinger, D., Stehle, T. & Gerok, W. (1985) Glutamine metabolism in isolated perfused rat liver. The transamination pathway. Biol. Chem. Hoppe-Seyler 366, 517-536

26 Kaiser, S. Gerok, W. & Häussinger, D. (1988) Ammonia and glutamine metabolism in human liver slices: new aspects on the pathogenesis of hyperammonemia in chronic liver disease. Eur. J. Clin. Invest. 18, in press

27 Häussinger, D., Gerok, W. & Sies, H. (1984) Hepatic role in pH regulation: role of the intercellular glutamine cycle. Trends Biochem. Sci. 9, 300-302

28 McGivan, J.D., Bradford, M.N., Verhoeven, A.J. & Meijer, A.J. (1984) Liver glutaminase. In: Glutamine metabolism in mammalian tissues. (Häussinger, D., Sies, H., eds.) pp. 122-137. Springer Verlag Heidelberg.

29 Verhoeven, A.J., van Iwaarden, J.F., Joseph, S.K. & Meijer, A.J. (1983) Control of rat liver glutaminase by ammonia and pH. Eur. J. Biochem. 133, 241-244

30 Häussinger, D. & Sies, H. (1979) Hepatic glutamine metabolism under the influence of the portal ammonia concentration in the perfused rat liver. Eur. J. Biochem. 101, 179-184

31 Häussinger, D. Gerok, W. & Sies, H. (1983) Regulation of flux through glutaminase and glutamine synthetase in isolated perfused rat liver. Biochim. Biophys. Acta 755, 272-278

32 McGivan, J.D. & Bradford, M.N. (1983) Characteristics of the activation of glutaminase by ammonia in sonicated rat-liver mitochondria. Biochim. Biophys. Acta 759, 296-302

33 Lenzen, C., Soboll, S., Sies, H. & Häussinger, D. (1987) pH control of hepatic glutamine degradation. Role of transport. Eur. J. Biochem. 166, 483-488

34 Meijer, A.J., Lof, C., Ramos, I.C. & Verhoeven, A.J. (1985) Control of ureogenesis. Eur. J. Biochem. 148, 189-196

35 Cohen, N.S., Kyan, F.S., Kyan, S.S., Cheung, C.W. & Raijman, L. (1985) The apparent K_m of ammonia for carbamoylphosphate synthetase (ammonia) in situ. Biochem. J. 229, 205-211

36 Meijer, A.J. (1985) Channeling of ammonia from glutaminase into carbamoylphosphate synthetase in rat liver mitochondria. Febs Lett. 191, 249-251

37 Häussinger, D. (1986) Urea synthesis and CO_2/HCO_3^- compartmentation in isolated perfused rat liver. Biol. Chem. Hoppe-Seyler 367, 741-750

38 Meijer, A.J. & Verhoeven, A.J. (1986) Regulation of hepatic glutamine metabolism. Biochem. Soc. Trans. 14, 1001-1004

39 Kovacevic, Z. & McGivan, J.D. (1984) Glutamine transport across biological membranes. In: Häussinger, D. & Sies, H., eds. "Glutamine metabolism in mammalian tissues". pp. 47-58, Springer Verlag Heidelberg.

40 Kilberg, M.S., Handlogten, M.E. & Christensen, H.N. (1980) Characteristics of an amino acid transport system in rat liver for glutamine, asparagine, histidine and closely related analogues. J. Biol. Chem. 255, 4011-4019

41 Fafournoux, P., Demigne, C., Remesy, C. & LeCam, A. (1983) Bidirectional transport of glutamine across the cell membrane in rat liver. Biochem. J. 216, 401-408

42 Häussinger, D., Soboll, S., Meijer, A.J., Gerok, W., Tager, J.M. & Sies, H. (1985) Role of plasma membrane transport in hepatic glutamine metabolism. Eur. J. Biochem. 152, 597-603

43 Kovacevic, Z. & Bajin, K. (1982) Kinetics of glutamine efflux from liver mitochondria loaded with the ^{14}C-labeled substrate. Biochim. Biophys. Acta 687, 291-295

44 Remesy, C., Morand, C., Demigne, C. & Fafournoux, P. (1988) Control of hepatic utilization of glutamine by transport processes or cellular metabolism in rats fed a high protein diet. J. Nutr. 118, 569-578

45 Welbourne, T.C. (1986) Hepatic glutaminase flux regulation of glutamine homeostasis: studies in vivo. Biol. Chem. Hoppe-Seyler 367, 301-305

46 Atkinson, D.E. & Camien, N.M. (1982) The role of urea synthesis in the removal of metabolic bicarbonate and the regulation of blood pH. Curr. Top. Cell. Reg. 21, 261-302

47 Bourke, E. & Atkinson, D.E. (1988) The development of acid-base concepts and their application to mammalian pH homeostasis. In: Häussinger, D., ed. pH Homeostasis, pp. 163-179, Academic Press London

48 Oliver, J., Koelz, A.M., Costello, J. & Bourke, E. (1977) Acid-base induced alterations in glutamine metabolism and ureogenesis in perfused muscle and liver of the rat. Eur. J. Clin. Invest. 7, 445-449

49 Bean, E.S. & Atkinson, D.E. (1984) Regulation of the rate of urea synthesis in liver by extracellular pH: a major factor in pH homeostasis in mammals. J. Biol. Chem. 259, 1552-1559

50 Remesy, C., Demigne, C. & Fafournoux, P. (1986) Control of ammonia distribution ratio across the liver cell membrane and of ureogenesis by extracellular pH. Eur. J. Biochem. 158, 283-288

51 Häussinger, D., Meijer, A.J., Gerok, W. & Sies, H. (1988) Hepatic nitrogen metabolism and acid-base homeostasis. In: Häussinger, D., ed. pH Homeostasis. pp. 337-377. Academic Press London, San Diego.

52 Häussinger, D. & Gerok, W. (1988) Urea synthesis and pH regulation. ISI Atlas of Science Biochemistry 1, 65-701

53 Boon, L. & Meijer A.J. (1988) Control by pH of urea synthesis in isolated rat hepatocytes. Eur. J. Biochem. 172, 465-469

54 Kamemoto, E.S. & Atkinson, D.E. (1985) Modulation of the activity of rat liver acetylglutamate synthase by pH and arginine concentration. Arch. Biochem. Biophys. 243, 100-107

55 Häussinger, D., Gerok, W. & Sies, H. (1986) The effect of urea synthesis on extracellular pH in isolated perfused rat liver. Biochem. J. 236, 261-265

ON THE CELLULAR MECHANISM OF ACTION OF DIPHTHERIA TOXIN

Cesare Montecucco, Emanuele Papini, Giampietro Schiavo, Dorianna Sandona', and Rino Rappuoli[*]

Centro C.N.R. Biomembrane e Istituto di Patologia Generale, Università di Padova, Via Loredan 16, 35131 Padova, Italy and [*]Centro Ricerche Sclavo S.p.a., Via Fiorentina 1, 53100 Siena, Italy

Introduction

A large group of bacterial protein toxins of paramount importance in human and animal pathology share the property of having an intracellular target. The cellular mechanism of action of these toxins can be conveniently divided into three steps: 1) binding to the cell surface, 2) membrane translocation, 3) modification of the cytoplasmic target (Middlebrook & Dorland, 1984).

The most studied of bacterial toxins with intracellular action is diphtheria toxin (DT), produced by strains of Corynebacterium diphtheriae lysogenic for a corynephage carrying the tox gene. This protein toxin is responsible for most clinical symptoms of diphtheria (Pappenheimer, 1977). As schematically shown in fig. 1, DT is secreted as a single chain of 58,342 daltons (Ratti et al., 1983; Greenfield et al., 1983), that is cleaved by tissue proteases or in vitro by trypsin at an exposed loop, giving rise to two chains : B (DT-B, 37,194 daltons), responsible for cell binding and A (DT-A, 21,164 daltons) which is able to ADP-ribosylate specifically elongation factor 2, thereby blocking cellular protein synthesis and causing cell death (Pappenheimer, 1977).

From binding curves of iodine-labelled DT it clearly appears that on sensitive cells, such as Vero cells, there are both high affinity, saturable, sites and low affinity, unsaturable sites. The association constant of DT to the high affinity binding sites is around 10^9 l/M and their number varies from the hundreds or thousands of HELA and CHO cells to the hundred of thousands of Vero cells (Boquet & Pappenheimer, 1976; Chang & Neville, 1978; Middlebrook et al., 1978; Didsbury et al., 1983). There is evidence that a protein is involved in the saturable high affinity binding site (Cieplak et al., 1987; Mekada et al., 1988). The nature of the low affinity binding site has not been identified, however there is evidence that the negatively charged phosphate head group of phospholipids is involved (Moehhring & Crispell, 1974; Alving et al., 1980; Olsnes et al., 1985).

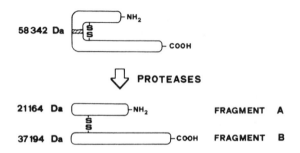

Fig. 1. Schematic structure of DT. The single (top) and di-chain
(bottom) forms of diphtheria toxin are shown together with their M.W..
The single chain DT is inactive and fragment A displays its enzymatic
activity only after protease nicking and reduction of the single
interchain disulfide bridge.

We have investigated the lipid interaction of DT with different
techniques. One approach was based on photolabelling of DT with
liposomes doped with trace amounts of the two photoreactive
phosphatidylcholines depicted in fig. 1. PC I, which is tritium
labelled, reacts only with those regions of a protein interacting with
the polar head group regions of the phospholipid bilayer, while PC II,
labelled with ^{14}C, probes deeper hydrophobic regions of the membrane and
reacts only with those part of a protein exposed to lipids inside the
membrane (Bisson & Montecucco, 1985; Montecucco, 1988). The labelling
reaction of these probes with a membrane interacting protein is
triggered by illumination with non-protein damaging radiations. This
labelling is rather unspecific and results in a covalent attachement of
the radioactive reagent to the protein. The analysis of the distribution
of radioactivity on the protein provides relevant information on its
membrane interaction (Bisson & Montecucco, 1985). Essentially this
membrane photolabelling is a semi-quantitative assay of the interaction
of toxins with phospholipids at two different levels of the lipid
bilayer (Montecucco, 1988).

Lipid interaction of DT at neutral pH

Fig. 2 shows that when DT is incubated with photoactive asolectin
liposomes at neutral pH and illuminated, both toxin subunits are
labelled with the surface probe PC I, but not with the deeper one PC II.
With egg lecithin liposomes, devoid of a net charge, DT-associated
radioactivity was barely detectable.

These observations indicate that DT at neutral pH interacts with
negatively charged lipid bilayers and that this interaction is confined
at the surface level without involving the fatty acid tails of
phospholipids. Unfortunately this approach does not allow one to
extimate an association constant of DT to phospholipids, that appears to

be a weak one (Alving et al., 1980). However, the surface phospholipid interaction of DT may be relevant for the binding of the toxin to cells if combined with that to a protein acceptor.

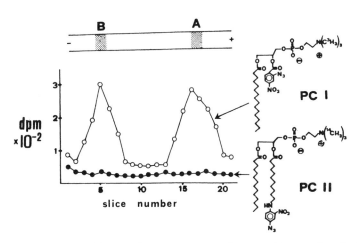

Fig. 2. Membrane labelling of DT at neutral pH. DT was incubated at pH 7.4 with asolectin vesicles doped with trace amounts of PC I and PC II, whose formulae are shown in the right side. The mixture was illuminated and electrophoresed. After staining (pattern shown in the top panel), the gel was sliced and counted and the radioactivity associated to each slice is shown. DT is labelled only with the surface probe PC I and not with the deeper reagent PC II indicating a surface interaction of the toxin with the lipid bilayer.

It was suggested that lipid adsorpion may be the first step in the association of DT to cells and that the lipid-bound toxin moves then laterally on the two-dimensional membrane solvent until it strikes its protein acceptor. At this point the cell binding of DT to the cell is described by the product of the lipid binding constant Klipid and of the protein binding constant Kprotein:

$$Kcell = Klipid \times Kprotein.$$

In this model the presence of an unsaturable unspecific binding both to DT-sensitive and DT-insensitive cells (from murine origine) is accounted for as due to DT adsorption to negatively charged phospholipids. The high affinity saturable site is present only on DT-sensitive cells and it is due to the additional interaction with a cellular protein, involved in a yet unknown function, that happens to have a certain degree of surface complementarity with DT. This dual mode of binding was first put forward to account for the highly specific binding of clostridial neurotoxins to nerve cells (Montecucco, 1986). Here it provides an explanation for the fact that both phospholipase C and protease treatment protects cells from DT intoxication (Mohering &

Crispell, 1974; Olsnes et al., 1985). It also takes into consideration the surprising fact that cells present on their surface a structure capable of binding DT with such high affinity without an evolutive pressure behind. Rather we have to admit that an evolutive pressure was on DT to improve its ability to parasitize a cell structure. A simple and rapid way to achieve this goal is that of fusing the rather common-in-nature phosphate interaction with that of binding a protein: the two single interactions are both rather weak, but, because they occur on the same cell, they give rise by multiplication to a strong "high affinity" interaction (Papini et al., 1987 a).

Low pH DT-lipid interaction

There is clear indication that DT penetrates into the cell cytoplasm from a low pH intracellular compartment (Olsnes & Sandvig, 1988). It has been shown that DT enters into endosome-like compartments after receptor-mediated endocytosis (Moya et al., 1985; Morris et al., 1985). It appears that a transmembrane pH gradient is sufficient to induce the traslocation of DT across the membrane (Sandvig et al., 1986), but the process is not at all clear on the basis of our present understanding of protein structure and function: in fact it is difficult to explain how a water soluble protein can cross the hydrophobic membrane barrier. On the other hand it is important to understand at the molecular level this passage because it is common to other biological events occurring in several physiological and pathological processes such as complement and perforins membrane damage, virus cell penetration and membrane

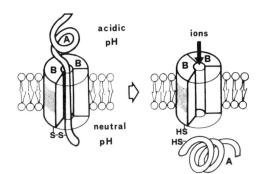

Fig. 3. Tunnel model for the membrane translocation of DT. A schematic drawing of the model put forward by Kagan et al. (1981) to explain the membrane translocation of DT. Fragment B is supposed to form at low pH a transmembrane hydrophilic pore large enough to allow the passage of the A chain in an extended form as well as ions.

translocation of cytoplasmically synthesized proteins into mitochondria and chloroplasts (White et al., 1983 ; Zimmerman & Meyer, 1986; Young et al., 1986; Eilers & Schatz, 1988).

It has been shown that DT at low pH changes conformation and becomes able to form voltage dependent ion channels in planar lipid bilayers (Donovan et al., 1981; Kagan et al., 1981; Hoch et al., 1985; Blewitt et al., 1985). On this basis, a *tunnel* model has been suggested for the translocation of DT across membranes. As depicted in fig. 3, this model envisages that fragment B of DT at low pH becomes hydrophobic and inserts into the lipid bilayer where it forms a hydrophilic pore large enough to accomodate the A chain in an extended form (Kagan et al., 1981; Hoch et al., 1985). DT-A is supposed to cross the membrane without entering in contact with lipids; a corollary of this hypothesis is that the ion channel formed by DT-B is functional to the A chain translocation into the cytosol and hence its study will contribute to understand the molecular aspect of the process itself.

Several studies have been devoted to try to understand the role of low pH in the membrane interaction of DT and to provide support to the idea that it is the driving force of the translocation of DT-A in the cytoplasm. A first requirement is that the pH dependence of the phenomenon overlaps the range of pHs actually measured in the endosomes of living cells, that has never been observed to reach values below 5 (Mellman et al., 1986). In planar lipid bilayers channels are detected only below pH 5 (Donovan et al., 1981).

To monitor the low pH lipid interaction of DT, we used different techniques such as light scattering, fluorescence energy transfer, fluorescence quenching and hydrophobic photolabelling and we found a remarkable agreement among the different methods showing that membrane binding and insertion of DT is essentially complete at pH 5 (Montecucco et al., 1985; Papini et al., 1987 b and c) as shown in fig. 4.

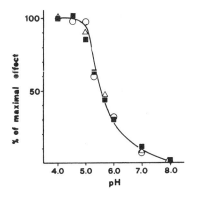

Fig. 4. pH dependence of the lipid interaction of DT. The change of the interaction of DT with phospholipid vesicles as detected with hydrophobic photolabelling with PC II (○), with light scattering (■) and with fluorescence energy transfer (△) is shown.

These results indicate that indeed, as pH is lowered, DT changes conformation with exposure of hydrophobic surfaces that enables it to penetrate into the lipid bilayer and that the phenomenon occurs in a range of pHs fully compatible with the acidification of the endosomal lumen.

Fig. 5 shows that DT, when inserted in the bilayer of negatively charged liposomes, is able to mediate the adhesion and promote the fusion of liposomes.

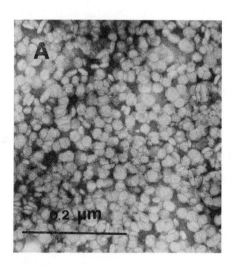 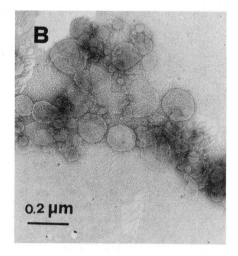

Fig. 5. DT-induced liposome fusion at acidic pHs. DT was incubated with liposomes of dipalmitoyllecithin containing 5% of dipalmitoyl-phosphatidic acid at pH 7.4 (A) or pH was lowered to 4.5 (B). A drop of the suspension was stained with phosphotungstic acid on a electron microscope copper grid and a picture was taken.

All these experiments were performed on model membrane systems and hence one is left with the doubt that they may be not fully relevant to the the *in vivo* process. Hence we performed a series of experiments to determine if some of the results hold true also for cellular systems (Papini et al., 1987 c; Papini et al., 1988). We took advantage of a protocol devised to by-pass the endocytosis step and to induce the cell penetration of DT from the external medium by exposing cells to a short acid pulse (Draper & Simon, 1980; Olsnes & Sandvig, 1980; Sandvig & Olsnes, 1981). Fig. 6 shows the pH dependence we found for the inhibition of cell protein synthesis of Vero cells that is closely similar to that of fig. 4, obtained with model systems.

Another set of experiments was aimed at determining if DT channels are also formed on cells at low pH. It was found that the toxin forms on Vero and CHO cells ion channels and are rather specific for monovalent cations (Papini et al., 1988). These channels open at low pH and close at neutral pH and appear to have a low conductance: a property unexpected for a large hydrophilic pore as suggested in the *tunnel* model.

Most approaches are however unable to discriminate the role of the two toxin fragments. This limitation does not hold for membrane photolabelling. As shown in fig. 7, at low pH both DT chains are labelled by both the surface and the deeper phospholipid probes thus indicating that the low pH-induced toxin interaction with the hydrophobic part of the membrane is mediated not only by the B chain, as proposed by the tunnel model, but also by the A chain. Fig. 7 also shows that DT interacts much more extensively with negatively charged liposomes than with lipid bilayers with no net charge (shaded peaks). The most likely explanation for this finding is that acidification of DT

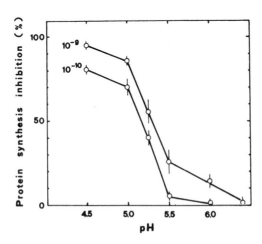

Fig. 6. pH dependence of the intoxication of Vero cells with DT. Different concentrations of DT were incubated at 4 C with Vero cells and the toxin was induced to enter into cells by a 5 min incubation in a 37 C buffer of different pHs. The amount of DT entered into cells is estimated by determining the block of protein synthesis and the effect is reported as percentage with respect to the control taken as 100 %.

preadsorbed onto a negative liposome surface causes its insertion into the bilayer with little self aggregation. On the contrary, in the presence of liposomes with no net charge, the toxin is mainly dissolved in the water phase and self aggregation, due to the matching of the newly exposed DT hydrophobic surfaces, predominates over lipid interaction.

Another point to be considered is that the membrane penetrating form of DT has a net positive charge and these charges have to be masked to decrease the energetic cost of the process. A simple way to achieve this goal is to form ionic couples with the phospholipid neagative head groups. As suggested before (Montecucco et al., 1985), the membrane inserting form of DT may actually be a phospholipid-protein complex where phospholipids are bound to the toxin via ionic couples.

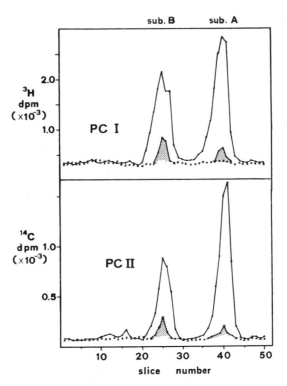

Fig. 7. Membrane labelling of DT with PC I and PC II. DT was incubated with photoactive liposomes made either of asolectin (non-shaded areas) or of egg lecithin (shaded areas), the pH was lowered to 4.5 and the mixtures were illuminated. DT was subjected to SDS-PAGE, aand after staining and destaining the gels were sliced and counted for radioactivity. Both DT subunits are heavily labelled at the surface and the inner core of the lipid bilayer in the presence of asolectin liposomes, while they are weakly labelled in the presence of egg lecithin liposomes.

Not all the results discussed here can be accomodated in the *tunnel* model for the traslocation of DT across membranes. In particular the model does not account for the fact that chain A interacts hydrophobically with phospholipids at low pH and for the low conductance of the DT channels both on cells and on planar lipid bilayers.

Fig. 8 shows a *cleft* model that considers all the presently available experimental data (Bisson & Montecucco, 1987; Papini et al.,

1988). It is proposed that DT can exist in two major conformations: *neutral* above pH 6 and *acid* below pH 6. While the *neutral* conformation binds strongly

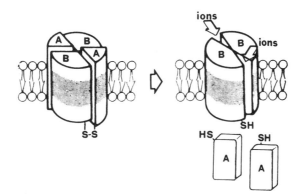

Fig. 8. Schematic drawing of a cleft model for the membrane translocation of DT. DT is supposed to exist in a *neutral* water soluble conformation (not shown here) that binds to the surface of the lipid bilayer and to its protein acceptor. At acid pHs DT converts to an *acid* conformation that detaches from the protein acceptor and, being characterized by large hydrophobic surfaces (shaded areas), penetrates into the lipid bilayer (left panel) with the interchain disulfide bond exposed to the reducing agents of the cytoplasm. Both chains interact with lipids with their hydrophobic surfaces (shaded areas), while their hydrophilic surfaces are facing each other. Hence during membrane translocation hydrophilic residues are not exposed to the hydrophobic membrane core thus reducing the energetic cost of the process. When facing the neutral pH of the cytoplasm, the disulfide bond is broken and chain A reacquires its *neutral* water soluble conformation. It leaves the membrane and it reaches and ADP-ribosylates it cytoplasmic specific substrate elongation factor two thereby blocking protein synthesis.

to its membrane acceptor, the *acid* form shows very little binding and detaches from it as transferrin does inside endosomes. The DT *acid* conformation is characterized by hydrophobic surfaces contributed both by fragment B and fragment A and consequently both subunits penetrate into the lipid bilayer and expose their hydrophobic surfaces to the hydrocarbon chains of lipids. During insertion into the hydrophobic core of the lipid bilayer it is likely that some external positive charges of both the A and B chains are neutralized and shielded from the surrounding hydrophobic solvent by formation of ionic couples with negative polar head groups of phospholipids. Areas of overall hydrophilicity of both the A and the B chains are supposed to face each other in such a way as to be protected from the contact with lipids. The hydrophilic part of the enzymic A chain is nested into an hydrophilic *cleft* formed by hydrophilic surfaces of chain B. Thus an effective matching of hydrophilic and hydrophobic interactions is realized that decreases the energetic cost of the process of translocation as compared with the *tunnel* model where the hydrophilic pore of fragment B has to allow the sliding passage of the 68 hydrophobic residues (out of 193) present in the A chain. This aspect of the *tunnel* model is hard to be accepted : it is difficult to envisage the kind of forces involved in the passage of aminoacid residues of different chemical properties and bulkness inside the same molecular funnel.

In principle, DT can insert with its interchain disulfide bridge facing either the cytoplasm (trans face) or the endosomal lumen (cis face). Only in the former case the disulfide bond can be quickly reduced by glutathion or other cytoplasmic reducing agents such reducing enzymes. When facing the neutral cytoplasmic pH DT-A regains its *neutral* water soluble conformation and thus leaves the membrane and reaches its specific cytoplasmic substrate elongation factor 2. The reversible membrane insertion of DT-A is quite a crucial aspect of the model and it is rather important that it has been proven experimentally to occur (Montecucco et al., 1985).

After chain A has left, the hydrophilic *cleft* of chain B is expected to reduce its size to minimize the amount of surface exposed to lipids. However the presence of a transmembrane alignement of hydrophilic residues would give rise to an ion channel as that we have observed in cells (Papini et al., 1988). Such a feature accounts for the low conductance of DT channels (few picoSiemens) better than a fragment B *tunnel*, that having to accomodate the size of a polypeptide chain, it is expected to have a conductance similar to or higher (hundreds of pS) than that of known pore-forming proteins (Colombini, 1980; Menestrina, 1986).

REFERENCES

- Alving, C.R., Iglewski, B.H., Urban, K.A., Moss, J., Richards, R.L. & Sadoff, J.C. (1980) Proc. Natl. Acad. Sci. U.S.A. 77, 1986-1990
- Bisson, R. & Montecucco, C. (1981) Biochem. J. 193, 757-763
- Bisson, R. & Montecucco, C. (1987) Trends Biochem. Sci. 12, 187-188
- Bisson, R. & Montecucco, C. (1985) in "Progress in Protein-Lipid Interactions", pp. 259-287, Watts, A & De Pont J.J.H.H.M. Eds., Elsevier Science Publishers, Amsterdam
- Cieplak, W., Gaudin, H.M. & Eidels, L. (1987) J. Biol. Chem. 262, 13246-13253
- Colombini, M. (1980) J. Membr. Biol. 53, 79-84
- Didsbury, J.R., Moehring, J.M. & Moehring, T.J. (1983) Moll. Cell. Biol. 3, 1283-1294
- Donovan, J.J., Simon, M.I., Draper, R.K.,& Montal, M. (1981) Proc. Natl. Acad. Sci. U.S.A. 78, 172-176
- Draper, R.K. & Simon, M.T. (1980) J. Cell Biol. 87, 849-854
- Greenfield, L., Bjorn, M.J., Horn, G., Fong, D., Buck, G.A., Collier, R.J. & Kaplan, D.D. (1983) Proc. Natl. Acad. Sci. U.S.A. 80, 6853-6857

- Hoch, D.H., Romero-Mira, M., Ehrlich, B., Finkelstein, A., DasGupta, B.R. & Simpson, L.L. (1985) Proc. Natl. Acad. Sci. U.S.A. **82**, 1692-1696
- Kagan, B.L., Finkelstein, A. & Colombini, M. (1981) Proc. Natl. Acad. Sci. U.S.A. **78**, 4950-4954
- Mekada, E., Okada, Y. & Uchida, T. (1988) J. Cell Biol. **107**, 511-519
- Menestrina, G. (1986) J. Membr. Biol. **90**, 177-190
- Middlebrook, J.L. & Dorland, R.B. (1984) Microbiol. Rev. **48**, 199-221
- Moehring, T.J. & Crispell, J.P. (1974) Biochem. Biophys. Res. Commun. **60**, 1446-1452
- Montecucco, C. (1986) Trends Biochem. Sci.11, 314-317
- Montecucco, C. (1988) Methods Enzymol. **165**, 347-357
- Morris, R.E., Gersten, A.S., Bonventre, P. & Salinger, C.B. (1985) Infect. Immun. **50**, 721-727
- Moya, M., Dautry-Varsat, A., Goud, B., Louvard, D. & Boquet, P. (1985) J. Cell Biol. **101**, 548-559
- Olsnes, S., Carvajal, E., Sundan, A. & Sandvig, K. (1985) Biochim. Biophys. Acta **846**, 334-341
- Olsnes, S. & Sandvig, K. (1988) in "Immunotoxins", Fraenkel, A.E. Ed., pp. 39-73, Kluwer Academic Publ., New York
- Papini, E., Colonna, R., Schiavo, G., Cusinato, F., Tomasi, M., Rappuoli, R. & Montecucco, C. (1987 c) F.E.B.S. Lett. **215**, 73-78
- Papini, E., Colonna, R., Cusinato, F., Montecucco, C., Tomasi, M. & Rappuoli, R. (1987 b) Eur. J. Biochem. **169**, 629-635
- Papini, E., Schiavo, G., Tomasi, M., Colombatti, M., Rappuoli, R. & Montecucco, C. (1987 c) Eur. J. Biochem. **169**, 637-644
- Papini, E., Sandona', D., Rappuoli, R. & Montecucco, C. (1988) E.M.B.O. J., in press
- Pappenheimer, A.M.Jr. (1977) Annu. Rev. Biochem. **46**, 69-94
- Ratti, G., Rappuoli, R. & Giannini, G. (1983) Nucl. Acids Res. **11**, 6589-6595
- Sandvig, K. & Olsnes, S. (1980) J. Cell Biol. **87**, 828-832
- Sandvig, K. & Olsnes, S. (1981) J. Biol. Chem. **256**, 9068-9076
- Sandvig, K., Tonnessen, T.I., Sandl, O. & Olsnes, S. (1986) J. Biol. Chem. **261**, 11639-11644
- Simpson, L.L. (1986) Annu. Rev. Pharmacol. Toxycol. **26**, 427-453

ORGANIZATION AND EVOLUTION OF MITOCHONDRIAL DNA IN METAZOA

Cecilia Saccone and Elisabetta Sbisa'

Centro di Studio sui Mitocondri e Metabolismo
Energetico, C.N.R. Bari e Dipartimento di
Biochimica e Biologia Molecolare, Universita'
di Bari, 70126 Bari, Italy

SIZE, SHAPE AND GENE CONTENT

The most relevant and peculiar feature of mitochondrial
(mt) DNA in animal cells is its compact gene organization.
Genes coding for transfer (t), ribosomal (r) and messenger (m)
RNAs are often contiguous, sometimes slightly overlap or are
separated by only a few nucleotides. One main non-coding
region is usually present in the genome. Its length, largely
variable, can be reduced, as in sea urchins, to a hundred
nucleotides (Attardi, 1981a, b; 1985; Cantatore and Saccone,
1987 for review). As a consequence of such economic gene
organization the size of mitochondrial DNA in metazoa is
extremely reduced compared with that of other eukaryotes.
Among the mitochondrial genomes completely sequenced so far,
whose length is shown in Table 1, the smallest size pertains
to Ascaris mt DNA with 14,284 base pairs (bp), the largest to
a frog, Xenopus laevis with 17,553 bp. More recently numerous
cases in which the length of mt DNA exceeds the normal, rather
constant average, size found in animal cells have been
described in the literature (Moritz et al., 1987 for review).
Such length variations can span an almost two-fold size range
(14-28 kb) or even more (39 kb as in sea scallop, Snyder et
al., 1987). However it seems that in such cases there is
always a duplication of some genomic regions. Length
variations are more frequent in invertebrates and in
poikilotermic vertebrates and can be distributed within
(heteroplasmy) and between individuals and thus may be
generated very rapidly. In general length differences are
confined to the control region but cases in which length
variations appear to be dispersed and/or include structural
genes have also been described. Deletion of mt genome tracts
have been reported, particularly in heteroplasmic animals.
Large deletions of the mt genome have been described in mice
(Boursot et al., 1987). Very recently it has been
demonstrated that heteroplasmy for mt DNA does occur in man
(Holt et al., 1988) and that defects of the mitochondrial

genome may be associated with maternally inherited human diseases. Both tissue specific sequence deletions or duplications and point mutations can be responsible for a variety of clinical disorders (Wallace et al., 1988).

Table 1. Nucleotide Composition for Mitochondrial Genomes of Metazoa

Species	Length	Percentage Nucleotide Composition			
		A	C	G	T
Ascaris suum	14,284	22.2	7.6	20.4	49.8
Drosophila yakuba	16,019	39.4	12.2	9.3	39.1
Strongylocentrotus p.	15,650	28.7	22.7	19.4	30.2
Paracentrotus lividus	15,700	30.8	22.5	17.2	29.5
Xenopus laevis	17,553	33.0	23.5	13.5	30.0
Mus musculus	16,295	34.5	24.4	12.4	28.7
Rattus norvegicus	16,298	34.1	26.2	12.5	27.2
Bos taurus	16,338	33.4	25.9	13.4	27.3
Homo sapiens	16,569	31.0	31.2	13.1	24.7

In general the nucleotide composition is A>T>C>G except in man (C>A>T>G) and in Ascaris (T>A>G>C). In Drosophila and in sea urchins the percentages of A and T are similar. The percentage of T decreases from 49.8 (Ascaris) to 24.7 (Homo Sapiens), whereas that of C increases from 7.6 (Ascaris) to 31.2 (Homo Sapiens). The percentage of G is the lowest in all the genomes except in Ascaris. The data reflect the base composition of one strand of the genome.

Small duplications and deletions can be explained by replication slippage mechanisms. However mechanisms other than intermolecular recombination which seems to be absent in animal mitochondria, are probably responsible for larger changes in the mt DNA sequence. Moritz and Brown (personal comunication) have shown that in lizard mt DNA several junction points of the duplications and deletions so far mapped, appear to align whith tRNAs. The Authors suggest that tRNAs might provide, through their secondary structure, a recognition signal for the enzymes responsible for the rearrangement. This observation confirms the peculiar role of tRNAs in mt evolution which will be discussed below. The extreme compacteness of the mt genome in animal cells has been explained by assuming a strong positive selection for a small genome size. Not only recombination but also repair are lacking in animal mitochondria and thus one can wonder what the strategy used by mt DNA for maintaining the fidelity of the genetic message is. Palumbi and Wilson (personal communication) have proposed a model according to which the replication rate of certain mt DNA molecules can replace the role of repair enzymes. In other words, under certain conditions, a replication race between cellular genomes might take place so as to ensure the constant presence of good templates. The winners will be the fastest replicating molecules which will predominate in the population.

As far as the shape of animal mt DNA is concerned, the circular double stranded structure appears to be a constant rule. To our knowledge there is as yet only one documented exception: the linear shape of mt DNA in Hydra which is probably not an artifact due to DNA isolation procedure (Warrior and Gall, 1985).

The gene content of mt genome is another "quasi" constant feature in animal cells. Again we have only one documented exception Ascaris, (Wolstenholme et al., 1987) which lacks one of the ATPase subunits, the so called subunit 8. It is well known that the gene content of mt DNA in metazoa is not dramatically different from those of other eukaryotes. It consists of the following:

i) a set of 22 tRNAs which, owing to a more extended U:N whobble, are able to recognize a degenerate triplet code slightly different from the universal genetic code.

ii) two ribosomal RNA species, smaller than their cytoplasmic counterparts and having (at least in vertebrates) a sedimentation coefficient of 12S and 16S.

iii) messenger RNA species for 13 proteins (in Ascaris 12) which are subunits of enzyme complexes present in the inner mitochondrial membranes, in particular: 3 subunits of cytochrome oxidase (CO), 2 subunits (in Ascaris 1) of ATPase (A), seven subunits of NADH dehydrogenase (ND) and cytochrome b (b). In some cases a messenger RNA may contain the information for two proteins in different reading frames. In vertebrates there are eleven mRNAs for thirteen proteins since the reading frames of ND4L and ND4 as well as those of ATPase 8 and ATPase 6 are partially overlapping and are translated from a single transcript.

In mitochondrial DNA the genetic code is slightly different from the universal one. Moreover, it is also varies from philum to philum. The deviations from universality of the mt genetic code in various species are shown in Table 2. Another peculiar aspect of mt protein coding genes is that several codons can act as initiators of protein synthesis. In addition to AUG, the other three codons of the some family namely AUA, AUC, AUU, and in a few cases also GUG, have been found at the 5' termini of mRNAs (Gadaleta et al., 1988) Recently, a detailed comparative analysis of the mRNA coding genes in mammals (Gadaleta et al., 1989) and echinoderma (Cantatore et al., 1989) has revealed that the use of the initiator codons appears to follow a well defined rule. In particular, we have observed that when the mRNA contains untranslated nucleotides at the 5' end, AUG is used as the initiator, whereas when the messenger starts with the first codon, other unusual initiators may be present at its 5' termini. In our opinion this indicates that AUG is necessary to put the messenger in the right frame when there are 5' untranslated nucleotides. The rule seems to be confirmed in the mt genomes of other animal mt DNAs with a few exceptions which in any case demonstrate the importance of sequencing not only the genes, but also the corresponding mRNAs and proteins.

Table 2. Deviation from the Universal Code in Mitochondria

Codons	Nuclear Code	Mitochondrial Code			
		Yeast	Drosophila	Vertebrates	
		Neurospora	Echinoderma		
AUA	Ile	Met	Ile	Met	Ile	Met
CUN	Leu	Thr	Leu	Leu	Leu	Leu
AGR	Arg	Arg	Arg	Ser	Ser	Ter
UGA	Ter	Trp	Trp	Trp	Trp	Trp
AAA	Lys	Lys	Lys	Lys	Asn	Lys

GENE ORGANIZATION

In constrast with the constant gene content of animal mt genome, the gene order and organization varies strikingly from philum to philum. Gene rearrangement consists of a different distribution of genes between the two strands (polarity inversion), gene transposition and also, at least in one case (Ascaris), a gene loss (Woltenholme et al., 1985). In Fig. 1 the mt gene organization in four different phila, namely sea urchin, vertebrates, Ascaris and Drosophila, is shown. The most remarkable feature is the lack of correlation of tRNA distribution in the four genomes. In several phila, as in vertebrates, tRNA genes are scattered among genes coding for proteins where they can act as punctuation marks in primary transcripts. In others, like sea urchins, 15 of the 22 tRNA genes are in a cluster that could have a role in the initiation and regulation of replication and transcription (Brown et al., 1986; Cantatore et al., 1989). Numerous observations by different Authors clearly indicate that mt tRNAs must be considered mobile elements.

Events of duplication and remoulding of tRNA genes during the evolutionary rearrangement of mitochondrial genomes have been suggested by us on the basis of comparative analysis carried out on the tRNA genes of Paracentrotus lividus. In particular we observed:

i) a strong sequence resemblance between the CUN and UUR leucine tRNA genes.

ii) an altered location of the CUN gene in mt DNA of Paracentrotus with respect to the location of the same gene in vertebrates.

iii) the persistance of a 72 bp sequence containing a trace of the old CUN gene at its original location, which now codes for an extra amino acid sequence at the amino termini of ND5, a subunit of NADH dehydrogenase.

On the basis of these findings we suggested that during the evolution of sea urchins, a transfer RNA gene lost its function and became part of a protein coding gene. This loss was accompained by the gain of a new tRNA with that specificity through duplication and divergence from a tRNA gene specific for another group of codons, namely leucine UUR tRNA gene. (Cantatore et al., 1987). The new urchin CUN gene is near the UUR gene, both residing in the cluster of 15 tRNA genes surrounding the noncoding region.

Starting from our observation (Brown et al., 1986;

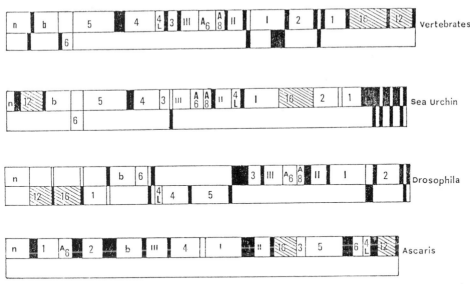

Fig. 1. Gene organization in metazoa mt DNA. n=main non-coding region; b= cytochrome b; 1, 2, 3, 4, 4L ,5, 6= NADH deydrogenase subunits; I, II, III= citochrome oxidase subunits; A6, A8= ATPase subunits; 12, 16= rRNAs; black boxes= tRNAs. The gene distribution between the two strands is indicated.

Saccone et al., 1987a) that tRNA like structures, probably acting as primers, may be necessary for DNA synthesis, we suggested that duplication of the tRNA gene could occur through such a mechanism: if tRNAs could themselves act as primers for mt DNA synthesis and sometimes fail to be removed from the newly synthesized DNA strand, the tRNA would became a template in the next round of duplication. The result would be the incorporation of a tRNA gene at the replication origin. If this mechanism takes place it is clear that tRNA genes can change their location indipendently of other functional products.

As we have pointed out in the previus section, the finding that tRNAs are present at the end of duplications and deletions, also supports the hypothesis that tRNA genes are involved in gene rearrangement. If we assume that tRNA genes can move independently of other mitochondrial genes, we can more easily interpret the differences in gene order between phila. Only a few events are needed to interrelate the various gene orders. An unrooted tree giving one possible explanation for the differences in gene order found among DNAs of the four animal phila depicted in Fig. 1 is reported in Cantatore et al. (1987). In general we can say that the gene order of each philum is more similar to that of vertebrates than to each other. This implies that the vertebrate lineage lost the capacity to rearrange before the other metazoa did and thus its present organization is more similar to that of the common ancestor. Other evidence based on the gene organization in Echinodermata indicate that between Asteroids and Echinoids, which are separated by about 400-500 milion years, there is only one major rearrangement event: the region included between the tRNA cluster and the 16S RNA was inverted.

In general it seems that in the evolution of metazoa the capacity to rearrange and/or to recombine was lost rather early, probably contemporaneously with the reduction in size. In this regard the mt genome of animals behaves exactly in the opposite way with respect to the mt genome of plants where we find a very large genome (up to ~2,400 Kb in some cucurbits) (Ward et al., 1981) and an extremely high intermolecular recombination capacity (Palmer and Herbon, 1988). The recent observations, reported in the previus section, indicating the occurrence of sequence duplication in philogenetically diverse animal mt genomes, suggest on the other hand that mt function may not be disrupted by at least some rearrangement mechanisms and thus the potentiality for gene rearrangement may still be present. As pointed out by Moritz et al. (1987), combined mechanisms of duplication and deletion, can be responsible for gene rearrangement according to the following scheme:

AB -> duplication -> ABAB -> deletion -> BA

BASE SUBSTITUTION RATE

The evolution of mt genome in various organisms has been throughly studied by using endonuclease restriction enzymes. These studies have provided important information on mitochondrial gene evolution but do not allow a precise measurement of the mt evolutionary dymamics. A much clearer and deeper insight into the sequence variation and the evolution of animal mt DNA was achieved with the determination of the nucleotide sequences of complete mitochondrial genomes or pieces of them.

It is well known that the evolutionary rate of biological molecules depends on both the mutation rate and the mutation fixation rate (Wilson et al., 1977). The latter value is determined by the functional constraints of the molecules. Since the mitochondrial genome codes for products having different functions, this thus implies that the evolution of the mitochondrial genome as a whole is the result of various evolutionary processes, having different dynamics, each of which must be considered separately. Among mt DNA products, mRNA genes are the most suitable for analysis in a quantitative manner. In the evolution of protein-coding genes we can distinguish between silent substitutions, a quasi-neutral process, and replacements, a process which is linked to the functional constraints of the protein. For rRNA and tRNA genes, whose functional constraints cannot be precisely determined, it is more difficult to quantitatively evaluate their evolutionary rate.

Reliable estimates of the evolutionary effective substitution rates of the four nucleotides strongly depend both on a trustworthy determination of the divergence time and on a correct calculation of the nucleotide substitution number. Several methods have been suggested for the correct estimation of the evolutionary substitution rate of homologous genes and, even though such methods have different bases, there is a wide consensus that the substitution process is stochastic in nature. The majority of the methods for calculating the rate of silent substitutions are derived from the widely used Jukes and Cantor formula (1969). However this formula seems somewhat unsatisfactory since it assumes that each nucleotide can go into the other three with the same probability, a situation which has been repeatedly proven to be unrealistic. Moreover, all the modifications apported to the Jukes and Cantor formula, as well as other methods that have appeared in the literature are always based upon a number of aprioristic assumptions which are often experimentally untenable (Kimura, 1981; Gojobori et al., 1982; Brown et al., 1982). It was precisely to measure gene evolution in a quantitative way that a few years ago we proposed a new method, the stationary Markov model (Lanave et al., 1984, 1985; Preparata and Saccone, 1987). Giving as imput the divergence time between two species, the method allows one to calculate:

i) the divergence time between any other couples of species, i.e. to construct phylogenetic trees.

ii) the substitution rate of each nucleotide going into the other three (substitution rate matrix).

iii) the average evolutionary rates for the genes under consideration.

The prerequisite for the application of our model is that the frequencies qi (i=A,C,G,T) of the 4 bases at the same codon positions (first, second, or third) in the homologous genes under comparison, must be constant within statistical fluctuations. Only when these conditions can one reconstruct in a quantitative way the dynamics of the process under consideration, from the structure of the rate matrix. According to our "base-drift hypothesis" (Preparata and Saccone, 1987), homologous genes mantain the same average base composition of the common ancestor (stationarity condition) when the divergence time (T) is not too large and/or when the dynamics of the evolutionary process are not too fast. In other words, the stationarity condition depends on: V.T, the product of the velocity of the evolutionary process (V) and the divergence time (T). Loss of stationarity can also be due to directional mutation pressure (Sueoka, 1988).

Applying the above mentioned method we have calculated for the mammalian mitochondrial genes the matrix of base substitution rates at synonymous sites (Cantatore and Saccone, 1987). It turned out that C<->T transitions are one order of magnitude higher that A<->G transitions or A<->Y transversions. The transversion rate G<->Y is practically zero. The avoidance in the use of G in neutral positions of mitochondrial genome seems to be a peculiar feature of the majority of eukaryotic organisms with the exception of plants (Pepe et al., 1983; Saccone et al., 1987b).The reasons for this strategy is still unknown. We are tempted to speculate that it can be linked to some metabolic characteristics of animal and lower eukaryote mitochondria, such as the absence of repair enzymes for a particularly strong damage to guanine.

Table 3 reports the average base substitution rates of several protein coding genes, linked together to form a "supergene", in the nuclei and mitochondria of mammals. The values of replacement rates (first and second positions) are similar, within the statistical fluctuations. So far it has been widely reported that mitochondrial genes evolve faster than nuclear genes. The results shown here and others previously published (Lanave ét al., 1985; Saccone et al., 1987b), clearly demonstrate that in animal cells the replacement rates of mt protein coding genes are comparable to those of many nuclear genes. In contrast, the mt silent substitution rate (3rd silent position) highly exceeds that in nuclei. Thus it can be argued that the faster silent substitution rate of mitochondrial genes depends on looser functional constraints on codon strategy, which in turn is presumably linked to the transcription and traslation apparatus of the organelles. However, it cannot be excluded that the higher mutation rate of mitochondrial genes does not proportionally affect the rate of replacement since mitochondrial proteins posses higher functional constraints.

THE NON-CODING REGION

A peculiar feature of metazoa mtDNA is the presence of a unique main "non coding" region which, in the light of the

Table 3. Mean Base Substitution Rate (10-9 Subs/Site.Year.)

	Mitochondrial Super-Gene		
	1st pos	2nd pos	3rd sil pos
Artiodactyls-Rodents	1.7±0.2	0.8±0.1	7.8±2.0
Human-Rodents	1.9±0.2	0.9±0.1	8.0±2.7 (*)
Artiodactyls-Human	2.1±0.2	1.0±0.1	8.5±3.0 (*)

	Nuclear Super-Gene		
	1st pos	2nd pos	3rd sil pos
Artiodactyls-Rodents	1.5±0.2	1.1±0.2	2.8±0.4
Human-Rodents	1.3±0.2	1.0±0.2	2.3±0.4
Artiodactyls-Human	0.9±0.1	0.7±0.1	1.7±0.2

(*) = Stationarity condition not verified

extremely economic gene organization, should contain the major regulatory elements for the replication and expression of the mt genome. Thus the study of its structure and evolution appears particularly interesting. In contrast to the high degree of conservation of other parts of the genome, this region evolves very rapidly displaying, according to species, great variability of length and base composition (Table 4). It ranges from 121-132 base pairs (bp) in sea urchins to 879-1,122 bp in mammals, to 2,135 bp in anphibia and 1,029-5,100 bp in Drosophila.

In invertebrates the structure and the evolution of the main non coding region has not yet been characterized as well as it has in vertebrates. In Ascaris and in Drosophila, owing to an extremely high A+T content, this part of the genome is called A+T rich region.

In Ascaris a main A-T region, 899 bp long, lies between the Ser-tRNA and the Asn-tRNA genes. A smaller non-coding sequence of 117 bp, which may be folded into a stem and loop structure, is located between the ND4 and the COI genes (Wolstenholme et al., 1987). In nematode worms all the genes are coded for by one strand but the sites and the mechanisms of replication and trascription are still unknown.

In Drosophila the A-T region is extremely polimorphyc. It varies in sequence and length according different species (1,029 bp in D. virilis and 5,100 bp in D. melanogaster) (Fauron and Wolstenholme, 1980a), in individuals of one species (Fauron and Wolstenholme 1980b) and in the different mtDNA molecules of a single fly (Solignac et al. 1983). By comparing the A-T region of D. virilis (1,029 bp) and D. yakuba (1,077 bp) two conserved sequences were identified (Clary and Wolstenholme, 1987). One of these sequences, 49 bp long and having 78% homology in the two species, is located near the Ile-tRNA gene and probably contains the promoter for the transcription of one strand. The second sequence, 276/274 bp long and having 87.4% homology, probably contains the replication origin as revealed by studies using the electron microscope (Goddard and Wolstenholme 1980). In this region a sequence of 70 bp, 80% homologous between the two species, can form a hairpin structure similar to the origin of L strand replication in vertebrates.

Table 4. Nucleotide Composition of the Main Non-Coding Region

Species	Length	Percentage Nucleotide Composition			
		A	C	G	T
Ascaris suum	899				
Drosophila yakuba	1077	50	5	3	42
Paracentrotus lividus	132	15	22	28	35
Xenopus laevis	2135	39	18	10	33
Mus musculus	879	34	25	12	29
Rattus norvegicus	898	33	25	12	30
Bos taurus	910	33	24	14	29
Homo sapiens	1122	30	33	14	23

The main non coding region in Metazoa shows great variability in length and base composition. It ranges from 123 b in Paracentrotus to 2135 in Xenopus. In mammals this region, like the whole genome, is characterized by a low G content (10-14%). However in the central domain (data not shown) the percentage of G increase up to 20%, a value slightly higher than that found in mt rRNA (18%), tRNA (15%) and mRNA (12%) (Saccone et al., 1987a). In Ascaris and Drosophila the A+T content is extremely high. The precise base composition of the non-coding region in Ascaris has not been published.

In echinoids the main non coding region is located in the tRNA gene cluster and is limited to only 121 bp in

S. purpuratus (Jacobs et al. 1988) and to 132 bp in P. lividus (Cantatore et al., 1989). This region is the most likely candidate for the site of replication origin because comparative analysis revealed that, even though it is in a condensed space, it contains signals similar to those present in vertebrate D-loop. In particular there is sequence homology with the conserved sequence blocks (CSBs) which are responsible for the transition from the RNA primer (starting from the LSP) to the nascent H DNA strand (as discussed below), but in an opposite orientation and a stem and loop structure similar to the origin of replication of L strand in vertebrates. These features suggest a novel unique model of replication in sea urchin mitochondria.

In vertebrates the main non-coding region is called the D-loop because a new synthesized strand (800-1,000 b in mammals), corresponding to the start of the H strand replication, creates a triple stranded structure called displacement (D) loop. This region, spanning the Phe-tRNA and Pro-tRNA genes, contains the origin of the replication of the H strand and the promoters for both the H (HSP) and L (LSP) strands. The synthesis of the two strands is asymmetrical: starting with a transcriptional event from the LSP, the replication of the H strand proceeds clockwise in a unidirectional manner. When the H strand synthesis is two-thirds complete, the origin site of the L strand is exposed as a single stranded template and its replication starts in a hairpin structure (30 b) rich in T surrounded by a tRNA gene cluster in the opposite direction (Clayton, 1982).

Although the D-loop containing region represents the most rapidly evolving part of the whole genome, a detailed comparative analysis of vertebrate mtDNAs has revealed a number of interesting properties well conserved in evolution. According to our observations (Brown et al., 1986; Saccone et al., 1987a) the region can be divided into three domains on the basis of both base composition and degree of conservation. The left and right domains, which have a low G content and contain the 5' and the 3' D-loop ends, respectively, are highly variable in both base sequence and length. They do, however, contain thermodynamically stable tRNA-like secondary structures which include the conserved sequence blocks, CSBs, firstly described by Walberg and Clayton (1981) and the termination associated sequences, TAS, (Doda et al., 1981) which could be regulatory signals for the start and the stop sites of the third D-loop strand synthesis (Fig. 2).

The left domain, immediately adjacent to the Phe-tRNA gene is probably the most important functional part of the region, containing the origin of the replication of the H strand and the two promoters HSP and LSP.

The central domain, about 200 bp long, is extremely well conserved in primary structure and characterized by a low A content and the highest G content of the whole L-strand DNA. A peculiar feature of the mtDNA is the asymmetric GC distribution between the L strand (very poor in G) and the H strand (very poor in C) which is probably responsible for the codon strategies of mRNAs in which the use of G at the third codon position is very low (Cantatore and Saccone, 1987). Thus, the presence in the D-loop containing region of a stretch rich in G and well conserved in the evolution is indicative of a functional constraint. Indeed we found that in all four mammals there is a potential reading frame (ORF). The ORFs however vary in length in the four species and show significant similarities when the aminoacids are compared using the functional alphabet (Saccone et al., 1987a).

The right domain, spanning from the middle region to the Pro-tRNA gene, has the highest A content and the lowest G content of the D-loop region. Since it contains the 3' D-loop end, it is at this level where it will be decided whether or not the synthesis of a new DNA molecule must take place.

In contrast to the high variability displayed by the D-loop containing region in distantly related mammals, by comparing closely related species (Rattus norvegicus, Rattus rattus) we found that this region does not diverge more than typical protein genes (Brown et al. 1986) and that deletions and insertions occur approximately equally as frequently as base substitutions. The features of this part of the mt genome where peculiar mechanisms (such as polymarase slippage and stuttering, possible intra-strand recombination and others) are taking place, may explain the different evolutionary behaviour we observed intra or inter species. It is likely that homogenization effects of the type postulated by Dover (1984) due to DNA turnover might be responsible for the unexpected high degree of conservation of the D-loop primary structure in different mt DNA types of the same species (Rattus norvegicus) and in closely related species (Rattus norvegicus versus Rattus rattus). It should remembered that this genomic tract must coevolve together with nuclear coded products such as polymerase enzymes and transcription factors. This makes the study of problems of concerted evolution within this region particularly attractive.

Fig. 2. Structural organization of the D-loop containing region in vertebrates spanning the Phe-tRNA and the Pro-tRNA genes. The central part (heavy line) is the most conserved region in mammals. HSP= heavy strand promoter; LSP= light strand promoter; CSB= conserved sequence blocks; OH= replication origin of H strand; TAS= termination associated sequences.

In the attempt to better understand the functional role of the D-loop we have carried out studies on its expression in rat liver. We have identified stable H and L transcripts in the D-loop region and, by using suitable ad hoc constructed riboprobes, we have been able to distinguish the main features of the transcriptional events occurring in both strands. In particular we detected long H transcripts which encompass the whole D-loop and stop a few bases upstream the HSP and a population of RNA species which terminate at level of the 3' end of the last coded gene (Thr-tRNA) which corresponds to the 3' termini of the nascent H strand. The processed RNA species contain polyA, suggesting a strict association between cleavage and polyadenilation. The pattern of the L transcripts shows, besides a long transcript which starting from the LSP covers the whole D-loop, RNA species which are actively processed at the level of the CSBs. These data appear particularly important since they show that:

i) the main non coding region of rat mtDNA is completely and symmetrically transcribed.

ii) there is in mitochondria the contemporaneous presence of both L and H transcripts.

iii) at the level of CSBs and TASs there are multiple transcription processing (termination ?) sites.

The symmetric transcription of the D-loop containing region of rat mtDNA and in particular the presence of stable transcripts complementary to the putative replication primer RNA species suggest that mechanisms mediated by interaction between complementary RNA molecules (antisense RNAs) might play a role in the regulation of mtDNA replication and expression. These studies together with the observations previously reported on the structure and evolution of this genomic segment are open to further experimental investigation.

ACKNOWLEDGEMENTS

This work has been supported by: Progetto Strategico Genoma Umano and MPI 40% Italy.

REFERENCES

Attardi, G., 1981a, Organization and expression of the mammalian mitochondrial genome: a lesson in economy, TIBS, 6:86.
Attardi, G., 1981b, Organization and expression of the mammalian mitochondrial genome: a lesson in economy, TIBS, 6:100.
Attardi, G., 1985, Animal mitochondrial DNA:an extreme example of genetic economy, Int. Rev. Cytol., 93:93.
Boursot, P., Yonekawa, H., Bonhomme, F., 1987, Heteroplasmy in mice with deletion of a large coding region of mitochondrial DNA, Mol. Biol. Evol., 46:55.
Brown, W. M., Prager, E. M., Wang, A. and Wilson, A. C.,

1982, Mitochondrial DNA sequences of primates: tempo and mode of evolution, J. Mol. Evol., 18:225.

Brown, G. G., Gadaleta, G., Pepe, G., Saccone, C. and Sbisa', E., 1986, Structural conservation and variation in the D-loop containing region of vertebrate mitochondrial DNA, J. Mol. Biol., 192:503.

Cantatore, P. and Saccone, C., 1987, Organization, structure and evolution of mammalian mitochondrial genes, Int. Rev. Cytol., 108:149.

Cantatore, P., Gadaleta, M. N., Roberti, M., Saccone, C. and Wilson, A. C., 1987, Duplication and remoulding of tRNA genes during evolutionary rearrangement of mitochondrial genomes, Nature, 329:853.

Cantatore, P., Roberti, M., Rainaldi, G., Gadaleta, M. N. and Saccone, C., 1989, The complete nucleotide sequence, the gene organization and the genetic code of the mitochondrial genome of Paracentrotus lividus, J. Biol. Chem., in press.

Clary, D. O. and Wolstenholme D. R., 1987, Drosophila mitochondrial DNA: conserved sequences in the A+T-rich region and supporting evidence for a secondary structure model of a small ribosomal RNA, J. Mol. Evol., 25:116.

Clayton, D. A., 1982, Replication of animal mitochondrial DNA, Cell, 28:693.

Doda, J. N., Wright, C. T. and Clayton, D. A., 1981, Elongation of displacement-loop strands in human and mouse mitochondrial DNA is arrested near specific template sequences, Proc. Natl. Acad. Sci. U.S.A., 78:6116.

Dover, G. A. and Flavell, R. B., 1984, Molecular coevolution: DNA divergence and the maintenance of function, Cell, 38:622.

Fauron, M. C. R. and Wolstenholme, D. R., 1980a, Extensive diversity among Drosophila species with respect to nucleotide sequence within the adenine + thymine-rich region of mitochondrial DNA molecules, Nucleic Acids Res., 8:2439.

Fauron, M. C. R. and Wolstenholme, D. R., 1980b, Intraspecific diversity of nucleotide sequences within the adenine and thymine-rich region of mitochondrial DNA molecules of Drosophila mauritiana, Drosophila melanogaster, and Drosophila simulans, Nucleic Acids Res., 8:5391

Gadaleta, G., Pepe, G., De Candia, G., Quagliariello, C., Sbisa', E. and Saccone, C., 1988, Nucleotide sequence of rat mitochondrial NADH dehydrogenase subunit 1. GTG, a new initiator codon in vertebrate mitochondrial genome, Nucleic Acids Res., 16:6233.

Gadaleta, G., Pepe, G., De Candia, G., Quagliariello, C., Sbisa', E. and Saccone, C., 1989, The complete nucleotide sequence of the Rattus Norvegicus mitochondrial genome: cryptic signals revealed by comparative analysis between vertebrates, J. Mol. Evol., in press

Goddard, J. M. and Wolstenholme, D. R., 1980, Origin and direction of replication in mitochondrial DNA molecules from the genus Drosophila. Nucleic Acids Res., 8:741.

Gojobori, T., Ishii, K. and Nei, M., 1982, Estimation of
    average number of nucleotide substitutions when the
    rate of substitution varies with nucleotide, J. Mol.
    Evol., 18:414.
Jukes, T. H. and Cantor, C. R., 1969, Evolution of protein
    molecules in: "Mammalian Protein Metabolism" III, H.
    N. Munro, ed., Academic Press, New York.
Holt, I. J., Harding, A. E. and Morgan-Hughes, J. A.,
    1988, Deletions of muscle mitochondrial DNA in patients
    with mitochondrial myopathies, Nature, 331:717.
Jacobs, H. T., Elliott, D., Math, V. B. and Farquharson,
    A., 1988, Nucleotide sequence and gene organization of
    sea urchin mitochondrial DNA, J. Mol. Biol.,
    202:185.
Kimura, M., 1981, Estimation of evolutionary distances between
    homologus nucleotide sequences, Proc. Natl. Acad.
    Sci. U.S.A., 78:454.
Lanave, C., Preparata, G., Saccone, C. and Serio, G., 1984, A
    new method for calculating evolutionary substitution
    rates, J. Mol. Evol., 20:86.
Lanave, C., Preparata, G. and Saccone, C., 1985, Mammalian
    genes as clocks?, J. Mol. Evol., 21:346.
Moritz, C., Dowling, T. E. and Brown, W. M., 1987,
    Evolution of animal mitochondrial DNA: relevance for
    population biology and systematics, Ann. Rev. Ecol.
    Syst., 18:269.
Palmer, J. D. and Herbon L. A., 1989, Plant mitochondrial
    DNA evolves rapidly in structure, but slowly in
    sequence, J. Mol. Evol., 28:87.
Pepe, G., Holtrop, M., Gadaleta, G., Kroon, A.M., Cantatore,
    P., Gallerani, R., De Benedetto, C., Quagliariello, C.,
    Sbisa', E. and Saccone, C., 1983, Non random patterns
    of nucleotide substitution and codon strategy in the
    mammalian mitochondrial genes coding for identified and
    unidentified reading frames, Bioch. Int., 6:553.
Preparata, G. and Saccone, C., 1987, A simple quantitative
    model of the molecular clock, J. Mol. Evol., 26:7.
Saccone, C., Attimonelli, M. and Sbisa', E., 1987a,
    Structural elements highly preserved during the
    evolution of the D-loop région in vertebrate
    mitochondrial DNA, J. Mol. Evol., 26:205.
Saccone, C., Attimonelli, M., Lanave, C., Gallerani, R. and
    Pesole G., 1987b, The evolution of mitochondrially
    coded cytochrome genes: a quantitative estimate, in:
    "Cytochrome Systems", S. Papa, B. Chance and L.
    Ernster, eds., Plenum Publishing Corporation, New York.
Snyder, M., Fraser, A. R., La Roche, J., Gartner-Kepkay, K.E.
    and Zouros, E., 1987, Atypical mitochondrial DNA from
    the deep-sea scallop Placopecten Magellanicus, Proc.
    Natl. Acad. Sci. U.S.A., 84:7595.
Solignac, M., Monnerot, M., Mounolou, J. C., 1983,
    Mitochondrial DNA heteroplasmy in Drosophila
    mauritiana, Proc. Natl. Acad. Sci. U.S.A., 80:6942.
Sueoka, N., 1988, Directional mutation pressure neutral
    molecular evolution, Proc. Natl. Acad. Sci. USA,
    85:2653
Walberg, M. W. and Clayton, D. A., 1981, Sequence and
    properties of the human KB cell and mouse L cell
    D-loop regions of mitochondrial DNA, Nucleic Acids

Res., 9:5411.

Wallace, D. C., Singh, G., Lott, M. T., Hodge, J. A.,
    Schurr, T. G., Lezza, A. M. S., Elsas II, L. J.
    and Nikoskelainen, E. K., 1988, Mitochondrial DNA
    mutation associated with Leber's hereditary optic
    neuropathy, Science, 242:1427.

Ward, B. L., Anderson, R. S. and Bendich, A. J., 1981, The
    mitochondrial genome is large and variable in a
    family of plants (Cucurbitaceae), Cell, 25:793.

Warrior, R. and Gall, J., 1985, The mitochondrial DNA of
    Hydra attenuata and Hydra littoralis consists of two
    linear molecules, Arch. Sc. Geneve, 439:445

Wilson, A. C., Carlson, S. S. and White, T. J., 1977,
    Biochemical evolution, Annu. Rev. Biochem., 46:573.

Wolstenholme, D. R., 1987, Bizarre tRNAs inferred from DNA
    sequences of mitochondrial genomes of nematode worms,
    Proc. Natl. Acad. Sci. U.S.A., 84:1324.

Wolstenholme, D. R., Clary, D. O., Macfarlane, J. L.,
    Wahleithner, J. A. and Wilcox, L., 1985, Organization
    and evolution of invertebrate mitochondrial genomes
    in: "Achievements and Perspectives of Mitochondrial
    Resarch", II, E. Quagliariello, E. C. Slater, F.
    Palmieri, C. Saccone and A. M. Kroon, eds., Elsevier
    Amsterdam.

TISSUE-SPECIFIC EXPRESSION OF NUCLEAR GENES

FOR MITOCHONDRIAL ENZYMES

Bernhard Kadenbach, Andrea Schlerf, Thomas Mengel,
Ludger Hengst[1], Xinan Cao[2], Guntram Suske[2],
C. Eckerskorn* and F. Lottspeich*

Fachbereich Chemie, Biochemie, Philipps-Universität
D-3550 Marburg, Fed. Rep. Germany

[1]Present address: Max-Planck-Institut für
biophysikalische Chemie, Abtlg. Molekulare Genetik
D-3400 Göttingen, Fed. Rep. Germany

[2]Present address: Institut für Molekularbiologie und
Tumorforschung, Philipps-Universität
D-3550 Marburg, Fed. Rep. Germany

*Max-Planck-Institut für Biochemie, Genzentrum
D-8033 Martinsried, Fed. Rep. Germany

INTRODUCTION

One of the most fascinating aspects of life is its innumerable
diversification in shape and function. Nevertheless all living species are
constructed by the same chemical principle, due to a common evolutionary
origin. In fact the basic reactions of cellular energy synthesis by oxidative phosphorylation in mitochondria, once established early during
evolution, remained unchanged up to the appearance of men.

In contrast the metabolism of life under anaerobic or oxygen limited
conditions evolved into multiple different pathways in order to generate
energy and chemical intermediates from different environmental sources.
Thus in bacteria a multitude of enzymatic reactions are found which do not
occur in higher organisms. Although the basic chemical reactions of energy
synthesis in multicellular organisms remained unchanged, an enormous amplification of their regulatory accessories occured during evolution. We are
just beginning to envision the multitude of genes involved in regulation of
ATP-synthesis, which are under the control of tissue-specific, developmental
and hormonal signals.

The mitochondrial genome

The unchanged principle of oxidative metabolism during evolution of
differentiated organisms is examplified at the mitochondrial genome. In
eucaryotic cells mitochondrial DNA codes for only a few proteins which,
however, are essential components of enzyme complexes involved in oxidative
phosphorylation. But whereas mitochondrial DNA of plants has a size of

208-2500 kb (e.g. 570 kb in maize), that of yeast has about 78 kb and the human mitochondrial DNA comprises only 16.569 kb. Thus the mammalian mitochondrial genome reflects a strikingly economical use of genetic material, containing very few intergenic nucleotides and no detectable introns (1-3). Mitochondrial DNA of mammals codes for 13 proteins: 7 subunits of NADH dehydrogenase (complex I), 1 subunit (cytochrome b) of ubiquinone cytochrome c oxidoreductase (complex III), 3 subunits of cytochrome c oxidase (complex IV) and 2 subunits of ATP synthase (complex V) (4). All of these enzyme complexes of the inner mitochondrial membrane, and only these 4 enzyme complexes, are involved in energy transduction for the synthesis of mitochondrial ATP by oxidative phosphorylation.

The surprising result of recent observations demonstrate, however, that mitochondrial DNA and thus ATP synthesis by mitochondrial respiration is not essential for the survival of eucaryotic cells. The growth of yeast under anaerobic conditions and even without a mitochondrial genome (petite mutants) is well known. Recently Desjardins et al. (5) and Attardi and co-worker (6) could demonstrate, that also mammalian cells can grow in single cell culture without having mitochondrial DNA. These cells do still have mitochondria, but they are lacking the 4 functional enzyme complexes of oxidative phosphorylation. Instead they synthesize ATP by glycolysis, accompanied by the formation of lactic acid, yielding only 2 moles of ATP as compared to 38 moles of ATP obtained by full oxidation of one mole of glucose.

Life of higher organisms is dependent on regulated respiration

Nevertheless higher life on earth is essential dependent on oxygen, i.e. by cellular respiration. Thus all eucaryotic organisms contain mitochondria, including fungi, algae and plants, but the detailed structure of respiratory enzyme complexes varies. Apparently not only the total capacity of energy synthesis is essential for a normal cellular function, but also its regulatory properties. During the last ten years many cases of human mitochondrial myopathy have been described, some of the fatal infantile type, which are based on a partial defective cytochrome c oxidase, the terminal enzyme of the mitochondrial respiratory chain (see DiMauro et al. (7) for review). Therefore it appears that life of multicellular differentiated organisms is only compatible with a sufficient and regulated supply of energy. Energy metabolism in bacteria continues as long as substrates are available. In contrast the rate of energy metabolism in higher organisms is not only depending on the availability of nutrients, but also on specific signals. Thus evolution of energy metabolism was accompanied with an increase in regulatory complexity (8).

Cytochrome c oxidase (COX) - a highly regulated enzyme complex

The increase of regulatory complexity during evolution is most evidently demonstrated at COX, the oxygen acceptor of the mitochondrial respiratory chain, which controls (at least in part (9,10)) the rate of cellular respiration and thus the amount of ATP synthesis. The unchanged catalytic principle of this enzyme during evolution from procaryotes to men, follows from the amino acid sequence homology of subunits I-III between Paracoccus denitrificans, yeast and mammals (11-13). But whereas the bacterial enzyme contains only these 3 subunits, additional subunits occur in eucaryotic cells: 5-6 in unicellular fungi, 8 in fish and 10 in mammals (14). The additional subunits are encoded on nuclear DNA, in contrast to the mitochondria-encoded catalytic subunits. They were suggested to have a regulatory function (8,15) and to represent receptors for intracellular allosteric effectors like nucleotides, ions, second messengers etc., which upon binding change the conformation of the enzyme complex and thus the rate of respiration and/or efficiency of energy transduction (16).

The regulatory function of nucleus-encoded subunits of COX could be established recently, at least in principal, by the following observations: a) Covalently bound 8-nitreno-ATP changes the kinetics of reconstituted bovine heart COX (17); b) intraliposomal ATP increases and ADP decreases the K_m for cytochrome c of the reconstituted enzyme (18); these effects are not obtained with the reconstituted Paracoccus enzyme (19) which lacks the nucleus-encoded subunits; c) extramitochondrial ATP changes the redox state of cytochrome aa₃ in yeast mitochondria (20); d) extraliposomal ATP and phosphate change the respiratory control index of the reconstituted enzyme (21) and e) intraliposomal phosphate changes the K_m and V_{max} for cytochrome c of the reconstituted enzyme from bovine heart and liver (22).

Tissue-specific isozymes of cytochrome c oxidase

A further indication for a regulatory function of nucleus-encoded subunits came from the discovery of different tissue-specific forms for the same subunit (8,23,24) as deduced from different apparent molecular weights (8,25,26), N-terminal amino acid sequences (8,24,27), immunological reactivities (28,29) and presence of SH-groups (30). An influence of different isoforms of the subunits on the different catalytic properties of COX from liver and heart was concluded from different kinetics (22,31), and different participation of nucleus-encoded subunits in the binding domain for cytochrome c (32,33), which is located at the mitochondria-encoded subunit II. In addition to tissue-specific also developmental-specific isoforms of the same subunit were concluded to occur in the corresponding tissue of fetal and adult rats, based on different immunological reactivities (29).

A different combination of subunit isoforms was found in COX from brown fat tissue of cold-adapted rats (34). In fig. 1 the N-terminal amino acid sequences of subunits VIa and VIII from rat liver, heart and brown fat tissue are compared. They were determined from protein bands of SDS-PAGE gels as described (35). The amino acid sequences of subunits VIa and VIII

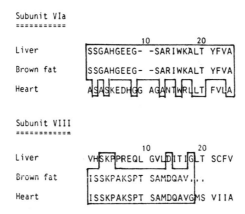

Fig. 1. Comparison of the N-terminal amino acid sequences of subunits VIa and VIII from rat liver, cold-adapted brown fat tissue and heart. The sequences of COX subunit VIa from liver and heart (49) and of subunit VIII from liver (52) were deduced from cDNA sequences, the other sequences were determined by solid-phase sequencing of separated gel bands (35).

are different in liver and heart. Surprisingly the brown fat isozyme differs from both the liver and the heart isozyme, because it contains the "liver-type" subunit VIa and the "heart-type" subunit VIII. This result clearly demonstrates that different isoforms of the same subunit can be exchanged in the enzyme complex, which appears to be always composed of 13 different subunits, from which the catalytic, mitochondria-encoded subunits I-III are identical in all tissues of an organism. Assuming only two expressed genes for each of the 10 nucleus-encoded subunits to occur in mammalian organisms, a total of 2^{10} = 1024 different isozymes of COX are theoretically possible. From immunological data it could be concluded that the "heart-type" subunit VIII occurs mainly in the heart (29). The expression of this isoform also in brown fat tissue might indicate its role in nonshivering thermogenesis, the main function of brown fat tissue (36,37). In fact it was found by Puchalski et al. (38) that noradrenaline increases in Djungarian hamster the organ blood flow not only in brown fat tissue but also in heart. In no other tissue an increase of blood flow was observed, but instead a decrease was found in kidney and small intestine. The authors suggested that the heart muscle may be a potential source of heat during noradrenaline-stimulated nonshivering thermogenesis (38).

Genes for nucleus-encoded subunits of mammalian cytochrome c oxidase

Subunit IV. A cDNA clone for bovine heart subunit IV comprising the precursor sequence and 82 of the 147 amino acids of the mature protein (39) was isolated by Lomax et al. (40), using synthetic oligonucleotides for screening a cDNA library. The corresponding chromosomal gene and a processed pseudogene for bovine subunit IV were isolated from a genomic library (41). Because the genomic Southern analysis revealed only 2-4 hybridizing restriction fragments in the bovine genome, the authors suggested that only one gene for subunit IV may be expressed in bovine. This conclusion appeared to be further supported by Northern blot hybridization using the full-length cDNA probe for subunit IV of human liver (42). With total RNA from HeLa cells, human muscle and liver the same sized transcript hybridized with the probe.

From immunological reactivities of subunit-specific antisera to rat liver COX subunits with the enzyme from different rat tissues, however, a large difference between subunit IV of adult liver and of fetal skeletal muscle was found (29). A less pronounced difference was found between adult liver and fetal heart or liver. Thus different isoforms for subunit IV appear to occur not between the enzyme of different organs but between the enzyme of different developmental stages. Interestingly two different genes for subunit V of yeast COX, the analog of mammalian subunit IV, were described by Cumsky et al. (43,45). One form (Va) was shown to be expressed under normal, oxygen saturated conditions, whereas the other gene is turned on only under oxygen-limited conditions (46). In mammals the fetal stage may represent an oxygen limited condition.

Subunit Va. A full-length cDNA clone specifying subunit Va of human COX was described by Rizzuto et al. (47). The clone was isolated from a human endothelial cell cDNA library and was identical to a clone derived from a human fetal muscle cDNA library. Northern blot hybridization with RNA from human muscle, liver and brain resulted in the same hybridizing band. The authors concluded that no isoforms for subunit Va may occur in human. This protein appears to be highly conserved during evolution, because 95 % homology was found between the human and bovine subunit Va.

Subunit Vb. The cDNA sequences encoding subunit Vb of human and part. of bovine COX were presented by Zeviani et al. (48). The human cDNA encodes a 31 amino acid long presequence and the deduced amino acid sequence shows 85 % homology to that of the (incomplete) bovine cDNA. Again no indication for different tissue-specific genes for subunit Vb could be obtained: identical hybridizing bands were found with total RNA from human muscle, liver and brain, and identical cDNA sequences were determined in clones isolated from cDNA libraries from human endothelial cells, fetal muscle and adult muscle.

Subunit VIa. The first description of two different tissue-specific genes for a nucleus-encoded subunit of COX was presented by Schlerf et al. (49). The two cDNAs were isolated from cDNA libraries of rat heart and liver in λgt11 using monospecific antisera to subunit VIa. The cDNA from the heart library displays a TAA stop codon in frame 18 bp 5' upstream of the first methionine codon, thus excluding a cleavable leader sequence for rat heart subunit VIa.

The two deduced amino acid sequences show only 50 % homology, indicating a long independent evolution of the two genes. While the heart-type gene is only expressed in heart and muscle, the liver-type gene is expressed in brain, kidney, liver and heart but very little in muscle, as shown by Northern blot analysis. The reason for the expression of both genes in rat heart is not known. Subunit VIa was shown to participate in the binding domain for cytochrome c in a tissue-specific manner (32,33). Subunits VIa, VIIa and VIII of pig liver and kidney but not of heart and muscle could be protected by cytochrome c against labelling with radioactive glycine ethyl-ester in the presence of a water soluble carbodiimide (EDC). As shown in fig. 2, the amino acid sequence of the liver-type subunit VIa contains 6 additional acidic amino acids as compared to the heart-type subunit VIa, suggesting their involvement in the binding of the basic cytochrome c.

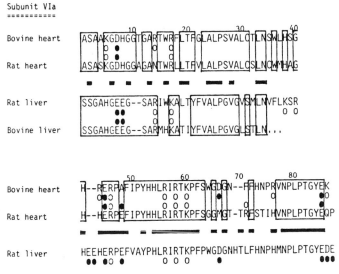

Fig. 2. Comparison of the complete amino acid sequences of COX subunit VIa from rat liver and heart with the complete sequence of bovine heart and the N-terminal sequence of bovine liver. Bovine heart (50) and liver (27) sequences are derived from protein sequencing, the rat sequences from cDNAs (49). The thick bar indicates homologous amino acids between rat liver and heart.

147

From Northern analysis, using the full-length liver-type cDNA as a probe, it is concluded that the two genes do not hybridize (50 % formamide at 42°C for 4 h), and therefore cannot be used for screening for the other "isogene". Thus the above described unsuccessfulness in finding isogenes for subunits IV, Va and Vb could be due to the small homology in nucleotide sequences of corresponding isogenes.

In fig. 2 are also presented the amino acid sequences of bovine heart and liver subunit VIa as determined by Meinecke and Buse (50) and Yanamura et al. (27). Clearly a much higher homology between the two liver (82 %) or heart sequences (74 %) is obtained as compared between liver and heart of the rat (49 %) or bovine (41 %), supporting the previous conclusion: tissue-specificity overrides species-specificity in cytochrome c oxidase polypeptides (28).

Subunit VIc. The first but incomplete gene (cDNA) for a nucleus-encoded subunit of mammalian cytochrome c oxidase was described by Parimoo et al. (51) for subunit VIc from rat liver. A cDNA clone containing the full length coding region of rat liver subunit VIc was published by Suske et al. (52). This clone contained a TAG stop codon in frame 18 bp 5' upstream of the first methionine codon, which was substantiated in the genomic sequence (53), indicating the absence of a cleavable leader sequence for this subunit. By genomic Southern blot analysis 8-9 hybridizing fragments were obtained, suggesting the occurrence of multiple genes for subunit VIc in the rat genome.

A rat genomic library was screened with the cDNA for subunit VIc as a probe, and two clones, encoding the functional and a processed gene, have been isolated and characterized by Suske et al. (53). Whereas the functional gene is composed of 4 exons and 3 introns and contains the identical sequence of the cDNA, the intronless processed gene shows 88 % homology with the rat liver cDNA. This retroposon is flanked by a 13 bp direct repeat and codes for a full-length subunit VIc protein containing 16 mutated out of 76 total amino acids. Because in the 5' region TATA- and CAAT-like structures are found, there exists the possibility that this gene could be expressed.

From a rat heart cDNA library a clone was isolated containing an insert which differed from the cDNA for subunit VIc of the rat liver library in a different 5' region and two mutations (54). One mutation affected the start methionine codon (ATG → ATC), resulting in a shortened open reading frame beginning with the 12th amino acid (Met), the other mutation changed a codon of the reading frame (Val → Ala). By using a probe comprising the different 5' region of the cDNA from heart in a Northern blot, a weak hybridization signal was found with RNA from heart but not from liver. The band had the same size as that obtained using the full length cDNA probe (55). Thus it cannot be excluded that this cDNA represents a gene for a protein which is slightly different from subunit VIc and is expressed in rat heart, although only to a small extent.

Subunit VIIa. A cDNA clone for bovine heart COX subunit VIIa was isolated by Lomax et al. (56), but not further characterized. In fig. 3 are compared the N-terminal amino acid sequences of COX subunits VIIa from rat and bovine, heart and liver. Out of 37 N-terminal amino acids of subunit VIIa bovine heart and liver 6 are different. In contrast no difference is found between 19 N-terminal amino acids of rat heart and liver subunit VIIa. Thus it cannot be excluded that the tissue-specific expression of COX subunits may differ in different species.

```
Subunit VIIa
============

                                      10         20         30
Bovine heart     FENRVAEKQKLFQEDNGLPVHLKGGATDNILYRVTMTL...

Rat heart        FENKVPEKQKLFQEDNGMP...
                 ══════════════════════

Rat liver        FENKVPEKQKLFQEDNGMPVHLKGGT...

Bovine liver     FENKVPEKQKLFQEDNGIPVHLKGGIADNILYRVTM...
```

Fig. 3. Comparison of N-terminal amino acid
sequences of COX subunit VIIa from bovine
and rat heart and liver. All sequences
were determined by protein sequencing:
bovine heart (57), bovine liver (27) and
rat liver and heart from SDS-PAGE protein
bands as described in (35). The thick bar
indicates homologous amino acids between
rat liver and heart.

Subunit VIIc. Lomax et al. (56) isolated from a bovine heart cDNA
library two different clones coding for the full-length precursor protein
of COX subunit VIIc. Only in the 5' region, encoding the presequence of
the protein, the two cDNAs are different, resulting in two different presequences of 16 and 21 amino acids length, respectively. The authors conclude that the two cDNAs are the result of alternative splicing of one and
the same primary transcript.

Subunit VIII. A cDNA clone encoding the partial sequence of rat liver
COX subunit VIII was isolated by Suske et al. (52) from a rat hepatoma cDNA
library with a specific antiserum to subunit VIII. This probe was used to
screen a rat genomic library for chromosomal genes. Three clones were
characterized and found to contain pseudogenes (processed retroposons) (54).
In fig. 4 the cDNA for rat liver COX subunit VIII is aligned with a processed retroposon (clone 5). The latter is characterized by i) flanking
direct repeats of 9 bp, ii) the absence of introns and iii) a poly (A)$^+$
tail at the 3' end. This retroposon codes for COX subunit VIII differing
from the mature rat liver protein in only one amino acid. Because in the
5' region TATA- and CAAT-like structures are found, it cannot be excluded
that this processed retroposon is expressed in the rat. This contrasts the
other two processed retroposons, which, due to multiple mutations, cannot
be expressed and therefore represent nonfunctional pseudogenes (54).

The amino acid sequence of subunit VIII deduced from the cDNA of rat
liver is compared in fig. 5 with the amino acid sequences of subunit VIII
from rat heart and bovine heart and liver, determined by protein sequencing.
Again a lower homology is found between the sequences from different tissues
of the same organism (rat liver/heart = 38 %; bovine liver/heart = 42 %)
than between the sequences from the same tissue of different species (rat/
bovine heart = 59 %; rat/bovine liver = 73 %.

The large difference between corresponding subunits VIa and VIII from
liver and heart suggests that these subunits have an as yet unknown different function in the catalytic and/or regulatory activity of COX in liver
and heart.

149

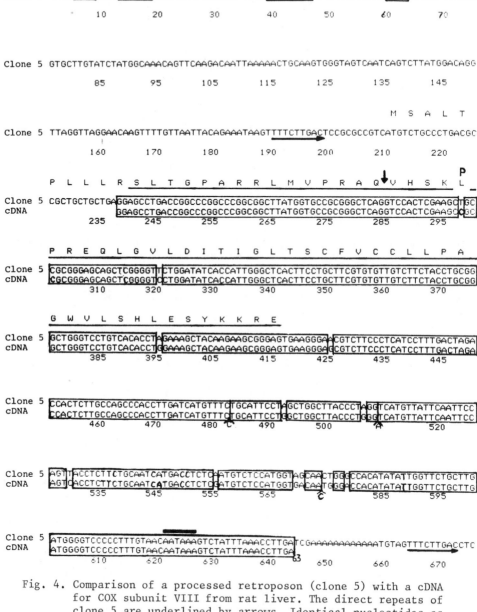

Fig. 4. Comparison of a processed retroposon (clone 5) with a cDNA
for COX subunit VIII from rat liver. The direct repeats of
clone 5 are underlined by arrows. Identical nucleotides as
well as promotor-like structures in the 5'-region of clone
5 are boxed. The indicated deduced amino acid sequence of
clone 5 differs only in one amino acid (L instead of P) from
that of the cDNA (underlined). The N-terminal end of the
mature protein is indicated by a vertical arrow. The poly-
adenylation signal is indicated by a thick horizontal bar.

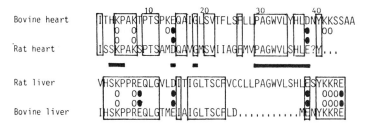

Fig. 5. Comparison of the complete amino acid sequences of COX subunit VIII from bovine heart and rat liver with partial sequences of rat heart and bovine liver. The rat liver sequence was deduced from the cDNA sequence (see fig. 4), the bovine heart (58), and bovine liver (27) sequences were determined by protein sequencing, the rat heart sequence from SDS-PAGE protein bands as described in (35). The thick bar indicates homologous amino acids between rat liver and heart. : Acid amino acids; o: basic amino acids.

Tissue-specific genes of other mitochondrial enzymes

ADP/ATP carrier. A cDNA clone encoding the full-length ADP/ATP carrier protein was isolated by Battini et al. (59) from a SV-40-transformed human fibroblast cDNA library. The expression of this gene in cultured cells was shown to depend on growth factors. From an adult human liver cDNA library two further clones coding for the human translocase have been isolated (60). The amino acid sequences of these cDNAs show either only one amino acid substitution or 91 % homology compared with that from human fibroblasts (59). The majority of nucleotide substitutions (82 %) are silent. A different cDNA clone for the ADP/ATP translocator was isolated from a human skeletal muscle cDNA library (61 %). Its protein sequence showed only 88 % homology with the human fibroblast translocator. Two different expressed ADP/ATP carrier genes from bovine (T_1 and T_2) have been described by Walker et al. (62). Like the human genes, they are expressed tissue-specific: T_1 predominates in heart and T_2 in smooth muscle. The corresponding human genes are expressed predominantly in various tissues and cell lines (T_1), in liver (T_2) and HL60 cells (T_3) (63). A similar chromosomal gene structure has been found for the human genes T_1 and T_2, each being composed of 4 exons and 3 introns (63).

ATP synthase. Two genes encoding the bovine mitochondrial ATP synthase proteolipid have been isolated by Gay and Walker from a cDNA library (64). Whereas the two genes (P_1 and P_2) contain completely different noncoding regions and different presequences of 61 and 68 amino acids length, the mature proteins are identical. Nevertheless the two genes are expressed in a tissue-specific manner. A very similar gene structure has been determined for the corresponding two human genes (P_1 and P_2), which are composed of 4 exons and 3 introns. The intron sizes, however, vary remarkably (P_1 = 915, 706 and 320 bp; P_2 = 2627, 529 and 3723 bp, respectively) (63).

A cDNA clone comprising the full-length protein sequence of the α-subunit of the ATP synthase from bovine heart was isolated by Walker et al. (63). Another cDNA, derived from a bovine liver cDNA library, encodes for a slightly different protein, but differs extensively in its DNA sequence. The two cDNAs are suggested to encode tissue-specific isoforms of the α-subunit.

The physiological meaning of tissue-specific isoforms of mitochondrial proteins

The specific physiological functions of various isoforms for the same mitochondrial carrier, enzyme or enzyme subunit are yet unknown. This is not surprising because also the function of most of the small subunits of respiratory chain complexes is still unknown. It may be assumed that the same functional principle required the evolution of additional subunits, as well as tissue-, hormone- and developmental-specific isoforms for the same subunit of mitochondrial enzyme complexes. Since all protein components of mitochondrial oxidative phosphorylation serve the same ultimate function, namely the generation of ATP, we may ask, why so many protein components are required in mammals but only a few in unicellular organisms?

As suggested previously (8,15,16), the optimization of energy supply may have driven the evolution of oxidative phosphorylation. This optimization may include the responsiveness of an organism, and of its various cells and tissues individually, to numerous environmental factors such as temperature, nutritional state, season (hibernation), mental excitement, immune response, relaxation (sleep) etc. For a variable and tissue-specific response of an enzyme complex to the same hormone, second messenger or intracellular metabolite, different proteins (or subunits) are required in different tissues. The need for such a complicated regulation of energy supply may become evident from the fact, that any manifestation of life, either in chemical or physical terms, requires energy. But the form of required energy varies, being of low or high intensity, of short or long duration, or being formed at low or high efficiency, accompanied with the production of large or small amounts of heat, respectively.

This functional argument for a tissue-specific regulation of energy supply, however, cannot explain the tissue-specific expression of e. g. two different genes for the same mature proteolipid of ATP-synthase (64) or two ADP/ATP-carriers differing in only one amino acid (59).

Evolution of genes involved in oxidative phosphorylation

The basic principle of evolution, namely gene mutation/duplication and natural selection, involves a slow and dynamic change of genes, including a large number of successive small variations. Thus the large number of genes for the same functional protein, which differs in only a single or up to 60 % of its amino acids may indicate different steps of independently evolving genes. In addition functional constraints determine the rate of amino acid exchange. This gene evolution, however, concerns mainly the regulatory properties of mitochondrial enzymes. In fact multiple genes, including functional processed retroposons and nonfunctional pseudogenes (54) have evolved for the regulatory protein components of respiratory chain complexes. For most mitochondrial enzymes or carriers different regulatory properties are manifestated by the expression of different isoforms, encoded on different genes. This is found e. g. for the ADP/ATP carrier, where 12 % difference in the deduced amino acid sequence was found between two genes (61), suggesting different regulatory properties for the same functional carrier protein.

With the membrane-bound enzyme complexes of oxidative phosphorylation, in particular with cytochrome c oxidase, nature has formed a new level of complexity. By exchanging multiple isoforms for the same subunit, the number of isozymes does not correspond to the sum of isogenes but to the product of isogenes and subunit number. Thus 2^{10} = 1024 different isozymes for cytochrome c oxidase are theoretically possible, if two expressed genes would occur for each subunit (see above). By this principle much less coding capacity is required for the same amount of regulatory complexity. In contrast the separate and independent evolution of the mitochondrial genome into the smallest possible size, coding for only one or a few components of each of the 4 energy transducing enzyme complexes, as well as 22 tRNAs and two rRNAs (1), excludes further diversification of these genes. Thus the basic principle of ATP-synthesis remained unchanged during evolution of eucaryotic cells, whereas the evolution of regulatory functions represents a currently ongoing process.

Much work still has to be done to identify all of the genes for cytochrome c oxidase subunits in mammals, which might vary in different species.

Acknowledgements. Roswitha Roller-Müller is gratefully acknowledged for typing the manuscript. This paper was supported by the Deutsche Forschungsgemeinschaft (Ka 192/17-2) and Fonds der Chemischen Industrie.

REFERENCES

1. S. Anderson, A. T. Bankier, B. G. Barrell, M. H. L. de Bruijn, A. R. Coulson, J. Drouin, J. C. Eperon, D. P. Nierlich, B. A. Roe, F. Sanger, P. H. Schreier, A. J. H. Smith, R. Staden and J. G. Young, Sequence and organization of the human mitochondrial genome, Nature 290:457-465 (1981).
2. J. Montoya, D. Ojala and G. Attardi, Distinctive features of the 5'-terminal sequences of the human mitochondrial mRNAs, Nature 290: 465-470 (1981).
3. D. Ojala, J. Montoya and G. Attardi, tRNA punctuation model of RNA processing in human mitochondria, Nature 290:470-474 (1981).
4. A. Chomyn, P. Mariottini, M. W. J. Cleeter, C. J. Ragan, R. F. Doolittle, A.M. Yagi, Y. Hatefi and G. Attardi, Functional assignment of the products of the unidentified reading frames of human mitochondrial DNA, in: "Achievements and perspectives of mitochondrial research, Vol. II", E. Quagliariello, E. C. Slater, F. Palmieri, C. Saccone and A. M. Kroon, eds., Elsevier, Amsterdam, pp. 259-275 (1985).
5. P. Desjardins, J.-M. de Muys and R. Morais, An established avian fibroblast cell line without mitochondrial DNA, Somat. Cell Molec. Genet. 12:133-139 (1986).
6. G. Attardi and M. King, Repopulation of normal and mtDNA-less human cells with mtDNA of exogenous mitochondria, in: "Organelle Genomes and the Nucleus", 14th EMBO Annual Symposium 1988, Abstracts, pp. 1-2 (1988).
7. S. DiMauro, M. Zeviani, S. Servidei, E. Bonilla, A. F. Miranda, A. Prelle and E. A. Schon, Cytochrome oxidase deficiency: clinical and biochemical heterogeneity, Ann. N. Y. Acad. Sci. 488:19-32 (1986).
8. B. Kadenbach, Structure and evolution of the "Atmungsferment" cytochrome c oxidase, Angew. Chem. Int. Ed. Engl. 22:275-282 (1983).
9. M. Erecinska and D. F. Wilson, Regulation of cellular energy metabolism, J. Membr. Biol. 70:1-14 (1982).
10. J. M. Tager, R. J. A. Wanders, A. K. Groen, W. Kunz, R. Bohnensack, U. Küster, B. Letko, G. Böhme, J. Duszynski and L. Wojtczak, Control of mitochondrial respiration, FEBS Lett. 151: 1-9 (1983).
11. G. C. M. Steffens, G. Buse, W. Oppliger and B. Ludwig, Sequence homology of bacterial and mitochondrial cytochrome c oxidases. Partial sequence data of cytochrome c oxidase from Paracoccus denitrificans, Biochem. Biophys. Res. Commun. 116: 335-340 (1983).

12. B. Ludwig, Cytochrome c oxidases in prokaryotes, FEMS Microbiol. Rev. 46: 41-56 (1987).

13. M. Raitio, T. Jalli and M. Saraste, Isolation and analysis of the genes for cytochrome c oxidase in Paracoccus denitrificans, EMBO J. 6: 2825-2833 (1987).

14. B. Kadenbach, L. Kuhn-Nentwig and U. Büge, Evolution of a regulatory enzyme: Cytochrome c oxidase (complex IV). Curr. Top. Bioenerg. 15:113-161 (1987).

15. B. Kadenbach and P. Merle, On the function of multiple subunits of cytochrome c oxidase from higher eucaryotes. FEBS Lett. 135:1-11 (1981).

16. B. Kadenbach, Mini Review. Regulation of respiration and ATP-synthesis in higher organisms: Hypothesis. J. Bioenerg. Biomembr. 18: 39-54 (1986).

17. F.-J. Hüther and B. Kadenbach, Specific effects of ATP on the kinetics of reconstituted bovine heart cytochrome c oxidase. FEBS Lett. 207: 89-94 (1986).

18. F.-J. Hüther and B. Kadenbach, ADP increases the affinity for cytochrome c by interaction with the matrix side of bovine heart cytochrome c oxidase, Biochem. Biophys. Res. Commun. 147:1268-1275 (1987).

19. F.-J. Hüther and B. Kadenbach, Intraliposomal nucleotides change the kinetics of reconstituted cytochrome c oxidase from bovine heart but not from Paracoccus denitrificans. Biochem. Biophys. Res. Commun. 153:525-534 (1988).

20. M. Rigoulet, B. Guerin and M. Denis, Modification of flow-force relationships by external ATP in yeast mitochondria, Eur. J. Biochem. 168:275-279 (1987).

21. F. Malatesta, G. Antonini, P. Sarti and M. Brunori, Modulation of cytochrome oxidase activity by inorganic and organic phosphate, Biochem. J. 248:161-167 (1987).

22. U. Büge and B. Kadenbach, Influence of membrane lipids, buffer composition and proteases on the kinetics of reconstituted cytochrome c oxidase from bovine liver and heart, Eur. J. Biochem. 161:383-390 (1986).

23. P. Merle and B. Kadenbach, On the heterogeneity of vertebrate cytochrome c oxidase polypeptide chain composition. Hoppe-Seyler's Z. Physiol. Chem. 361:1257-1259 (1980).

24. B. Kadenbach, R. Hartmann, R. Glanville and G. Buse, Tissue-specific genes code for polypeptide VIa of beef liver and heart cytochrome c oxidase. FEBS Lett. 138:236-238 (1982).

25. B. Kadenbach, U. Büge, J. Jarausch and P. Merle, Composition and kinetic properties of cytochrome c oxidase from higher eucaryotes. in: "Vectorial Reactions in Electron and Ion Transport in Mitochondria and Bacteria", F. Palmieri et al., eds., Elsevier North-Holland Biomedical Press, Amsterdam, pp. 11-23 (1981).

26. R. Bisson and G. Schiavo, Two different forms of cytochrome c oxidase can be purified from the slime mold Dictyostelium discoideum, J. Biol. Chem. 261: 4373-4376 (1986).

27. W. Yanamura, Y.-Z. Zhang, S. Takamiya and R. A. Capaldi, Tissue-specific differences between heart and liver cytochrome c oxidase, Biochemistry 27:4909-4914 (1988).

28. J. Jarausch and B. Kadenbach, Tissue-specificity overrides species specificity in cytochrome c oxidase polypeptides, Hoppe-Seyler's Z. Physiol. Chem. 363:1133-1140 (1982).

29. L. Kuhn-Nentwig and B. Kadenbach, Isolation and properties of cytochrome c oxidase from rat liver and quantification of immunological differences between isozymes from various rat tissues with subunit-specific antisera. Eur. J. Biochem. 149:147-158 (1985).

30. A. Stroh and B. Kadenbach, Tissue-specific and species-specific distribution of SH-groups in cytochrome c oxidase subunits,

Eur. J. Biochem. 156:199-204 (1986).

31. P. Merle and B. Kadenbach, Kinetic and structural differences between cytochrome c oxidases from beef liver and heart, Eur. J. Biochem. 125:239-244 (1982).

32. B. Kadenbach and A. Stroh, Different reactivity of carboxylic groups of cytochrome c oxidase polypeptides from pig liver and heart. FEBS Lett. 173:374-380 (1984).

33. B. Kadenbach, A. STroh, M. Ungibauer, L. Kuhn-Nentwig, U. Büge and J. Jarausch, Isozymes of cytochrome c oxidase: characterization and isolation from different tissues. Methods Enzymol. 126:32-45 (1986).

34. B. Kadenbach, A. Stroh, A. Becker, C. Eckerskorn and F. Lottspeich, Brown fat cytochrome c oxidase is different from the isozymes of liver and heart, submitted.

35. C. Eckerskorn, W. Mewes, H. Goretzki and F. Lottspeich, A new siliconized-glass fiber as support for protein-chemical analysis of electroblotted proteins, Eur. J. Biochem. 176:509-519 (1988).

36. D. G. Nicholls and R. M. Locke, Thermogenic mechanism in brown fat, Physiol. Rev. 64:1-64 (1984).

37. J. Nedergaard and B. Cannon, Thermogenic mitochondria, in: "Bioenergetics", L. Ernster, ed., Elsevier Science Publishers B.V., Amsterdam, pp. 291-314 (1984).

38. W. Puchalski, H. Böckler, G. Heldmaier and L. Langfeld, Organ blood flow and brown adipose tissue oxygen consumption during noradrenaline-induced nonshivering thermogenesis in the Djungarian hamster, J. Exp. Zool. 242:263-271 (1987).

39. R. Sacher, G. J. Steffens and G. Buse, Studies on cytochrome c oxidase, VI. Polypeptide IV. The complete primary structure, Hoppe-Seyler's Z. Physiol. Chem. 360:1385-1392 (1979).

40. M. I. Lomax, N. J. Bachman, M. S. Nasoff, M. H. Caruthers and L. I. Grossman, Isolation and characterization of a cDNA clone for bovine cytochrome c oxidase subunit IV, Proc. Natl. Acad. Sci. 81:6295-6299 (1984).

41. N. J. Bachman, M. I. Lomax and L. I. Grossman, Two bovine genes for cytochrome c oxidase subunit IV: a processed pseudogene and an expressed gene, Gene 55:219-229 (1987).

42. M. Zeviani, M. Nakagawa, J. Herbert, M. I. Lomax, L. I. Grossman, A. A. Sherbany, A. F. Miranda, S. DiMauro and E. A. Schon, Isolation of a cDNA clone encoding subunit IV of human cytochrome c oxidase, Gene 55:205-217 (1987).

43. M. G. Cumsky, C. Ko, C. E. Trueblood and R. O. Poyton, Two nonidentical forms of subunit V are functional in yeast cytochrome c oxidase, Proc. Natl. Acad. Sci. USA 82:2235-2239 (1985).

44. M. G. Cumsky, C. E. Trueblood, C. Ko and R. O. Poyton, Structural analysis of two genes encoding divergent forms of yeast cytochrome c oxidase subunit V, Mol. Cell. Biol. 7:3511-3519 (1987).

45. C. E. Trueblood and R. O. Poyton, Differential Effectiveness of yeast cytochrome c oxidase subunit V gene results from differences in expression not function, Mol. Cell. Biol. 7:3520-3526 (1987).

46. R. O. Poyton, R. M. Wright, J. D. Trawick, C. E. Trueblood and B. Roecklein, Intergenomic signalling and the assembly of yeast cytochrome c oxidase, in: "Organelle Genomes and the Nucleus", 14th EMBO Annual Symposium 1988, Abstracts, pp. 49-50 (1988).

47. R. Rizzuto, H. Nakase, M. Zeviani, S. DiMauro and E. A. Schon, Subunit Va of human and bovine cytochrome c oxidase is highly conserved, submitted for publication.

48. M. Zeviani, S. Sakoda, A. A. Sherbany, H. Nakase, R. Rizzuto, C.E. Samitt, S. DiMauro and E. A. Schon, Sequence of cDNAs encoding subunit Vb of human and bovine cytochrome c oxidase, Gene 65:1-11 (1988).

49. A. Schlerf, M.Droste, M. Winter and B. Kadenbach, Characterization of two different genes (cDNA) for cytochrome c oxidase subunit VIa from heart and liver of the rat, EMBO J. 7:2387-2391 (1988).

50. L. Meinecke and G. Buse, Studies on cytochrome c oxidase, XII. Isolation and primary structure of polypeptide VIb from bovine heart, Biol. Chem. Hoppe-Seyler 366:687-694 (1985).

51. S. Parimoo, S. Seelan, S. Desai, G. Buse and G. Padmanaban, Construction of a cDNA clone for a nuclear-coded subunit of cytochrome c oxidase from rat liver, Biochem. Biophys. Res. Commun. 118:902-909 (1984).

52. G. Suske, T. Mengel, M. Cordingley and B. Kadenbach, Molecular cloning and further characterization of cDNAs for rat nuclear-encoded cytochrome c oxidase subunits VIc and VIII. Eur. J. Biochem. 168:233-237 (1987).

53. G. Suske, C. Enders, A. Schlerf and B. Kadenbach, Organization and nucleotide sequence of two chromosomal genes for rat cytochrome c oxidase subunit VIc: a structural and a processed gene, DNA 7:163-171 (1988).

54. X. Cao, L. Hengst, A. Schlerf, M. Droste, T. Mengel and B. Kadenbach, Complexity of nucleus-encoded genes of mammalian cytochrome c oxidase, Ann. N. Y. Acad. Sci., in press.

55. L. Hengst, Isolation and Charakterisierung von cDNA-Klonen für die Untereinheit VIc der Cytochrom c Oxidase aus Rattenherz- und Rattenleber-cDNA Genbänken, Diplomarbeit, Fachbereich Chemie, Philipps-Universität, Marburg (1988).

56. M. I. Lomax, L. I. Grossman, D. Scheuner and M. S. Aqua, Molecular genetics of mammalian cytochrome c oxidase, Fifth Eur. Bioenerg. Conference, Short Reports, Vol. 5:268 (1988).

57. L. Meinecke and G. Buse, Studies on cytochrome c oxidase, XIII. Amino acid sequence of the small membrane polypeptide VIIIc from bovine heart respiratory complex IV, Biol. Chem. Hoppe-Seyler 367:67-73 (1986).

58. L. Meinecke, G. J. Steffens and G. Buse, Studies on cytochrome c oxidase, X. Isolation and amino-acid sequence of polypeptide VIIIb, Hoppe-Seyler's Z. Physiol. Chem. 365:313-320 (1984).

59. R. Battini, S. Ferrari, L. Kaczmarek, B. Calabretta, S.-T. Chen and R. Baserga, Molecular cloning of a cDNA for a human ADP/ATP carrier which is growth-regulated, J. Biol. Chem. 262:4355-4359 (1987).

60. J. Houldsworth and G. Attardi, Two distinct genes for ADP/ATP translocase are expressed at the mRNA level in adult human liver, Proc. Natl. Acad. Sci. USA 85:377-381 (1988).

61. N. Neckelmann, K. Li, R. P. Wade, R. Shuster, D. C. Wallace, cDNA sequence of a human skeletal muscle ADP/ATP translocator: Lack of a leader polypeptide, divergence from a fibroblast translocator cDNA, and coevolution with mitochondrial DNA genes, Proc. Natl. Acad. Sci. USA 84:7580-7584 (1987).

62. J. E. Walker, A. L. Cozens, M. R. Deyer, I. M. Fearnley, S. J. Powell and M. J. Runswick, Chemica Scripta, 27B: 97-105 (1987).

63. J. E. Walker, A. L. Cozens, M. R. Dyer, I. M. Fearnley and M. J. Runswick, Multigene families for mitochondrial proteins in mammals: ATP synthase and transport proteins, in: "Molecular Basis of Biomembrane Transport", F. Palmieri and E. Quagliariello, eds., Elsevier, Amsterdam, pp. 209-216 (1988).

64. N. J. Gay and J. E. Walker, Two genes encoding the bovine mitochondrial ATP synthase proteolipid specify precursors with different import sequences and are expressed in a tissue-specific manner, EMBO J. 4:3519-3524 (1985).

THE PLASTID ENVELOPE MEMBRANES:

PURIFICATION, COMPOSITION AND ROLE IN PLASTID BIOGENESIS

Roland Douce, Claude Alban, Maryse A. Block and Jacques Joyard

Laboratoire de Physiologie Cellulaire Végétale, URA CNRS n° 194, Département de Recherche Fondamentale, Centre d'Etudes Nucléaires et Université Joseph Fourier, 85X, F-38041 Grenoble-Cédex, France

INTRODUCTION

All plastids (proplastids, leucoplasts, amyloplasts, chromoplasts, etioplasts and chloroplasts) are limited by a pair of outer membranes, known as the envelope. Together, these membranes provide a flexible boundary between the plastid and the surrounding cytosol. Studies with isolated intact chloroplasts have demonstrated the different permeability properties of each envelope membrane (Heber and Heldt, 1981). The outer envelope membrane is freely permeable to small molecules, due to a pore-forming protein (Flügge and Benz, 1984), whereas the inner envelope membrane is the functional border between the plastid stroma and the cytosol. Specific translocators regulate the flow of metabolites across the inner envelope membrane (Heber and Heldt, 1981). Furthermore, biochemical studies of purified envelope membranes (mostly from spinach and pea) have demonstrated the complexity of their enzymatic equipment which has a major role in plastid biogenesis: up to 20 different enzymes involved in the synthesis of plastid components such as galactolipids, phosphatidylglycerol and prenylquinones have been demonstrated and/or characterized in envelope membranes (Douce and Joyard, 1979 ; Douce et al, 1984).

In the first part of this article, we will provide pratical informations for the preparation of envelope membranes from chloroplasts and non-green plastids. The procedures described below are largely based (a) on the method to purify total envelope membranes from spinach chloroplasts originaly described by Douce et al (1973), and extended to non-green plastids (Alban et al, 1988) and (b) on the method we have developped for the separation of outer and inner envelope membranes (Block et al, 1983).

Abbreviations: MGDG, monogalactosyldiacylglycerol ; DGDG, digalactosyl-diacylglycerol ; SL, sulfolipid ; PC, phosphatidylcholine ; PG, phosphatidylglycerol ; PI, phosphatidylinositol ; PE, phosphatidylethanolamine ; DPG, diphosphatidylglycerol ; UDP-gal, UDP-galactose.

The last part of this article is a short review on the chemical composition (polypeptides, glycerolipids, pigments and prenylquinones) and on the role of envelope membranes in plastid biogenesis (synthesis of glycerolipids and prenylquinones).

I - PURIFICATION OF PLASTID ENVELOPE MEMBRANES

The key step in the purification of envelope membranes from chloroplasts or non-green plastids is actually the preparation of large amounts of intact and pure organelles.

A - PREPARATION OF INTACT AND PURIFIED PLASTIDS

ISOLATION OF CHLOROPLASTS FROM SPINACH (adapted from Douce and Joyard, 1982).

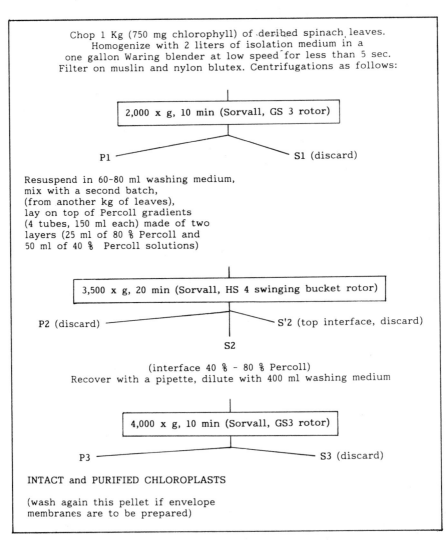

Chop 1 Kg (750 mg chlorophyll) of deribed spinach leaves.
Homogenize with 2 liters of isolation medium in a
one gallon Waring blender at low speed for less than 5 sec.
Filter on muslin and nylon blutex. Centrifugations as follows:

2,000 x g, 10 min (Sorvall, GS 3 rotor)

P1   S1 (discard)

Resuspend in 60-80 ml washing medium,
mix with a second batch,
(from another kg of leaves),
lay on top of Percoll gradients
(4 tubes, 150 ml each) made of two
layers (25 ml of 80 % Percoll and
50 ml of 40 % Percoll solutions)

3,500 x g, 20 min (Sorvall, HS 4 swinging bucket rotor)

P2 (discard)   S'2 (top interface, discard)

S2

(interface 40 % - 80 % Percoll)
Recover with a pipette, dilute with 400 ml washing medium

4,000 x g, 10 min (Sorvall, GS3 rotor)

P3   S3 (discard)

INTACT and PURIFIED CHLOROPLASTS

(wash again this pellet if envelope
membranes are to be prepared)

Figure 1. Procedure for isolation of intact and purified spinach chloroplasts.

Solutions

Isolation medium: mannitol (or sorbitol), 330 mM ; tetrasodium pyrophos-
    phate, 50 mM, pH 7.8 ; bovine serum albumin, 0.1 % (w/v)

Washing medium: mannitol (or sorbitol), 330 mM ; HEPES-KOH, 10 mM,
    pH 7.8

Percoll solutions: Percoll 40 % and 80 % (v/v) ; mannitol (or sorbitol)
    330 mM ; HEPES-KOH, 10 mM, pH 7.8

Materials

Nylon blutex (toile à bluter), 50 μm aperture (Tripette et Renaud,
    Sailly-Saillisel, Combles, France)

Muslin or cheesecloth or Miracloth, 50-60 cm large

1-gallon Waring Blender

Superspeed centrifuge, refrigerated (such as RC5 Sorvall, for instance)
    with a whole set of fixed angle (Sorvall SS-34 for 50 ml tubes
    and Sorvall GS3 for 500 ml bottles) and swinging bucket rotor (Sor-
    vall HS-4 for 250 ml bottles)

Procedure (All the steps are done between 0 and 4°C)

    The method used is described in figure 1. A key step in this
procedure is the blending of the leaves: homogenization should be limited
to a few (less than 5) sec.

    The final pellets of intact chloroplasts are then carefully resus-
pended in a minimum volume of washing medium, with gentle agitation.
From 2 kg of spinach leaves, intact and purified chloroplasts corresponding to 100-150 mg chlorophyll are routinely prepared.

ISOLATION OF NON-GREEN PLASTIDS FROM CAULIFLOWER BUDS (adapted from
                                     Journet and Douce, 1985)

Solutions

Isolation medium: mannitol, 300 mM ; tetrasodium pyrophosphate, 20 mM,
    pH 7.6 ; EDTA, 1 mM ; bovine serum albumin, 0.2 % (w/v ; cysteine,
    4 mM (cystein should be added just before the blending step)

Washing medium: sucrose, 300 mM ; MOPS-KOH, 10 mM, pH 7.2 ; EDTA, 1 mM ;
    bovine serum albumin, 0.1 % (w/v)

Percoll solution: Percoll, 35 % (v/v) ; sucrose, 300 mM ; MOPS-KOH,
    10 mM, pH 7.2 ; EDTA, 1 mM ; bovine serum albumin, 0.1 % (w/v)

Materials (see above)

Procedure (All the steps are done between 0 and 4°C)

    The major problem in the procedure described in figure 2 is that
it is not possible to get intact plastids devoid of any contaminating
extraplastidial structures by a single-step purification. The efficiency
of the two-step purification method outlined in figure 2 was clearly
demonstrated by Journet and Douce (1985).

Peel the top of cauliflower inflorescences (about 5 kg of tissue).
Homogenize in a one gallon Waring blender (1 kg of cut material for 2
liters of isolation medium) at low speed for less than 5 sec.
Filter on nylon blutex. Centrifugations as follows:

| 3,000 x g, 20 min (Sorvall, GS3 rotor) |

P1 ⎯⎯⎯⎯⎯⎯⎯⎯⎯⎯⎯ S1 (discard)

Resuspend each pellet in 10 ml washing medium,
filter on nylon blutex to separate the aggregates.

| 200 x g, 5 min (Sorvall, SS 34 rotor) |

P2 (discard carefully) ⎯⎯⎯⎯⎯⎯ S2

| 3,000 x g, 10 min (Sorvall, SS 34 rotor) |

P3 ⎯⎯⎯⎯⎯⎯⎯⎯⎯⎯⎯ S3 (discard)

WASHED PLASTIDS

Resuspend in 12 ml washing medium. Load (3 ml on
each tube) over 36 ml of 35 % Percoll solution.

| 40,000 x g, 30 min (Sorvall, SS 34 rotor) |

(this centrifugation results
in the formation of a continuous gradient)

P4 (discard) ⎯⎯⎯⎯⎯⎯⎯⎯ S'4 (top band, discard)

S4

(lower band)
Recover with pipette, dilute with washing medium
(4 vol. for 1 vol. of fraction). Load (4 ml on each tube)
over 36 ml of 35 % Percoll solution.

| 25,000 x g, 30 min (Sorvall, HB 4 swinging bucket rotor) |

(this centrifugation does not result
in the formation of a gradient)

S5 (lower band) ⎯⎯⎯⎯⎯⎯⎯ S'5 (top band, discard)

Recover with a pipette, dilute with washing medium
(10 vol. for 1 vol. of plastid suspension).

| 5,000 x g, 10 min (Sorvall, SS 34 rotor) |

P6 ⎯⎯⎯⎯⎯⎯⎯⎯⎯⎯⎯ S6 (discard)

INTACT and PURIFIED NON-GREEN PLASTIDS

Figure 2. Procedure for isolation of intact and purified non-green plastids from cauliflower buds.

The final yield of intact plastids is, however, very low (0.5-1 % of the total amounts of plastids present in the cauliflower tissue used). Starting from 1 kg of cut material, intact and purified plastids corresponding to approximately 10-15 mg protein are routinely prepared.

PURIFICATION OF AMYLOPLASTS FROM SYCAMORE CELLS (adapted from Journet et al, 1986).

Solutions

Culture medium: see Bligny and Leguay (1987)

Cell digestion mixture: mannitol, 0.5 M ; cellulase (Onozuka RS, Yakult Pharmaceutical Co, Nishimomiya, Japan), 1 % (w/v) ; pectolyase Y-23 (Seighin Pharmaceutical Co, Nishinomiya, Japan), 0.1 % (w/v) ; all these compounds are added to the cell culture medium, the final pH is adjusted to 5.7

Protoplasts washing medium: mannitol, 0.5 M ; in the cell culture medium

Suspension medium: mannitol, 0.5 M ; MOPS-NaOH, 20 mM, pH 7.5 ; EDTA, 2 mM ; bovine serum albumin, 0.1 % (w/v) ; spermidine, 0.4 mM ; β-mercaptoethanol, 7 mM ; PMSF, 1 mM ; leupeptine, 10 μM ; benzamidine-HCl, 1 mM ; ϵ-aminocaproic acid, 5 mM ; insoluble polyvinylpyrolidone (K c.a. 25, Serva)

Fractionation medium: mannitol, 0.5 M ; MOPS-NaOH, 10 mM, pH 7.5 ; EDTA, 2 mM ; PMSF, 1 mM ; leupeptine, 10 μM ; benzamidine-HCl, 1 mM ; ϵ-aminocaproic acid, 5 mM

Percoll solution: Percoll, 50 % (v/v) ; mannitol, 0.5 M ; MOPS-NaOH, 10 mM, pH 7.5 ; EDTA, 2 mM ; PMSF, 1 mM ; leupeptine, 10 μM ; benzamidine-HCl, 1 mM ; ϵ-aminocaproic acid, 5 mM

Materials

Shaker for digestion of cells at 25°C microscope. Other materials: see above.

Procedure

A very important step in the preparation of sycamore amyloplasts is the transfer of the cells to a sucrose-free medium about 15 hours prior to the extraction. This is essential to decrease the size of the starch grains to get reasonable yields of intact plastids. A description of the events which occur within isolated cells after their transfer to a sucrose-free medium is given by Journet et al (1986).

150-200 g of sycamore cells in exponential growth (3-5 days after the beginning of the culture) are transfered to a sucrose-free medium the evening prior to the experiment and are harvested at least 15 hours later. The cells are then washed twice in culture medium containing 0.5 M mannitol. Sycamore cells are suspended in 150 ml of cell digestion mixture and incubated with constant shaking (20 cycles/min) for 1 hour at 25°C. Release of protoplasts from cells is monitored using light microscope. The suspension is filtered and protoplasts are recovered as described by Journet et al (1986). The procedure for the gentle rupture of the protoplasts, the separation of the organelles and amyloplasts purification is described in Figure 3. From 150-200 g of sycamore cells, the yield of purified intact amyloplasts is about 1 mg protein.

PLASTID INTACTNESS AND PURITY

The procedures described above (Figures 1-3) allow the preparation of highly intact and pure plastids. Chloroplast integrity can be controlled by observation with a phase contrast microscope. The best physiological assay for chloroplast integrity is to assess the CO_2-dependent O_2 evolution using an oxygen electrode (Walker, 1980 ; Walker et al, 1987). The intactness of plastid preparations from cauliflower buds as well as from sycamore cells can be evaluated by measurement of gluconate 6-phosphate dehydrogenase (E.C. 1.1.1.44) latency in a quick and reliable spectrophotometric assay, as described by Journet and Douce (1985).

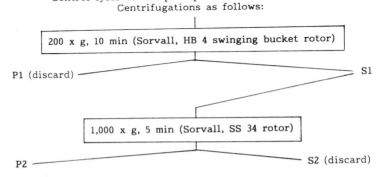

Sycamore protoplasts (from 150-200 g cells) are washed twice in fractionation medium. The suspension (100 ml) is loaded via the open end into the barrel of a 25 ml disposable syringe and protoplasts are forced out through a nylon blutex (20 µm). Repeat this step a second time. Control lysis of the protoplasts with a microscope. Centrifugations as follows:

200 x g, 10 min (Sorvall, HB 4 swinging bucket rotor)

P1 (discard) ___ S1

1,000 x g, 5 min (Sorvall, SS 34 rotor)

P2 ___ S2 (discard)

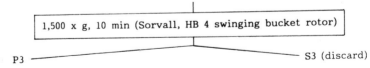

Resuspend carefully in fractionation medium, filter on miracloth to separate the aggregates. Load (2 ml) on top of 50 % Percoll solution (8 ml).

1,500 x g, 10 min (Sorvall, HB 4 swinging bucket rotor)

P3 ___ S3 (discard)

INTACT and PURIFIED AMYLOPLASTS

Figure 3. Procedure for isolation of intact and purified amyloplasts from sycamore protoplasts.

The methods used to assess plastid purity are valid for chloroplasts as well as for non-green plastids. The reader is refered to the review by Quail (1979) for details on convenient marker enzymes. In addition, it is convenient to determine whether the preparation contains PE and cardiolipin, which are absent from plastids and are two excellent markers for extraplastidial and inner mitochondrial membranes, respectively (Douce and Joyard, 1980).

B - PURIFICATION OF PLASTID ENVELOPE MEMBRANES

From chloroplasts, it is possible to prepare total envelope, containing both the outer and the inner envelope membrane, and membrane fractions enriched in outer or inner envelope membranes. To prepare total envelope membranes intact chloroplasts are broken by a gentle osmotic shock in a hypotonic medium (Douce et al, 1973). At this stage, the two envelope membranes fuse along their breaking edges and it is no longer possible to separate the outer from inner membrane. This was achieved by using an hypertonic incubation of the plastids to enlarge the space between the two membranes, followed by rupture of the chloroplasts using freeze-thawing (Cline et al, 1981, Keegstra and Yousif, 1986) or mechanical (Block et al, 1983) procedures.

PURIFICATION OF TOTAL ENVELOPE MEMBRANES

Solutions

Swelling medium: tricine-NaOH, 10 mM, pH 7.8 ; MgCl$_2$, 4 mM

Sucrose solutions for gradients: sucrose, 0.6 and 0.93 M (for step gradient centrifugation) or 0.4 and 1.2 M (for continuous density gradients) ; tricine-NaOH, 10 mM, pH 7.8 ; MgCl$_2$, 4 mM

Suspension medium: sucrose, 0.3 M ; tricine-NaOH, 10 mM, pH 7.8

Materials

Ultracentrifuge (such as Beckman L65 or Sorvall OTD 2) with swinging-out bucket rotors (Beckman rotors SW-27, 6 x 38 ml ; SW-40, 6 x 13 ml and SW-50, 6 x 5 ml, according to the amount of plastids used).

Procedures

PURIFICATION OF TOTAL ENVELOPE MEMBRANES FROM SPINACH CHLOROPLASTS (adapted from Douce et al, 1973)

Intact, purified chloroplasts (50-200 mg chlorophyll) prepared from Percoll gradients are suspended in 90 ml (final volume) of swelling medium. Under these conditions, water enters very quickly into the chloroplasts, the envelope membranes are unable to support the high pressure and burst in a few seconds with the liberation of stroma material. The suspension of broken chloroplasts is then layered on top of a discontinuous sucrose gradient (0.6 and 0.93 M) and centrifuged (Beckman SW-27 rotor) for 1 hour at 72,000 x g (Rmax), at 4°C. At the conclusion of this step, chloroplast components are clearly separated giving three distinguishable subfractions:
- a tightly-packed, dark-green pellet at the bottom of the tube (thylakoid subfraction) ;
- a yellow band at the interface of the two sucrose layers (envelope membranes subfraction) ;
- and a clear brown supernatant (soluble or stroma subfraction).

PROCEDURE FOR THE PURIFICATION OF ENVELOPE MEMBRANES FROM CAULIFLOWER BUD NON-GREEN PLASTIDS (adapted from Alban et al, 1988)

Intact cauliflower plastids (0.5 ml, 50-80 mg protein) are frozen for 30 min at -80°C, then thawned to room temperature, treated with 3 ml of swelling medium and homogenized smoothly with a Potter-Elvehjem apparatus with a loose fitting pestle, the envelope of almost 100 % plastids are ruptured, as judged by the lack of latency of gluconate 6-phosphate dehydrogenase (see above).

The broken plastid suspension (3.5 ml, 50-80 mg protein) thus obtained is layered on top of a linear sucrose gradient (total volume, 7.5 ml) made of 0.4 to 1.2 M sucrose, 10 mM tricine-NaOH (pH 7.8) and

4 mM MgCl$_2$. At the bottom of the gradient, a 2 ml sucrose (1.2 M) cushion, containing 10 mM tricine-NaOH (pH 7.8) and 4 mM MgCl$_2$, is layered prior to the sucrose gradient. Centrifugation (SW 40 rotor, Beckman) for 15 hours at 71,000 x g (Rmax) results in the separation of three fractions:
- a supernatant, containing the soluble material (stroma subfraction), on top of the tube ;
- a dense white starch pellet covered by a loose greenish pellet at the bottom of the tube ;
- a large yellow band approximately in the middle of the tube (envelope membrane subfraction).

The supernatant and the yellow band are removed successively from the top of the tube using a pasteur pipette. In this case, the yellow fraction is diluted with swelling medium devoid of MgCl$_2$ (final volume, 13 ml) and centrifuged (rotor SW 40, Beckman) for 1 hour at 218,000 x g (Rmax). The pellets obtained are suspended in a minimal volume of suspension medium by using a Potter-Elvehjem homogenizer.

PROCEDURE FOR THE PREPARATION OF ENVELOPE MEMBRANES FROM SYCAMORE AMYLO-PLASTS (adapted from Alban et al, 1988)

The differences with the method described above for cauliflower plastids are the following:
(a) a sample osmotic shock is sufficient for the disruption of the envelope membranes from sycamore amyloplasts ;
(b) the swelling medium as well as the different layers of the sucrose gradient are devoid of MgCl$_2$;
(c) protease inhibitors are added in all media used and at all the steps of the purification.
With sycamore amyloplasts, a starch pellet (devoid of membranes), is obtained at the bottom of the tube.

PREPARATION OF MEMBRANE FRACTIONS ENRICHED IN OUTER AND INNER CHLOROPLAST ENVELOPE MEMBRANES (adapted from Block et al, 1983)

Solutions

Hypertonic medium: mannitol, 0.6 M (final concentration, after addition of chloroplasts) ; tricine-NaOH, 10 mM, pH 7.8 ; MgCl$_2$, 4 mM

Dilution medium: tricine-NaOH, 10 mM, pH 7.8 ; MgCl$_2$, 4 mM

Sucrose gradients: sucrose, 1 M, 0.65 M and 0.4 M (for the different layers of the discontinuous gradient) ; tricine-NaOH, 10 mM, pH 7.8 ; MgCl$_2$, 4 mM

Materials

Yeda Press (Linca Scientific Instruments, Tel Aviv, Israel)

Centrifuge and Ultracentrifuge (see above)

Procedure (all steps are done between 0 and 4°C)

Intact chloroplasts (150-200 mg chlorophyll) are kept for 10 min in hypertonic medium (60 ml). After this treatment, the outer envelope membrane appears to be loosely attached to the inner membrane with large empty spaces in between. 30 ml of the chloroplast suspension are loaded into the chamber of a precooled (0-4°C) Yeda Press. Pressure is increased in the chamber using N_2 until 5 bars (applied pressure) are reached. The needle is gently and slowly opened so that the entire chloroplast suspension is extruded through the orifice at a flow rate of about 10 ml/min. The chloroplast suspensions thus obtained are centrifuged (SS 34 rotor, Sorvall) for 10 min at 12,000 x g. The greenish supernatants (2-5 mg protein/ml) are removed and the pellets (containing thylakoids and intact chloroplasts) are discarded. The supernatants are combined and diluted two times with the dilution medium (final volume: 80 ml). Aliquots of 13 ml of the suspension (2-5 µg chlorophyll ; 1-2 mg protein/ml) are layered on top of discontinuous sucrose gradients prepared with 8 ml of gradient solutions (1 M at the bottom, then 0.65 M and 0.4 M on top). Centrifugation (SW 27 rotor, Beckman) for 90 min at 90,000 x g (Rmax) results in the separation of membranes into two bands at the interface of the sucrose layers. The yellow band at the top interface (0.4 M/0.65 M) consists of outer envelope membrane (more than 90 % pure), whereas the yellow band at the lower interface (0.65 M/1 M) is strongly enriched in inner envelope membrane (about 80 % pure). Each band is diluted 4 times with dilution medium, and is centrifuged (SW 27 rotor, Beckman) for 1 hour at 113,000 x g. The pellets obtained are recovered and suspended in a minimum volume of suspension medium with a Potter-Elvehjem homogenizer.

YIELDS FOR ENVELOPE MEMBRANE PURIFICATION

TOTAL ENVELOPE PREPARATIONS

- about 8-10 mg total chloroplast envelope proteins from 2 kg spinach leaves ;

- about 1-1.5 mg envelope proteins from 4-5 kg cauliflower bud plastids ;

- about 0.2-0.3 mg envelope proteins from 150-200 g of isolated sycamore cells.

PREPARATION OF MEMBRANE FRACTIONS ENRICHED IN OUTER AND INNER ENVELOPE MEMBRANES

- about 0.5-1 mg outer envelope membrane and 1-2 mg of inner envelope membrane from 3-4 kg of spinach leaves.

II - CHEMICAL COMPOSITION OF THE PLASTID ENVELOPE MEMBRANES

A - POLYPEPTIDES

In chloroplasts as well as in cauliflower bud plastids, the polypeptides have Mr values less than 15,000 to more than 100,000. The polypeptide patterns are extremely complex, each fraction analyzed containing more than 100 distinct bands after staining with coomassie blue. Some polypeptides have apparently the same electrophoretic mobility when all envelope membrane profiles are analyzed. But a more precise analysis should be done. To determine whether the same polypeptides were present in envelope membranes from chloroplasts and non-green plastids, we have used antibodies raised against several chloroplast envelope polypeptides: E10, E24, E30 and E37 (Joyard et al, 1982, 1983). By immunoblotting experiments, we have demonstrated that all of them led to a reaction with envelope proteins from non-green plastids (cauliflower and sycamore). The most interesting result was obtained with the antibody to E30, since this polypeptide is involved in the phosphate/triose phosphate transport across the inner envelope membrane (Flügge and Heldt, 1976). Western blotting experiments with our antibody to spinach E30 clearly demonstrate that the major 28,000-dalton poly-peptide of the cauliflower envelope fraction and the phosphate trans-locator from spinach chloroplasts have closely related antigenic sites. The same experiment made using envelope membranes from sycamore amylo-plasts demonstrates that the major 30,000-dalton polypeptide also reacts with antibody against the spinach phosphate translocator. These observa-tions demonstrate that there are probably only limited differences between the sequence of the phosphate translocator in spinach, cauli-flower and sycamore, since good cross-reactivity of the antibody was observed between all these species, despite the differences in Mr.

B - GLYCEROLIPIDS

In contrast to extraplastidial membranes, plastid membranes (enve-lope and thylakoids) are characterized by a low phospholipid content and by the presence of glycolipids (Douce et al, 1973). The major plastid lipid compounds are galactolipids, which contain one (or two) galactose molecule(s) attached to the sn-3 position of the glycerol backbone, corresponding respectively to MGDG and DGDG. Interestingly, these glyce-rolipids are also the major components in cyanobacteria (Murata and Sato, 1983), which are supposed to have a common ancestor with chloro-plasts.

Galactolipids in thylakoid and envelope membranes contain a high amount of polyunsaturated fatty acids: up to 95 % (in some species) of the total fatty acids are linolenic acid. In non-green plastids, 18:3 is still a major component although appreciable amounts of 18:1 and 18:2 are present. Therefore, the most abundant molecular species of MGDG and DGDG have 18:3 at both sn-1 and sn-2 positions of the gly-cerol backbone. Some plants, such as pea, having almost only 18:3 in MGDG are called "18:3 plants". Other plants, such as spinach, contain important amounts of 16:3 in MGDG, they are called "16:3 plants" (Heinz, 1977). These two types are also found when non-green plastids are ana-lyzed: for instance, sycamore is a 18:3 plant whereas cauliflower is a typical 16:3 plant. The positional distribution of 16:3 in MGDG is highly specific: this fatty acid is only present at the sn-2 position of glycerol, and is almost excluded from sn-1 position. Therefore, two major structures are found in galactolipids, one with C18 fatty acids at both sn position and one with C18 and C16 fatty acids respecti-vely at sn-1 and sn-2 position. The first one is typical of "eukaryotic" lipids (such as PC) and the second one corresponds to a "prokaryotic" structure. These differences are probably due to galactolipid biosyn-thetic pathways (see below).

Plastid membranes sometimes contain galactolipids with 3 (Tri-GDG) and 4 (Tetra-GDG) galactoses. They are formed by an enzymatic galactose exchange between MGDG and DGDG, owing to a galactolipid:galactolipid galactosyltransferase (Van Besouw and Wintermans, 1978 ; Heemskerk and Wintermans, 1987) which catalyzes the following reactions:

$$2 \text{ MGDG} \rightarrow \text{DGDG} + \text{diacylglycerol}$$

$$\text{MGDG} + \text{DGDG} \rightarrow \text{Tri-GDG} + \text{diacylglycerol}$$

$$2 \text{ DGDG} \rightarrow \text{Tetra-GDG} + \text{diacylglycerol}$$

We have demonstrated (Dorne et al, 1982) that this enzyme (a) is located on the cytosolic side of the outer envelope membrane, (b) is susceptible to proteolytic digestion by thermolysin, a non-penetrant protease, (c) is present in all plastids analyzed so far (chloroplasts and non-green plastids). Consequently, in order to obtain a glycerolipid

Table I . **Glycerolipid composition of membranes from plastids and mitochondria.**

ORGANELLE	MGDG	DGDG	SL	PC	PG	PI	PE	DPG
MITOCHONDRIA								
- cauliflower buds	tr	tr	0	37	2	8	38	13
- mung bean hypocotyls								
total	0	0	0	36	1	2.5	46	14
inner membrane	*0*	*0*	*0*	*29*	*1*	*2*	*50*	*17*
outer membrane	*0*	*0*	*0*	*68*	*2*	*5*	*24*	*0*
- sycamore cells								
total	0	0	0	43	3	6	35	13
inner membrane	*0*	*0*	*0*	*41*	*2.5*	*5*	*37*	*14.5*
outer membrane	*0*	*0*	*0*	*54*	*4.5*	*11*	*30*	*0*
PLASTIDS								
- spinach chloroplasts								
thylakoids	52	26	6.5	4.5	9.5	1.5	0	0
inner envelope membrane	*49*	*30*	*5*	*6*	*8*	*1*	*0*	*0*
outer envelope membrane	*17*	*29*	*6*	*32*	*10*	*5*	*0*	*0*
total envelope	32	30	6	20	9	4	0	0
- pea etioplasts								
prothylakoids	42	35	6	9	5	2	0	0
envelope	34	31	6	17	5	4	0	0
- cauliflower bud non-green plastids								
envelope	31.5	27.5	6	20	9	4.5	1	0
- sycamore cells								
envelope	24	21	4	2	9	5.5	0.8	0

composition which could represent the in vivo situation within plastid membranes, the galactosyltransferase should be destroyed prior to fractionation of plastids.

The glycerolipid composition of envelope membranes from chloroplasts and non-green plastids is given in table 1. As expected after thermolysin treatment, no diacylglycerol was detected (whereas it represents about 10 % of the total envelope glycerolipids from non-treated plastids). The most striking feature is that the glycerolipid pattern is almost identical in envelope membranes from chloroplasts, etioplasts or other non-green plastids. The high amount of MGDG reflects the presence of the inner envelope membrane which has a glycerolipid composition close to that of thylakoids as shown in pea (Cline et al, 1981) or spinach (Block et al, 1983): in the inner envelope membrane and in thylakoids, MGDG is the major component. The presence of PC in envelope membranes reflects the presence of the outer envelope membrane: using phospholipase c digestion of intact chloroplasts, we have demonstrated that no PC is present in thylakoids. In contrast, PC represents about 30-35 % of the outer envelope glycerolipids, where it is concentrated in the outer leaflet of the membrane (Dorne et al, 1985). The major phospholipid in the inner envelope membrane and in thylakoids is PG, which is unique because of a $16:1_{trans}$ fatty acid at sn-2 position of the glycerol backbone. This phospholipid is therefore different from that found in extra-plastidial membranes. Finally, table 1 also confirm the observation that plastid membranes are devoid of PE which is a major component (together with PC) of mitochondrial, endoplasmic reticulum or other extra-plastidial membranes.

C - PIGMENTS AND PRENYLQUINONES

In contrast to thylakoids, envelope membranes from chloroplasts or non-green plastids are yellow, due to the presence of carotenoids and the absence of chlorophyll. Envelope membranes from chloroplasts and from amyloplasts present the same absorption spectrum, thus demonstrating that they have a very close pigment composition. Envelope membranes from chloroplasts and non-green plastids contain a higher proportion of violaxanthin (even the outer membrane), compared to thylakoids which are rich in β-carotene and xanthophyll (Douce et al, 1984).

Although devoid of the most conspicuous plastid pigment (chlorophyll) some non-green plastids contain chlorophyll precursors such as protochlorophyllide, however, nothing was known about the presence of such pigments in any chloroplast membrane. We have demonstrated that envelope membranes from mature spinach chloroplasts contain low amounts of pigments having the absorption and fluorescence spectroscopic properties, together with the behavior in polar/non polar solvents of protochlorophyllide and chlorophyllide (Pineau et al, 1986). It is not yet known whether the outer and the inner envelope membranes contain similar or different levels of these pigments. The low temperature fluorescence emission spectrum of total envelope membranes is strikingly different from that obtained with thylakoids. In addition, upon illumination in presence of NADPH, a decrease of the level of fluorescence at 636 nm together with a parallel increase of fluorescence level at 680 nm was observed, thus suggesting a possible phototransformation of protochlorophyllide into chlorophyllide (Pineau et al, 1986).

In addition to pigments, envelope membranes from spinach chloroplasts contain prenylquinones as genuine components. Both envelope membranes contain plastoquinone-9, α-tocopherol and phylloquinone (Soll et al, 1985). However, in contrast to thylakoids, the major prenylquinone in envelope membranes is α-tocopherol whereas it is plastoquinone-9 in thylakoids (Soll et al, 1985).

Together, these observations demonstrate that envelope membranes contain qualitatively the same pigments and prenylquinones as thylakoids. However, they are present in different amounts, probably due to their role within envelope membranes. Unfortunately, if the functions of prenylquinones and pigments are well established for thylakoids, nothing is known for their function in envelope membranes. The observation that these compounds are present in envelope membranes from non-green plastids and chloroplasts (Douce et al, 1984 ; Alban et al, 1988) is most interesting for studies to determine the specific functions of envelope pigments and prenylquinones.

III - ROLE OF ENVELOPE MEMBRANES IN PLASTID BIOGENESIS

A. GLYCEROLIPID BIOSYNTHESIS

The first observation that envelope membranes could be involved in the biosynthesis of plastid components was provided by Douce (1974): isolated envelope membranes purified from spinach chloroplasts are able to catalyze MGDG synthesis from diacylglycerol at high rates. Following the observations of Douce and Guillot-Salomon (1970) who established the presence of the enzymes from the Kornberg-Pricer pathway in chloroplasts and non-green plastids, a considerable body of data has been accumulated which clearly demonstrate that plastids are able to incorporate sn-glycerol 3-phosphate into lysophosphatidic acid, phosphatidic acid and diacylglycerol, and then into MGDG, sulfolipid or PG. Most of these data concern chloroplasts but we have recently demonstrated that non-green plastids, such as amyloplasts and other starch-containing plastids, behave like chloroplasts as far as glycerolipid biosynthesis by the so-called eukaryotic and prokaryotic pathways are concerned (Alban et al, 1988). Glycerolipid biosynthesis requires the assembly of three parts (Joyard and Douce, 1987): fatty acids, glycerol and a polar head group (galactose, for galactolipids ; sulfoquinovose, for sulfolipid ; and phosphorylglycerol, for PG).

FORMATION OF PHOSPHATIDIC ACID   (Figure 4)

The plastid Kornberg-Pricer pathway is unique in the sense that it is probably the only one to be closely associated with fatty acid synthesis, which is localized within the plastid stroma. All the enzymes of this pathway are closely associated with the envelope membranes (Joyard and Douce, 1977), and more precisely with the inner membrane (Block et al, 1983 ; Andrews et al, 1985).

The first enzyme of the pathway is a "soluble" enzyme which produces lysophosphatidic acid (lyso-PA):

sn-glycerol 3-phosphate + acyl-ACP →
                    1-acyl-sn-glycerol 3-phosphate + ACP

In fact, this enzyme is closely associated with the inner envelope membrane and lyso-PA is released directly into the membrane (Joyard and Douce, 1977). This enzyme is most interesting since it is specific for acyl-ACP thioesters (Frentzen et al, 1983). In addition, regardless of the thioester used, a striking specificity for sn-1 position of the glycerol is observed (Joyard et al, 1979 ; Frentzen et al, 1983). Finally, when a mixture of 16:0 and 18:1 thioesters is offered, 18:1 is preferably used (Frentzen et al, 1983). We have found that in non-green plastids from cauliflower buds and amyloplasts from sycamore cells, this enzyme was also soluble and presents the same specificities and selectivities as in mature chloroplasts (Alban et al, 1988). This enzyme has been purified from chloroplasts and several isoforms have been demonstrated in chloroplast stroma from several plant species (Bertrams and Heinz, 1981 ; Nishida et al, 1987).

The second enzyme, acyl-ACP:1-acyl-sn-glycerol 3-phosphate acyltransferase, catalyses the formation of phosphatidic acid:

1-acyl-sn-glycerol 3-phosphate + acyl-ACP →
                    1,2-diacyl-sn-glycerol 3-phosphate + ACP

This enzyme is firmly bound to the inner envelope membrane from chloroplasts (Joyard and Douce, 1977) and non-green plastids (Alban et al, 1988). Since lyso-PA used for this reaction is esterified at sn-1 position, the enzyme will direct fatty acids to the available sn-2 position (Joyard et al, 1979 ; Frentzen et al, 1983). However, the enzyme is highly specific for palmitic acid. Interestingly, the same results were obtained with non-green plastids purified from cauliflower buds and with amyloplasts from sycamore cells (Alban et al, 1988).

Therefore, the two acyltransferases have distinct specificities and selectivities for acylation of sn-glycerol 3-phosphate: together, they led to the formation of phosphatidic acid having 18:1 and 16:0 fatty acids respectively at sn-1 and sn-2 position of the glycerol backbone. This structure is typical for the so-called prokaryotic glycerolipids which are found in 16:3 plants (see above). In contrast, in 18:3 plants, plastid glycerolipids (MGDG) do not contain C16 fatty acids at sn-2 position, and therefore probably not derive directly from the envelope Kornberg-Pricer pathway, their origin is still under investigation.

FORMATION OF DIACYLGLYCEROL AND GALACTOLIPIDS   (Figures 4-6)

The phosphatidic acid formed in the inner envelope membrane can be used either for the synthesis of PG (Mudd et al, 1987) or for the formation of diacylglycerol (Joyard and Douce, 1977) which is the substrate for MGDG and sulfolipid biosynthesis.

Diacylglycerol biosynthesis occurs in the inner envelope membrane owing to a phosphatidate phosphatase (Joyard and Douce, 1977):

1,2-diacyl-sn-glycerol 3-phosphate → 1,2-diacylglycerol + Pi

This enzyme is unique: it is membrane-bound, active at alkaline pH and highly sensitive to cations (Joyard and Douce, 1979). In addition, 18:3 plants have a rather low phosphatidate phosphatase, in contrast to 16:3 plants (Heinz and Roughan, 1983). It is possible that the level of phosphatidate phosphatase activity in envelope membranes could be responsible for the difference observed between 18:3 and 16:3 plants at the level of MGDG synthesis. We have recently demonstrated that this was also true for non-green plastids. Chloroplast envelope membranes isolated from 16:3 and 18:3 plants have the same capacity to form MGDG when supplied with diacylglycerol and UDP-gal. However, very little MGDG is formed from sn-glycerol 3-phosphate in 18:3 plastids (or envelope membranes), although the biosynthesis of phosphatidic acid is very active, thus suggesting that the limiting step is indeed the formation of diacylglycerol (Heinz and Roughan, 1983).

The inner envelope membrane of spinach chloroplasts is characterized by the presence of a UDP-galactose:diacylglycerol galactosyltransferase (or MGDG synthase) which transfers a galactose from a water-soluble donor, UDP-gal, to an hydrophobic acceptor molecule, diacylglycerol, for synthesizing MGDG (Douce, 1974):

UDP-gal + 1,2-diacyl-sn-glycerol →
        1,2-diacyl-3-0-β-D-galactopyranosyl-sn-glycerol + UDP.

This enzyme is present in all plastids (chloroplasts and non-green plastids) analyzed so far. We have achieved a partial purification of this enzyme from spinach chloroplasts envelope membranes (Covès et al, 1986). CHAPS was used to solubilize envelope membranes and was rather efficient for MGDG synthase. After solubilization, we used hydroxyapatite chromatography to fractionate envelope proteins. A fraction containing most of the MGDG synthase activity was prepared. It contained less than 5 % of the envelope proteins and was almost entirely devoid of lipids. We have demonstrated a strong lipid requirement for the MGDG synthase. Acidic glycerolipids, and especially PG, were shown to be the best activators of the enzyme: the maximum activity thus obtained was as high as 10 μmol galactose incorporated into MGDG/h/mg protein. The preparation of the delipidated MGDG synthase-enriched fraction, together with the developement of optimal assay conditions allowed the determination of kinetic parameters for both UDP-gal and the hydrophobic substrate, diacylglycerol (Covès et al, 1988).

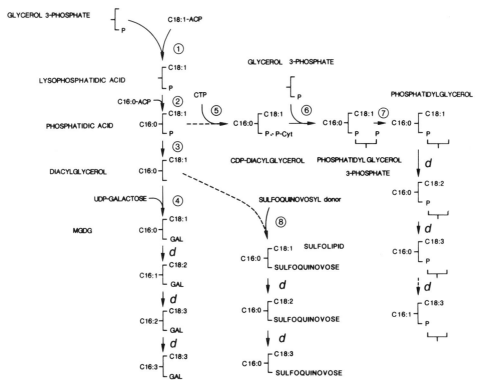

Figure 4. Pathways for glycerolipid biosynthesis in plastids.
1: acyl-ACP:sn-glycerol 3-phosphate acyltransferase;
2: acyl-ACP:monoacyl-sn-glycerol 3-phosphate acyltransferase;
3: phosphatidate phosphatase; 4: MGDG synthase; 5: phosphatidate cytidylyl transferase; 6: CDP-diacylglycerol:sn-glycerol
3-phosphate 3-phosphatidyltransferase; 7: phosphatitylglycero-
phosphatase; 8: sulfolipid biosynthesis; d: desaturases.

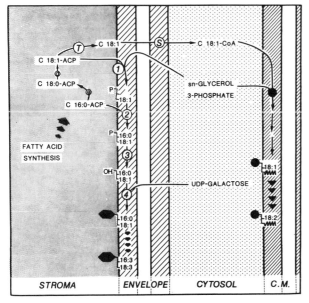

Figure 5. Role of the
chloroplast inner envelope
membrane from 16:3 plants
in MGDG biosynthesis.

171

The biosynthesis of DGDG is still a problem. Van Besouw and Wintermans (1978) and Heemskerk and Wintermans (1987) have proposed that this galactolipid could be synthesized owing to the galactolipid:galactolipid galactosyltransferase (see above § II-A).

FORMATION OF OTHER GLYCEROLIPIDS   (Figure 4)

Sulfolipid is a typical plastid component. This acidic glycolipid is synthesized within intact chloroplasts (Kleppinger-Sparace et al, 1985 ; Joyard et al, 1986). We have demonstrated that there is a competition between MGDG and sulfolipid synthesis at the level of diacylglycerol (Joyard et al, 1986), thus suggesting that envelope membranes are involved in sulfolipid biosynthesis. Unfortunately, the natural sulfoquinovosyl donor is still unknown, thus limiting experiments with isolated envelope membranes.

As discussed above, chloroplast PG is unique because it contains a $16:1_{trans}$ fatty acid at sn-2 position. PG synthesis in chloroplasts has been unsuccessful for a long time because the optimal experimental conditions for assay in the microsomal fraction were actually inhibitory for chloroplast PG synthesis. Mudd and coworkers (see for instance Mudd et al, 1987) have first demonstrated the ability of intact chloroplasts to synthesize PG and they have also demonstrated that all the enzymes involved in its formation are localized in the inner envelope membrane:

phosphatidic acid + CTP   →   CDP-diacylglycerol + PPi

CDP-diacylglycerol + sn-glycerol 3-phosphate   →
        3-sn-phosphatidyl-1'-sn-glycerol 3-phosphate + CMP

3-sn-phosphatidyl-1'-sn-glycerol 3-phosphate   →
                        phosphatidylglycerol + Pi

An additional proof for the demonstration of PG formation from phosphatidic acid synthesized via the envelope Kornberg-Pricer pathway was provided by a careful analyses of the nature and the position of the fatty acids on PG.

B - BIOSYNTHESIS OF PRENYLQUINONES

Prenylquinones, and especially plastoquinone-9, are very important compounds for photosynthesis. Since chloroplasts were able to synthesize prenylquinones, we have investigated whether the envelope membranes could be involved in this process.

Tocopherols are synthesized from homogentisic acid and a C20-prenyl unit to form 2-methyl-6-prenyl quinol. By a series of methylations and cyclization, the 2-methyl-6-prenylquinol gives rises successively to 2,3-dimethyl-6-prenylquinol, γ-tocopherol (or γ-tocotrienol) and finally α-tocopherol (α-tocotrienol). Plastoquinone-9 is synthesized from homogentisic acid and solanesyl-pyrophosphate to form 2-methyl-6-solanesylquinol, which is methylated and oxidized to form successively plastoquinol-9 and plastoquinone-9. All these enzymes have been localized in the inner envelope membrane of spinach chloroplasts (Soll et al, 1985). Consequently, the presence in envelope membranes of all these enzymes requires the massive transport of prenylquinones from envelope membranes to thylakoids. The mechanism for such a transport is not yet known.

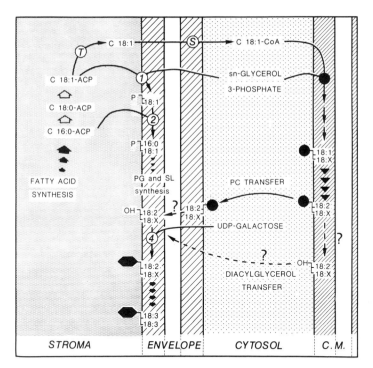

Figure 6. Glycerolipid biosynthesis in the chloroplast envelope membranes from 18:3 plants.

CONCLUSION

The data presented above demonstrate first that the two envelope membranes from chloroplasts and non-green plastids have a typical plastid composition: both envelope membranes contain galactolipids, sulfolipid, carotenoids and prenylquinones. Therefore, the outer envelope membrane has very little in common with the endoplasmic reticulum or the other extra-plastidial membranes. The only exception is the presence of PC in the cytosolic leaflet of the membrane. These observations could reflect the integration of the prokaryotic ancestor of plastids within the protoeukaryotic host. Indeed, the nature, the structure and the biosynthesis of the typical plastid glycolipids strongly support this hypothesis.

A second conclusion is that envelope membranes from all kind of plastids analyzed so far have almost identical chemical composition and functions. They are the only permanent membrane structure in all plastids, and the presence in envelope membranes of the enzymes involved in the biosynthesis of plastid compounds such as galactolipids, phosphatidylglycerol or sulfolipid (Figure 4) explain the flexibility of the transformation of proplastids into various plastid types. However, differentiations of all plastid types and their developmental transitions are probably associated with, or dependent on, marked changes in the specific enzymatic equipement of plastid envelope membranes. Nothing is presently known at this level, but this constitute a challenging goal for the next few years.

Finally, it is clear that envelope membranes are a major structure in plastid biogenesis. The contribution of envelope membranes, and especially the inner envelope membrane, to the biosynthesis of phosphatidic acid, diacylglycerol and MGDG is well established in plastids (chloroplasts and non-green plastids) from 16:3 plants, as shown in Figure 5. Unfortunately this is not as clear in 18:3 plants (Figure 6). The contribution of eukaryotic PC to the formation of the C18/C18 backbone is postulated (Roughan and Slack, 1982), but there is a missing link in the pathway: although it is possible to transfer PC to the outer envelope membrane, the existence of a phospholipase c type enzyme has not yet been demonstrated in envelope membranes. Such an enzyme would provide diacylglycerol for the MGDG synthase which is active in 16:3 and 18:3 plastids. Another possibility is that C18/C18 diacylglycerol, and not PC, is transfered from the extraplastidial membranes to the outer envelope membrane. This remains to be demonstrated.

REFERENCES

Alban, C., Joyard, J. and Douce, R. (1988) Plant Physiol. in press.

Andrews, J., Ohlrogge, J.B. and Keegstra, K. (1985) Plant Physiol. 78, 459-465.

Bertrams, M. and Heinz, E. (1981) Plant Physiol. 68, 653-657.

Bligny, R. and Leguay, J.J. (1987) Methods Enzymol. 148, 3-16.

Block, M.A., Dorne, A.-J., Joyard, J. and Douce, R. (1983) J. Biol. Chem. 258, 13281-13286.

Cline, K., Andrews, J., Mersey, B., Newcomb, E.H. and Keegstra, K. (1981) Proc. Natl. Acad. Sci. USA 78, 3595-3599

Covès, J., Block, M.A., Joyard, J. and Douce, R. (1986) FEBS Lett. 208, 401-406.

Covès, J., Joyard, J. and Douce, R. (1988) Proc. Natl. Acad. Sci. USA, 85, 4966-4970.

Dorne, A.-J., Joyard, J., Block, M.A. and Douce, R. (1985) J. Cell. Biol. 100, 1690-1697.

Douce, R. (1974) Science 183, 852-853.

Douce, R. and Guillot-Salomon, T. (1970) FEBS Lett. 11, 121-126.

Douce, R. and Joyard, J. (1979) Adv. Bot. Res. 7, 1-116

Douce, R. and Joyard, J. (1980) In the Biochemistry of Plants, vol. 4, (Stumpf P.K., ed) pp. 321-362.

Douce, R. and Joyard, J. (1982) In Methods in Chloroplast Molecular Biology (M. Edelman, R.B. Hallick and N-H. Chua, eds.) Elsevier Biomedical Press, Amsterdam, pp. 239-256.

Douce, R., Holtz, R.B. and Benson, A.A. (1973) J. Biol. Chem. 248, 7215-7222.

Douce, R., Block, M.A., Dorne, A.-J. and Joyard, J. (1984) In Subcellular Biochemistry, vol. 10, (D.B. Roodyn, ed.) pp. 1-86, Plenum Press, New york.

Flügge, U.I. and Heldt, H.W. (1976) FEBS Lett. 68, 259-262.

Flügge, U.I. and Benz, R. (1984) FEBS Lett. 68, 85-89.

Frentzen, M., Heinz, E., Mc Keon, T. and Stumpf, P.K. (1983) Eur. J. Biochem. 129, 629-639.

Heber, U. and Heldt, H.W. (1981) Ann. Rev. Plant Physiol. 32, 139-168.

Heemskerk, J.W.M. and Wintermans, J.F.G.M. (1987) Physiol. Plantarum 70, 558-568

Heinz, E. (1977) In Lipids and Lipid Polymers (M. Tevini and H.K. Lichtenthaler, eds.), pp. 102-120, Springer, Berlin.

Heinz, E. and Roughan, P.G. (1983) Plant Physiol. 72, 273-279.

Journet, E.P. and Douce, R. (1985) Plant Physiol. 79, 458-467.

Journet, E.P., Bligny, R. and Douce, R. (1986) J. Biol. Chem. 261, 3193-3199.

Joyard, J. and Douce, R. (1977) Biochim. Biophys. Acta 486, 273-285.

Joyard, J. and Douce, R. (1979) FEBS Lett. 102, 147-150.

Joyard, J., Grossman, A.R., Bartlett, S.G., Douce, R. and Chua, N.H. (1982) J. Biol. Chem. 257, 1095-1101.

Joyard, J., Billecocq, A., Bartlett, S.G., Block, M.A., Chua, N.H. and Douce, R. (1983) J. Biol. Chem. 258, 10000-10006.

Joyard, J. and Douce, R. (1987) The Biochemistry of Plants, vol. 9, lipids (P.K. Stumpf, ed.) pp. 215-274, Academic Press, New York.

Joyard, J., Chuzel, M. and Douce, R. (1979) in Advances in the Biochemistry and Physiology of Plant Lipids (L.A. Appelqvist and C. Liljenberg, eds.) pp. 181-186, Elsevier, Amsterdam.

Joyard, J., Blée, E. and Douce, R. (1986) Biochim. Biophys. Acta 879, 78-87.

Keegstra, K. and Youssif, A.E. (1986) Methods Enzymol. 118, 316-325.

Mudd, J.B., Andrews, J. and Sparace, S.A. (1987) Methods Enzymol. 148, 338-345.

Murata, N. and Sato, N. (1983) Plant Cell Physiol. 14, 133-138.

Neuburger, M., Journet, E.P., Bligny, R., Carde, J.P. and Douce, R. (1982) Arch. Biochem. Biophys. 217, 312-323.

Nishida, I. Frentzen, M., Ishizaki, O. and Murata, N. (1987) Plant Cell Physiol. 28, 1071-1079.

Pineau, B., Dubertret, G., Joyard, J. and Douce, R. (1986) J. Biol. Chem. 262, 9210-9215.

Quail, P.H. (1979) Ann. Rev. Plant Physiol. 30, 425-484.

Roughan, P.G. and Slack, C.R. (1982) Ann. Rev. Plant Physiol. 33, 97-132.

Soll, J., Schultz, G., Joyard, J., Douce, R. and Block, M.A. (1985) Arch. Biochem. Biophys. 238, 290-299.

Van Besouw, A. and Wintermans, J.F.G.M. (1978) Biochim. Biophys. Acta 529, 44-53.

Walker, D.A. (1980) Methods Enzymol. 69, 94-104.

Walker, D.A., Cerovic, Z.G. and Robinson, S.P. (1987) Methods Enzymol. 148, 145-157.

BIOGENESIS OF PEROXISOMES

H.F. Tabak and B. Distel

Laboratory of Biochemistry, University of Amsterdam
Meibergdreef 15
1105 AZ Amsterdam, The Netherlands

INTRODUCTION

All eukaryotic cells possess the same basic architectural design. They are divided in specific sub-compartments or organelles, each of which is equipped to carry out its particular function in biosynthetic or degradative processes. These different functions are each mediated by specific proteins and most of them are unique for a certain type of organelle. It requires that newly synthesised proteins are routed to the organelles or membrane structures to which they belong.

Studies on protein routing have revealed that relatively short amino acid stretches in proteins are responsible for routing to various subcellular compartments. Such routing signals can occur at the N-terminal end of proteins and after translocation of the organelle membrane this signal is often proteolytically removed. It is a feature typical for proteins translocated across the membranes of the endoplasmatic reticulum (ER)[1], mitochondria[2], and chloroplasts[3]. Sometimes the amino acid signal sequence is located internally in the protein and processing does not take place. This is the case with proteins that enter the nucleus[4]. Proteins that translocate the peroxisomal membrane are in general also not processed[5] but the identity and location of the routing signal within the protein is still somewhat controversial.

When results on protein routing obtained with cells from different organisms are compared, it is obvious that basic biochemical principles exist, again illustrating the uniformity in biochemical rules between eukaryotic organisms. It allows investigators to choose the best organism to study a particular problem, be it a mammal or yeast, and produce results of common interest. Comparison of data derived from studies on widely different organisms is especially useful to extract information on routing of proteins to peroxisomes.

Peroxisomes are mostly small spherical vesicles (0.2 - 1 μm) bounded by a single membrane with specialised functions depending on the cell type of organism in which they occur. They were originally defined as organelles containing an H_2O_2 producing oxidase and catalase but they can be equipped with a great variety of other enzymes, such as those catalysing the different steps in the fatty acid β-oxidation pathway[6,7]. Their indispensability to normal cellular metabolism is illustrated by the occurrence of inherited diseases in man which are due to deficiencies in peroxisomal function(s)[8].

Two major questions arise concerning the biogenesis of peroxisomes: i) where does the organelle originate from and ii) how are existing organelles equipped with their characteristic proteins. It has long been thought that peroxisomes arise by budding from the ER membrane. This was mainly based on morphological studies using the electron microscope, which often showed peroxisomes in close association with ER and even membrane continuities between peroxisome and ER membranes were claimed to be visible sometimes. More recent biochemical evidence favours the view that peroxisomes arise from pre-existing peroxisomes by division as is the case for other organelles such as mitochondria. Peroxisomal matrix proteins and, most importantly in this connection, membrane proteins are translated on free polyribosomes and subsequently routed to peroxisomes. They do not show any of the features typical for proteins traversing the ER such as possession of a cleavable N-terminal signal peptide or post-translational modifications such as carbohydrate groups. Particularly the post-translational insertion of proteins into the peroxisomal membrane has been taken as evidence for the autonomic growth and proliferation of peroxisomes. More recent electronmicroscopical studies carried out with fungi support the budding of daughter peroxisomes from a mother organelle[9]. A systematic account in favour of the division model is given by Fujiki and Lazarow[10].

PEROXISOMAL TARGETING SIGNALS

A question which is not completely solved is the mechanism of post-translational import of proteins into peroxisomes. Compared to the refined state of the art in the area of protein import into mitochondria or chloroplasts, the peroxisome field is still in statu nascendi. It may therefore be useful to try to review the current state of affairs and discuss the technical problems involved with the various experimental approaches applied. An overview is given in Table 1.

Import of proteins into other organelles has revealed that relatively short amino acid sequences are responsible for membrane targeting and a comparison of targeting signals between proteins belonging to the same organelle shows a certain conservation of features. Identification of these targeting signals is simplified, however, by the fact that the signal is removed from many precursor proteins upon entry into the organelle. Positive confirmation has been made possible in most cases by constructing in-frame gene fusions between a targeting signal and the body of a gene coding for a protein with another subcellular location. Often the addition of the targeting signal suffices to re-address the reporter protein to the new location specified by the targeting signal.

Identification of a peroxisomal targeting signal proved to be more difficult. The primary translation product is often similar to the mature protein, indicating that no processing takes place upon entry into the peroxisomes. The targeting signal may therefore be present somewhere along the entire length of the amino acid sequence. This makes it extremely difficult to deduce a possible signal from a comparison of amino acid sequences between various peroxisomal proteins. A fortunate case is afforded by import of proteins into glycosomes of Trypanosomes. Here glycolytic iso-enzymes exist, one present in the cytoplasm the other in the glycosome, that differ in their primary amino acid sequence only at a few positions. Most remarkable is a carboxy-terminal extension only present in the glycosomal version of phosphoglycerate kinase, implicating it as a potential targeting signal[11].

Some cell types can be transformed with genes coding for peroxisomal

Table 1. Identification of organelle import
          signals

In vivo

Methods   -computer analysis of gene sequences to
             detect conserved targeting signals
            -analysis of cells transformed with
             (mutant) genes

Criteria
-co-purification with target organelle
-localisation with immuno-fluorescence under LM
-localisation with immuno-gold decoration under EM
-sensitivity to selective energy depletion

Handicaps and pitfalls
-targeting sequences often short and difficult to
 trace due to absence of precursor processing
-peroxisomes from some organisms are difficult to
 isolate
-aggregated (mutant) proteins can co-purify with
 organelles
-under LM import indistinguishable from outside
 binding
-inhibitors may have side effects on other
 organelles (mitochondria)

In vitro

Method   -labelled primary translation product
            incubated with isolated organelles

Criteria
-selective binding to organelles (sedimentation)
-acquirement of protease resistance
-time-, temperature- and energy
 dependence/efficiency
-organelle specific
-saturable
-dependence (short?)on signal sequence

Handicaps and pitfalls
-no signal peptide removal to monitor succes of
 protein import
-quality of peroxisome preparation
     *leakiness or intrinsic property due to pores
     *requirement of cytosolic factors (chaperonins
     or heat shock like proteins)
-protease resistance
     *possible increase of protease resistance due to
     ligand binding or aggregation
     *choice of detergent: increase of protease
     sensitivity due to unfolding
-unnatural folding
     *aggregation of mutant proteins resulting in
     import incompetence
     *exposure of false import signals

proteins and, after expression, their subcellular location can be deter mined. This turned out to be particularly useful when heterologous genes are used coding for proteins that are absent in the untransformed cell. With specific antibodies the location of the protein product can be determined without interference of the cell's own proteins. Thus, when firefly luciferase, a protein located in the peroxisomes of the lantern organ of the firefly, was expressed in monkey kidney cells grown in tissue culture, fluorescent antibodies light up peroxisomes indicating luciferase binding to or import into these organelles[12]. By introduction of deleted versions of the luciferase gene into the cells, the targeting signal could be located in the carboxy-terminal end of the protein. Fusion of this carboxy-terminal end to the dihydrofolate reductase (DHFR) gene directs DHFR to the peroxisomes[13]. Finally, an amino acid substitution in the penultimate amino acid destroys the targeting signal. The essential amino acid sequence SER-LYS/HIS-LEU (SKL, see below) is also found in a number of other peroxisomal proteins and together the data strongly suggest that in many cases a peroxisomal targeting signal consists of a single short amino acid stretch located at the carboxy terminus of the protein[14]. Formally, however, the immuno-fluorescence technique used by the authors of this pioneering work does not distinguish between binding to and import into the organelles and it would be nice if the results could get final corroboration by using immuno-gold decoration under the EM. For a given protein this technique can distinguish between matrix location, membrane insertion or outside binding.

A similar approach was followed by transforming Saccharomyces cerevisiae with the Hansenula polymorpha gene coding for the peroxisomal protein alcohol oxidase. Immuno cytochemistry with antibody-gold particles showed that alcohol oxidase is imported into S. cerevisiae peroxisomes[15]. The imported protein remains enzymatically inactive, however, because subunits do not assemble to the octameric state, the active form of the enzyme in H. polymorpha. Here deletion studies to trace the targeting signal were less successful. Each mutational alteration led to severe protein insolubility, resulting in aggregation in the cytoplasm and no import could be detected (B. Distel, unpublished observations). Similar results have been obtained with mutant versions of dihydroxy acetone synthase, despite the fact that these studies were performed in the homologous H. polymorpha system[16]. Even in the successful studies that led to the identification of an import signal, unexplainable results were uncountered: i) insertion of a short linker sequence in various positions in the N-terminal half of the luciferase gene resulting in the in-frame insertion in the protein of a short oligopeptide, led to inability of the mutant protein versions to be imported although the import signal at the carboxy-terminal ends was still present[13]; ii) fusion of the carboxy terminal half of the human catalase gene to a reporter gene led to fusion proteins that were not imported in peroxisomes. The SKL import signal was functional only when the catalase part was reduced in size and was limited to the last 27 carboxy-terminal amino acids[14]. These examples point out the danger that certain mutant derivatives of proteins may turn out to be import-incompetent, possibly caused by permanent shielding or hiding of the targeting signal still present.

In the field of mitochondrial protein topogenesis the opposite has also been demonstrated. The cytosolic mouse DHFR contains a cryptic mitochon-drial targeting signal that can be exposed and activated by manipula-tions leading to improper protein folding[17].

Another and more difficult approach is based on reconstitution in vitro of peroxisomal import inspired by the success of import in vitro with other organelles such as mitochondria and chloroplasts. For peroxisomes, however, this line of experimentation has been less productive and conflicting results have been reported. This may be due to a number of reasons. First, peroxisomes are rather fragile and difficult to isolate. A substantial amount of peroxisomal enzymes can occur in soluble form in homogenates due to leakage or disruption of the organelles after cell breakage. Second, no established criteria exist to assess the quality of the sedimentable peroxisome fraction apart from enzyme latency. Most fractions contain organelles from which small molecules can leak out. Whether this is due to membrane damage or due to the existence of natural protein pores in the peroxisomal membrane[18] is still a matter of debate.

More recently it became obvious from studies on mitochondrial import that cytosolic factors are required for efficient import. These factors are supposed to act as unfoldases to prevent the protein from taking up too much structure that could result in an import-incompetent state. For instance, when DHFR together with its specific inhibitors methotrexate is incubated in vitro with mitochondria, import of DHFR is severely inhibited[19]. Methotrexate binds to DHFR and tightens the conformation of the native protein and this is taken as evidence that too much folding can indeed inhibit import. Peroxisomal fractions that have been used for some of the import studies thus far may have lacked such cytosolic factors. This may have negatively influenced the efficiency of the import process.

A number of criteria is available to establish whether import of a protein into mitochondria or chloroplasts has indeed taken place: i), the specific conversion of the precursor form of the protein to its smaller mature form due to signal peptide removal and ii), the transition of protease sensitivity to resistance. Furthermore, import is energy-dependent and by the subtle use of various inhibitors the import process can be dissected in discrete steps[20].

Criteria for import into peroxisomes are less well established. Almost the only technique applicable is to study the acquirement of protease resistance in a time- and temperature-dependent manner. This is a rather tricky procedure, however, since the protease-sensitive (outside) and resistant (inside) forms of the protein do not differ in size due to lack of processing upon entry into the peroxisome. Therefore, controls to judge the outcome of an import experiment in vitro are extremely demanding. For instance, to demonstrate that the imported protein is protected by the peroxisomal membrane, detergents are used to solubilise the membrane. Subsequent incubation with protease must show that this treatment renders the imported protein fully sensitive. The same effect would be obtained, however, if for instance ionic detergents are used in such controls, the protein may be partly unfolded making it more sensitive to proteolysis than the protein in the absence of detergent. Binding of the protein to membranes or ligands can also confer enhanced protease resistance. It can result in the misleading interpretation that resistance to protease was acquired due to import into the organelle.

Although evidence for the existence of an ATP-driven proton pump in the peroxisomal membrane has been published[21], indicating the possible involvement of a proton gradient in the protein-translocation step, this has not led to a useful criterium for import. This is among other things due to the lack of knowledge about the permeability of the peroxisomal membrane, as referred to earlier.

Import studies in vitro with deleted versions of the protein to be imported suffer from the same problems, such as the insolubility of mutated versions of the proteins, as import work in vivo (see above). In summary, although results have been published reporting success in vitro with import of proteins into peroxisomes, it is not yet completely clear how the possible pitfalls discussed were avoided in a number of cases. Strong support for results obtained in vitro is provided when studies in vivo confirm the observations in vitro. In two recent cases such confirmation was indeed obtained. First, when the last fifteen amino acids of rat acyl CoA oxidase, containing the Ser-Lys-Leu tripeptide at the extreme C terminus, were fused to the bacterial protein chloramphenicol acetyltransferase, the fusion protein was directed to peroxisomes in CV I monkey kidney cells as indicated by indirect immunofluorescence[13]. Fusion of the same C-terminal fragment to DHFR resulted in import of the chimeric protein in isolated rat peroxisomes (personal communication Fujiki and Hashimoto). The second example is that of yeast (Candida tropicalis) acyl CoA oxidase. In vitro studies with this protein suggest the existence of at least two targeting signals, one located at the N-terminus and one in the middle of the protein[22]. Similar results were obtained when deleted versions of the C. tropicalis acyl CoA oxidase gene were expressed in C. maltosa[23]. Cell fractionation studies with such transformed cells followed by protease treatment of the organelle fraction indicated that the primary targeting information is located internally and secondary information is located at the N-terminus of the protein. Efficient import is only obtained when both sequences are present. Interestingly, neither of the putative targeting sequences in the yeast protein contain the tripeptide Ser-Lys-Leu which is found in the peroxisomal targeting sequence of the rat acyl CoA oxidase. These results can be explained in two ways. One is that yeast and higher eukaryotes have diverged to such an extent that there is no longer structural and functional homology between the peroxisomal targeting sequences in these organisms. The other possibility is that there exist several (at least two) pathways in which proteins can enter peroxisomes, either in yeast or in higher eukaryotes. This last hypothesis is supported by the observation that overexpression in vivo of a truncated peroxisomal protein can inhibit the import of several, but not all other peroxisomal proteins[23].

Although it is claimed that firefly luciferase is imported into yeast peroxisomes, this observation has not yet been substantiated. In view of this it would be of great interest to see whether peroxisomal proteins of higher eukaryotes can be imported into yeast peroxisomes and vice versa. Obviously more work is required to further define peroxisomal targeting signals, but the first promising clues have been found.

MECHANISM OF PEROXISOMAL PROTEIN IMPORT

Only a few remarks on general aspects of peroxisomal protein import are possible, due to scarcety of available data. Peroxisomes can be induced rapidly to very high levels in a number of fungi when grown on certain substrates such as for instance methanol[24]. The two most abundant proteins in peroxisomes of such cells are alcohol oxidase (AO) and dihydroxy acetone synthase (DHAS)[25,26,27]. AO is an octameric enzyme consisting of 8 identical subunits. Pulse/chase experiments have shown that monomers are post-translationally imported into peroxisomes and the assembly of monomers takes place after arrival in the peroxisomal matrix[28]. Treatment of cells with the proton ionophore, carbonyl-cyanide m-chlorophenyl hydrazone (CCCP), interrupts this process: octamerisation is inhibited and AO together with some other proteins among which DHAS, is found in a complex bound to the external side of the peroxisomal membrane[29]. It suggests an inhibitor-induced entrapment of proteins in an intermediate stage of the import process. Its functional significance is supported by the observation that the intermediate complex also exists in cells that were not treated with the inhibitor and that the AO protein can be chased from the intermediate complex into octameric AO. The experiments suggest that certain peroxisomal proteins are coordinately imported in an energy-dependent way.

Other observations do not easily fit into this interpretation, however. Imanaka et al.[30] conclude from import studies in vitro that ionophores such as CCCP do not inhibit import, obviating the need for a membrane potential in protein translocation. They only observe a direct requirement for ATP in their system. The possible conflict between the existence of both a membrane potential and membrane pores has been indicated already. Moreover, the formation of the intermediate complex as observed by Bellion and Goodman is not a condition sine qua non for successful import. The AO gene, provided with a constitutive promoter derived from the S. cerevisiae phosphoglycerate kinase gene, was stably introduced into the genome of H. polymorpha[31]. The extra copy of the AO gene is expressed when cells are grown on glucose or ethanol: conditions under which the endogenous AO gene is completely silent, as are the other genes coding for proteins characteristic of the methanol-induced state. Nevertheless, the AO monomers are imported into the single peroxisome of the cell and assembled into enzymatically active octamers. Formation of an intermediate complex consisting of AO monomers and other proteins such as DHAS is therefore not an obligatory intermediate step in the import process.

These experiments also suggest that the peroxisomal membrane is competent to import proteins under all conditions of growth and that proteins characteristic of the methanol-induced state are not dependent on reconditioning of the highly adaptable organelle to permit their import. The existence of a changing repertoire of import receptors depending on growth conditions seems therefore rather unlikely.

CONCLUSION

The field of peroxisome biogenesis is still in its infancy, especially with regard to protein import and the mechanism of membrane translocation by proteins. This immaturity is reflected by the fact that conflicting data are reported. It is not easy to pinpoint specific reasons for these discrepancies. A number of possible pitfalls have been enumerated in the hope that it will help to critically find a way in this newly emerging field. Recent reports and optimism among researchers in this area gathered at a recent meeting held in Amsterdam, indicate that the first steps to maturity have been set, however.

REFERENCES

1. W.T. Wickner, and H.F. Lodish, Multiple mechanisms of protein insertion into and across membranes. Science 230: 400-407 (1985)
2. M.G. Douglas, M.T. McCammon, and A. Vassarotti, Targeting proteins into mitochondria. Microbiol. Rev. 50: 166-178 (1986)
3. G.W. Schmidt, and M.L. Mishkind, The transport of proteins into chloroplasts. Annu. Rev. Biochem. 55: 879-912 (1986)
4. C. Dingwall, and R.A. Laskey, Protein import into the cell nucleus. Annu. Rev. Cell Biol. 2: 367-390 (1986)
5. P. Borst, How proteins get into microbodies (peroxisomes, glyoxysomes, glycosomes). Biochem. Biophys. Acta 866: 179-203 (1986)
6. P.B. Lazarow, and C. de Duve, A fatty acyl-CoA oxidizing system in rat liver peroxisomes; enhancement by clofibrate, a hypolipidemic

drug. Proc. Natl. Acad. Sci. USA 73: 2043-2046 (1976)

7. S. Kawamoto, C. Nozaki, A. Tanaka, and S. Fukui, Fatty acid β-oxidation system in microbodies of n-alkane-grown Candida tropicalis. Eur. J. Biochem. 83: 609-613 (1978)

8. S. Goldfischer and J.K. Reddy, Peroxisomes (microbodies) in cell pathology. Int. Rev. Exp. Pathol. 26: 45-84 (1984)

9. M. Veenhuis, J.P. Van Dijken, S.A.F. Pilon, and W. Harder, Development of crystalline peroxisomes in methanol-grown cells of the yeast Hansenula polymorpha and its relation to environmental conditions. Arch. Microbiol. 117: 153-163 (1978)

10. P.B. Lazarow, and Y. Fujiki, Biogenesis of peroxisomes. Annu. Rev. Cell Biol. 1: 489-530 (1985)

11. B.W. Swinkels, R. Evers, and P. Borst, The topogeneic signal of the glycosomal (microbody) phosphoglycerate kinase of Crithidia fasciculata resides in a carboxy-terminal extension. EMBO J. 7: 1159-1165 (1988)

12. G.-A. Keller, S. Gould, M. Deluca, and S. Subramani, Firefly luciferase is targeted to peroxisomes in mammalian cells. Proc. Natl. Acad. Sci. USA 84:3264-3268 (1987)

13. S.J. Gould, G.-A. Keller, and S. Subramani, Identification of a peroxisomal targeting signal at the carboxyterminus of firefly luciferase. J. Cell. Biol. 105: 2923-2931 (1987)

14. S.J. Gould, G.-A. Keller, and S. Subramani, Identification of peroxisomal targeting signals located at the carboxy terminus of four peroxisomal proteins. J. Cell. Biol. 107: 897-906 (1988)

15. B. Distel, M. Veenhuis, and H.F. Tabak, Import of alcohol oxidase into peroxisomes of Saccharomyces cerevisiae. EMBO J. 6: 3111-3116 (1987)

16. H. Hansen, and R. Roggenkamp, Import of proteins into peroxisomes: use of gene fusions and deletions for the identification of functional domains. Abstr. 14th Int. Congres on Yeast Genetics and Molecular Biology, Helsinki, Finland.

17. E.C. Hurt, and G. Schatz, A cytosolic protein contains a cryptic mitochondrial targeting signal. Nature 325: 499-503 (1987)

18. G.P. Mannaerts, and P.P. van Veldhoven, Permeability of the peroxisomal membrane. In: Peroxisomes in Biology and Medicine, H.D. Fahimi and H. Sies, editors, Springer, Berlin, 169-176 (1987)

19. M. Eilers, and G. Schatz, Binding of a specific ligand inhibits import of a purified precursor protein into mitochondria. Nature 322: 228-232 (1986)

20. N. Pfanner, F.-U. Hartl, and W. Neupert, Import of proteins into mitochondria: a multiple-step process. Eur. J. Biochem. 175: 205-212 (1988)

21. A.C. Douma, M. Veenhuis, G.J. Sulter, and W. Harder, A proton-translocating adenosine triphosphatase is associated with the peroxisomal membrane of yeasts. Arch. Microbiol. 147: 42-47 (1987)

22. G.M. Small, L.J. Szabo and P.B. Lazarow, Acyl-CoA oxidase contains two targeting signals each of which can mediate protein import into peroxisomes. EMBO J. 7: 1167-1173 (1988)

23. T. Kamiryo, Y. Sakasegawa and H. Tan, Expression and transport of Candida tropicals peroxisomal Acyl-coenzym A oxidase in the yeast Candida maltosa. Agric. Biol. Chem. in press.

24. M. Veenhuis, J.P. van Dijken and W. Harder, The significance of peroxisomes in the metabolism of one-carbon compounds in yeast. In: Advances in Microbial Physiology, A.H. Rose, J. Gareth Morris and D.W. Tempest, editors, Acad. Press New York. 1-82 (1983)

25. M. Veenhuis, J.P. van Dijken, W. Harder and F. Mayer, Substructure of crystalline peroxisomes in methanol-grown Hansenula polymorpha: evidence for an in vivo crystal of alcohol oxidase. Molec. Cell Biol. 1: 949-957 (1981)

26. A.C. Douma, M. Veenhuis, W. de Koning, M. Evers and W. Harder, Dihydroxyacetone synthase is localized in the peroxisomal matrix of methanol-grown Hansenula polymorpha. Arch. Microbiol. 143: 237-243 (1985)

27. J.M. Goodman, Dihydroxyacetone synthase is an abundant constituent of the methanol-induced peroxisome of Candida boidinii. J. Biol. Chem. 260: 7108-7113 (1985)

28. J.M. Goodman, C.W. Scott, P.N. Donahue and J.P. Atherton, Alcohol oxidase assembles post-translationally into the peroxisome of Candida boidinii. J. Biol. Chem. 259: 8485-8493 (1984)

29. E. Bellion and J.M. Goodman, Proton ionophores prevent assembly of a peroxisomal protein. Cell 48: 165-173 (1987)

30. T. Imanaka, G.M. Small and P.B. Lazarow, Translocation of acyl-CoA oxidase into peroxisomes requires ATP hydrolysis but not a membrane potential. J. Cell Biol. 105: 2915-2922 (1987)

31. B. Distel, I. Van Der Ley and H.F. Tabak, Alcohol oxidase expressed under non-methylotrophic conditions is imported, assembled and enzymatically active in peroxisomes of Hansenula polymorpha. J. Cell. Biol. 107: 1669-1675 (1988)

BIOGENESIS AND EVOLUTIONARY ORIGIN OF PEROXISOMES

Fred R. Opperdoes and Paul A.M. Michels

Research Unit for Tropical Diseases
International Institute of Cellular and Molecular Pathology
B - 1200 Brussels, Belgium

INTRODUCTION

In 1965 de Duve coined the name peroxisome for the microbody-like organelles present in most eukaryotic cells. Microbodies, as a group, are quite heterogeneous and comprise a variety of organelles such as peroxisomes, glyoxysomes and glycosomes. They may have quite different functions within the eukaryotic cell (Lazarow and Fujiki, 1985; Huang et al, 1983; Opperdoes, 1987). Table 1 gives an overview of the principal metabolic pathways that have been found in microbodies. Despite such differences in function, it is now generally accepted that all these microbodies are members of one family of organelles. Morphologically, they have the same appearance. They are round or oval-shaped, but range in size from 0.2 to 1 micrometer. They are surrounded by a single membrane and have an electron-dense matrix, which sometimes contains a crystalloid inclusion (Fig. 1). They are found in almost all eukaryotic cells, such as protozoa, fungi, plants and animals. All microbodies have beta-oxidation as the common pathway. Peroxide metabolism, consisting of H_2O_2-producing oxidases and catalase, was originally thought to be the main characteristic of these organelles, since catalase is present in the microbodies of most organisms. However, this 'marker enzyme' may as well be absent as has been described in the case of the Euglenoids (Muller, 1975), the Trypanosomatids (Opperdoes, 1987) and some fungi (Kunau et al., 1987). Glyoxysomes, typical for germinating plants contain, in addition to the above mentioned pathways, enzymes of the glyoxylate cycle. Glycosomes, the microbodies typical of the Kinetoplastida - flagellated protozoa that comprise the parasitic trypanosomes, responsible for a number of important diseases of mankind - are highly specialised in glycolysis and contain the first seven enzymes of the Embden-Meyerhof pathway as well as two enzymes of glycerol metabolism. The peroxisomes of methylotrophic fungi and the fungi that grow on alkanes, are highly specialised in the oxidation of methanol and fatty acids, respectively.

Only primitive protozoa, sometimes referred to as the Archaezoa, lack microbodies. These organisms also lack other cell organelles such as mitochondria or chloroplasts (Cavalier-Smith, 1987). Microbodies, together with the other major cell organelles, probably all arose later in evolution.

Table 1. Major metabolic pathways associated with microbodies

Pathway	Peroxisomes Animals	Fungi	Plants	Glyoxysomes Plants	Glycosomes Trypanosomes
Peroxide metabolism	+	+	+	+	± [a]
β-Oxidation system	+	+	+	+	±
Ether-lipid biosynthesis	+				+
Glyoxylate cycle		±		+	
Pyrimidine biosynthesis					+
Purine salvage					±
CO_2-fixation					+
Glycolysis					±

[a] ± indicates that only part of the enzymes of the pathway are present or have been identified in microbodies

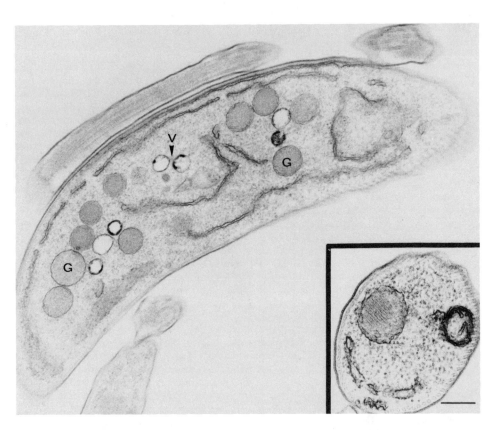

Figure 1. Thin section of the bloodstream form of *Trypanosoma brucei* showing two clusters of glycosomes (G) in the vicinity of rough endoplasmic reticulum and of membrane-bounded vacuoles (V). A unit membrane which surrounds the glycosome is clearly seen. Bar, 0.3 µm. Inset shows a cell with a glycosome containing a crystalloid core displaying a lamellar structure. Bar, 0.2 µm. Photograph kindly provided by Dr. Isabelle Coppens.

THE ORIGIN OF MICROBODIES

The evolutionary origin of microbodies, has been until now an enigma. For a long time it was thought that peroxisomes were formed in the cell by budding off from the endoplasmic reticulum (ER). This, however, has proven to be incorrect. Peroxisomes often are interconnected via a membrane system: the so called peroxisomal reticulum which is quite distinct from the ER (Fujiki et al, 1982). Therefore, it is highly unlikely that there exist direct connections between peroxisomes and ER.

Microbodies cannot be formed de novo. In yeast at least one peroxisome must be present and new peroxisomes always arise from pre-existing organelles by enlargement and subsequent fission into two unequally sized daughter organelles (Lazarow and Fujiki, 1985). In this respect microbodies resemble mitochondria and chloroplasts, two other cell organelles that multiply by binary division. It is generally accepted that the latter two organelles entered the primitive eukaryotic cell as endosymbionts, then gradually lost the majority of their DNA after they transferred a great number of their genes to the host nucleus. Microbodies, however, do not contain any DNA (Lazarow and Fujiki, 1985; Opperdoes, 1987). They are absent from the very primitive eukaryotes and were acquired by the eukaryotic cell presumably at about the same time in evolution as the mitochondria and chloroplasts. It is, therefore, well possible that the ancestral microbody, like mitochondria and plastids, also originated from an endosymbiont, as soon as the primitive eukaryotic cell had acquired the capacity to phagocytize bacteria. The fact that microbodies do not have any DNA, whereas mitochondria and chloroplasts still do, cannot be taken as an argument against this hypothesis. Microbodies would have had ample time to transfer all their genes towards the host nucleus. Moreover, there may not have been any need for retention of genes involved in the synthesis of such complicated integral membrane protein systems, as is the case with e.g. the components of the mitochondrial respiratory chain.

The fact that microbodies are surrounded by one membrane, while mitochondria and chloroplasts have a double membrane, also does not necessarily argue against an endosymbiotic origin. In the case the mitochondrial outermembrane is a remnant from the phagosomal membrane, the ancestral microbody could easily have escaped from the phagosome into the cytosol, as nowadays still occurs with *Trypanosoma cruzi*, when it invades mammalian cells. If, on the other hand, the mitochondrial outermembrane is of true prokaryotic origin, then the absence of a second membrane from the microbody can be explained if the ancestral microbody that invaded a primitive eukaryote was a Gram-positive bacterium (Cavalier-Smith, 1987). Both ancestors of mitochondria and chloroplasts are believed to have been Gram-negative bacteria.

Under such a scenario it is easy to envisage how the enzymes of a complete pathway, such as the nine enzymes of glycolysis in *Trypanosoma brucei* or the entire beta-oxidation pathway, ended up inside an organelle. They simply were already there.

As alternative, microbodies could have been an invention of the eukaryotic cell itself, as part of an intracellular membrane system, that started to live an independent life. Enzymes belonging to entire metabolic pathways subsequently should have been transferred, one by one, to the newly formed organelle. It is rather difficult to understand how this could have happened without doing any serious harm to the cell.

In the case the first scenario is true there should still be some microbody enzymes that originally were present in the endosymbiont and that have retained their prokaryotic characteristics. Amino-acid sequence analysis of microbody proteins might thus reveal such an endosymbiotic ancestry. By now many of such sequences have become available, but in most cases there is no cytosolic counterpart available for comparison. In the case of the glycosomal enzymes such an analysis could however be made. The glycolytic enzymes from trypanosomes have a slightly, but significantly, higher degree of similarity with the homologous enzymes from other eukaryotic organisms than from prokaryotes (Osinga et al, 1985; Michels et al, 1986; Swinkels et al, 1986). The eukaryotic character of the glycosomal enzymes becomes even more apparent when specific regions of these proteins are considered: e.g. the C-terminal part of triosephosphate isomerase (Borst, 1986) or the so-called S-loop region of glyceraldehyde-phosphate dehydrogenase (Michels, unpublished, Fig. 2). When the glycosomal sequences are used to reconstruct their phylogeny, it appears that the Trypanosomatidae branched off already very early from the main line of eukaryotic evolution, well before the separation between the other Kingdoms took place. Figure 3 shows such an analysis for glyceraldehyde-phosphate dehydrogenase. Similar analyses, using phosphoglycerate kinase and triosephosphate isomerase, led to an identical conclusion (not shown).

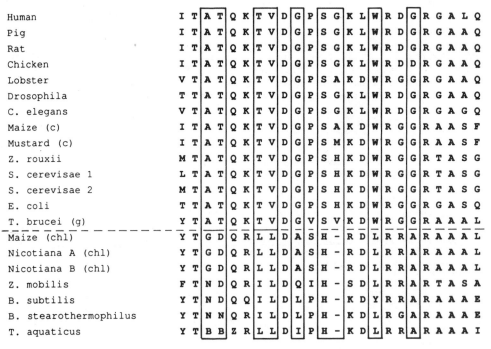

Human	I T	A T	Q K	T V	D G P S G K L	W R D	G	R G A L Q
Pig	I T	A T	Q K	T V	D G P S G K L	W R D	G	R G A A Q
Rat	I T	A T	Q K	T V	D G P S G K L	W R D	G	R G A A Q
Chicken	I T	A T	Q K	T V	D G P S G K L	W R D	D	R G A A Q
Lobster	V T	A T	Q K	T V	D G P S A K D	W R G	G	R G A A Q
Drosophila	T T	A T	Q K	T V	D G P S G K L	W R D	G	R G A A Q
C. elegans	V T	A T	Q K	T V	D G P S G K L	W R D	G	R G A G Q
Maize (c)	I T	A T	Q K	T V	D G P S A K D	W R G	G	R A A S F
Mustard (c)	I T	A T	Q K	T V	D G P S M K D	W R G	G	R A A S F
Z. rouxii	M T	A T	Q K	T V	D G P S H K D	W R G	G	R T A S G
S. cerevisae 1	L T	A T	Q K	T V	D G P S H K D	W R G	G	R T A S G
S. cerevisae 2	M T	A T	Q K	T V	D G P S H K D	W R G	G	R T A S G
E. coli	T T	A T	Q K	T V	D G P S H K D	W R G	G	R G A S Q
T. brucei (g)	Y T	A T	Q K	T V	D G V S V K D	W R G	G	R A A A L
Maize (chl)	Y T	G D	Q R	L L	D A S H - R D L	R R	A	R A A A L
Nicotiana A (chl)	Y T	G D	Q R	L L	D A S H - R D L	R R	A	R A A A L
Nicotiana B (chl)	Y T	G D	Q R	L L	D A S H - R D L	R R	A	R A A A L
Z. mobilis	F T	N D	Q R	I L	D Q I H - S D L	R R	A	R T A S A
B. subtilis	Y T	N D	Q Q	I L	D L P H - K D Y	R R	A	R A A A E
B. stearothermophilus	Y T	N N	Q R	I L	D L P H - K D L	R G	A	R A A A E
T. aquaticus	Y T	B B	Z R	L L	D I P H - K D L	R R	A	R A A A I

Figure 2. Sequence comparison of the S-loop region of glyceraldehyde-phosphate dehydrogenases. The S-loop regions involved in subunit-subunit interactions of 21 sequences of both prokaryotic and eukaryotic origin were compared. Boxed are the residues which are highly conserved. The horizontal division separates the eukaryotic from the prokaryotic sequences.

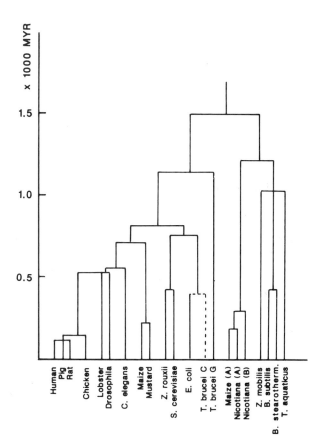

Figure 3. Phylogenetic tree showing the evolutionary relationships between various glyceraldehyde-phosphate dehydrogenase from both prokaryotic and eukaryotic origin.

The two glyceraldehyde-phosphate dehydrogenase isoenzymes from *T. brucei* are remarkably different with only 50-60% sequence identity between them (M. Marchand and P. Michels, preliminary results). This indicates that the two isoenzymes are not more related to each other than any glyceraldehyde-phosphate dehydrogenases from a prokaryote and eukaryote would be. This indicates that the two trypanosome isoenzymes must have a separate origin. It is of interest to note that the cytosolic isoenzyme is closely related to the *E. coli* enzyme (about 80% sequence identity, Michels, unpublished), however this may not be interpreted as an indication for a prokaryotic origin of this trypanosome isoenzyme, since amongst all the known bacterial glyceraldehyde-phosphate dehydrogenase sequences, the *E. coli* enzyme itself behaves anomalously in that it is the only prokaryotic sequence with typical eukaryotic character (Branlant and Branlant, 1985). This suggests that *E. coli*, or its ancestor, may have gained its gene for glyceraldehyde-phosphate dehydrogenase from a eukaryotic host by an event of horizontal gene transfer. Similarly the Trypanosomatid ancestor could have obtained an additional gene by horizontal transfer of DNA, either from an endosymbiont, or by other means.

So far, our observations do not lend support to a prokaryotic ancestry of the glycosome and this would apply also to the other members of the microbody-family. However, as yet such analyses may not be used as proof for a true eukaryotic nature of these organelles. At this stage it cannot be excluded that, in the case of the glycosome, although originally of prokaryotic origin, the original enzymes were gradually replaced by their counterparts that were already present in the eukaryotic cell. Clearly much more sequences have to be analyzed before a definite answer can be given as to the evolutionary origin of microbodies.

THE BIOGENESIS OF MICROBODIES

The general picture of microbody biogenesis that has emerged over the last few years is as follows (see Lazarow and Fujiki, 1985; Borst, 1986 and Opperdoes, 1988, for reviews). Microbodies do not have their own genetic machinery. DNA is absent from microbodies and no evidence has been found for the presence of ribosomes. Microbody proteins are thus encoded in the nucleus and several genes have indeed been localised on nuclear chromosomes. All microbody proteins studied so far, whether present in the matrix or in the microbody membrane, are synthesized on free polysomes in the cytosol; most of them at their mature size. Import into the organelle occurs post-translationally and is highly efficient. Cytosolic half-lives as short as 1 to 15 minutes have been measured. Contrary to the situation with the majority of the chloroplast, mitochondrial and ER proteins, import of microbody proteins does not require any proteolytic processing (reviewed by Lazarow and Fujiki, 1985; Borst, 1986 and Opperdoes, 1987, 1988). There is also no evidence for the requirement of any secondary modification, such as glycosylation, or phosphorylation.

TOPOGENIC SIGNALS

The glycosomal proteins, owing to the fact that they have well studied cytosolic counterparts, were the first microbody enzymes for which a topogenic signal was proposed (Wierenga et al., 1987). Comparison of the purified glycosomal enzymes with their respective cytosolic counterparts from other organisms has revealed that in general the glycosomal proteins have a larger subunit size (Misset et al, 1986), while a comparison of amino-acid sequences showed that this larger size results from the presence of glycosome-specific insertions and extensions (Wierenga et al., 1987). Such extensions have both been found at the N- as well as at the C-termini of glycosomal proteins and are not removed upon import into the glycosome (Hart et al., 1987). Another intriguing observation was that several of these unique sequence elements are responsible for the high overall positive charge that characterises most of the glycosomal proteins. Subsequent modelling of three of the glycosomal proteins into the three-dimensional structures of their respective homologous enzymes from other organisms, revealed the presence of so-called "hot spots" of positive charge, 4 nm apart, on the surface of each of these proteins. It was then suggested that these unique "hot spots" could function as the topogenic signal (Wierenga et al., 1987). Subsequent in vitro uptake experiments, using deletions and/or mutations of trypanosomal phosphoglycerate kinase (Dovey et al., 1988 and C.C. Wang, personal communication) and aldolase (Opperdoes and Michels, unpublished) support the idea that insertions with positively charged residues could be involved in the binding to, and the subsequent passage through, the glycosomal membrane. However, a recent comparison by Swinkels et al. (1988) of the cytosolic and glycosomal sequences of phosphoglycerate kinase of a closely related Trypanosomatid, revealed that apart from a long C-terminal extension, the two isoenzymes were 99.5 % homologous. Moreover, there was no difference at all in overall charge between the two proteins. Swinkels et al. (1988), therefore, concluded that the topogenic signal for phosphoglycerate kinase had to reside in the C-terminus.

Gould et al. (1987,1988) and subsequently Miyazawa et al. (1989) identified a topogenic signal responsible for the targeting of microbody enzymes to insect and mammalian peroxisomes. By deletion and fusion experiments Gould and coworkers located the targeting signal for several peroxisomal proteins within the last 15 amino-acids of the carboxy-terminus. This peptide was both necessary, and sufficient, for the proper targeting of luciferase from firefly peroxisomes and fusion proteins, into mammalian peroxisomes. This demonstrates that topogenic signals for uptake into microbodies are conserved between insects and mammals. Working with an in vitro import assay and various truncated and mutant cDNAs, Miyazawa et al. (1989) subsequently showed that in the case of rat-liver acyl-CoA oxidase the C-terminal part of the protein, is also responsible for the uptake, by mammalian peroxisomes, of both the oxidase and a fusion protein. They succeeded in narrowing down the signal sequence to the last 5 C-terminal residues. Comparison of a number of peroxisomal C-terminal sequences then revealed that many peroxisomal enzymes end with SKL, or a similar sequence (Fig. 4). Some flexibility however seems to be allowed in that the basic amino acid lysine (K) can be replaced by a histidine (H), but not by an asparagine (Gould et al, 1988) and possibly the serine (S) by other small amino acids, such as alanine.

Figure 4. Comparison of some C-terminal sequences of microbody enzymes. Modified from Miyazawa et al. (1989)

Acyl CoA oxidase (rat)	K H L K P L Q S K L
Bifunctional enzyme (rat)	S L A G P H G S K L
Luciferase (firefly)	K A K K G G K S K L
Uricase (soybean)	A S L S R L W S K L
Malate synthase (cucumber)	I H H P R E L S K L
D-amino-acid oxidase (pig)	N L L T M P P S H L
Glucosephosphate isomerase (T.brucei)	I N M F N E L S H L
Glyceraldehyde-P dehydrogenase (T.brucei)	M A A R D R A A K L

Probably a prerequisite for the SKL sequence is that it is well exposed to the outside of the protein. Although by itself sufficient for import, Gould et al. (1987) showed that deletions towards the N-terminal part of luciferase could prevent import, probably through an altered folding of the modified protein. Furthermore, a location at the very C-terminus is not an absolute requirement. Human catalase has an internal SHL sequence in a C-terminal peptide that also has been shown to function as a signal sequence (Gould et al, 1988).

An SKL sequence near or at the C-terminus cannot be the only signal responsible for import of proteins into microbodies since not all microbody proteins contain this sequence. Moreover, Small et al. (1988) have shown that in the case of acyl-CoA oxidase of the yeast *Candida*, the C-terminus is not necessary for import into peroxisomes. On the contrary, an N-terminal, as well as an internal peptide, exhibited targeting properties in an in vitro import assay. Neither of these peptides contain the SKL, or a related sequence (Small et al., 1988). This suggest that there might be more than one topogenic signal responsible for the targeting of proteins to microbodies.

The glycosomal enzymes glucosephosphate isomerase (M. Marchand and P. Michels, unpublished observation) and glyceraldehyde-phosphate dehydrogenase also have each a variant of the SKL sequence at their respective C-termini (Fig. 4). Although there is as yet no indication that these sequences function as a signal peptide, it is interesting to note that the AKL sequence of the glycosomal glyceraldehyde-phosphate dehydrogenase is present in a three amino-acid extension at the C-terminus. A chloroplast enzyme is the only other one, so far, with an extension. Modelling of the glycosomal extension in the three-dimensional crystal structure of the homologous enzyme from *Bacillus* reveals that it is located at the outside of the protein. In the case of other glycosomal enzymes, for which the sequence has become available, only for phosphoglycerate kinase a C-terminal extension of 20 and 39 amino acids, respectively (Swinkels et al., 1988), was identified as a putative C-terminal signal peptide (see above). Neither this peptide, nor the C-termini of the other glycosomal proteins have any detectable similarity with the SKL-like sequences discussed above. Therefore, the true topogenic signal or signals responsible for the targeting of glycosomal proteins remain to be identified.

Microbodies are the only organelles for which the targeting signal, as identified so far, is at the same time three amino acids short and also located at the carboxy terminus. The signal sequences of chloroplast, mitochondrial and ER proteins are invariably located at their N-termini. They are much longer and also much less conserved. In the latter signal peptides secondary structure and consequently the physical properties, rather than a precise sequence of amino acids seems to be important. The signal sequence of microbody proteins on the contrary is the shortest identified thus far, and a precise sequence therefore seems to be required.

CONCLUSION

A large body of information has been gathered on the function and the biogenesis of microbodies in a variety of organisms. Although it is tempting to assume, based on the fact that microbodies cannot be made de novo and multiply by division, that this organelle entered the cell as an endosymbiont, thus far there is no real evidence for a prokaryotic origin.

Microbody enzymes are made in the cytosol and are then rapidly transferred to the organelle. So far one type of topogenic signal has been identified. It is a C-terminally located SKL or related sequence. Not all microbody proteins contain such a sequence and therefore at least one other signal sequence should be responsible for the topogenesis of the remainder of the proteins. Such a signal(s), however, remains to be identified.

ACKNOWLEDGEMENT

Supported by the UNDP/World Bank/WHO Special Programme for Research and Training in Tropical Diseases.

REFERENCES

Borst, P., 1986, Biochim. Biophys. Acta, 866:179.

Branlant, G. and Branlant, C., 1985, Eur. J. Biochem., 150:61.

Cavalier-Smith, T., 1987, Ann. N. Y. Acad. Sci., 503:55.

de Duve, C., 1965, J. Cell Biol., 27:25A.

Dovey, H.F., Parsons, M. and Wang, C.C., 1988, Proc. Natl. Acad. Sci., 85:2598.

Fujiki, Y., Fowler, S., Shio, H., Hubbard, A.L. and Lazarow, P.B., 1982, J. Cell Biol., 93:103.

Gould, S.J., Keller, G.A. and Subramani, S., 1987, J. Cell Biol., 105:2923.

Gould, S.J., Keller, G.A. and Subramani, S., 1988, J. Cell Biol., 107:897.

Hart, D.T., Baudhuin, P., Opperdoes, F.R. and de Duve, C., 1987, EMBO J. 6:1403.

Huang, A.H.C., Trelease, R.N. and Moore Jr., T.S., 1983, Plant Peroxisomes, Academic Press, New York and London, pp 252.

Kunau, W.-H., Kionka, C., Ledebar, A., Mateblowski, M., Moreno de la Garza, M., Schultz-Borchard, U., Thieringer, R. and Veenhuis, M., 1987, in: "Peroxisomes in Biology and Medicine", Fahimi, D. and Sies, H., eds, Springer Verlag, Heidelberg, 128-140.

Lazarow, P.B. and Fujiki, Y., 1985, Annu. Rev. Cell Biol., 1:489.

Michels, P.A.M., Poliszczak, A., Osinga, K., Misset, O., Van Beeumen, J. Wierenga, R.K., Borst, P. and Opperdoes, F.R., 1986, EMBO J., 5:1049.

Misset, O., Bos, O.J.M. and Opperdoes, F.R., 1986, Eur. J. Biochem., 157:441.

Miyazawa, S., Osumi, T., Hashimoto, T., Ohno, K., Miura, S. and Fujiki, Y., 1989, Mol. Cell. Biol., 9:83.

Muller, M., 1975, Annu. Rev. Microbiol., 29:467.

Opperdoes, F.R., 1987, Annu. Rev. Microbiol., 41:127.

Opperdoes, F.R., 1988, Trends Biochem. Sci., 13:255.

Osinga, K.A., Swinkels, B.W., Gibson, W.C., Borst.,P., Veeneman, G.H., Van Boom, J.H., Michels, P.A.M. and Opperdoes, F.R., 1985, EMBO J. 4:3811.

Small, G.M., Szabo, L.J. and Lazarow, P.B., 1988, EMBO J., 7:1167.

Swinkels, B.W., Evers, R. and Borst, P., 1988, EMBO J., 7:1139.

Swinkels, B.W., Gibson, W.C. Osinga, K., Kramer, R., Veeneman, G.H., Van Boom, J.H. and Borst, P., 1986, EMBO J., 5:1291.

Wierenga, R.K., Swinkels, B.W., Michels, P.A.M., Misset, O., Van Beeumen, J., Gibson, W.C., Postma, J.P.M., Borst, P., Opperdoes, F.R. and Hol, W.G.J., 1987, EMBO J., 6:215.

FUNCTIONAL AND IMMUNOLOGICAL CHARACTERIZATION OF MITOCHONDRIAL F_oF_1 ATP-SYNTHASE

Ferruccio Guerrieri[1], Jan Kopecky[2] and Franco Zanotti[1]

[1]Institute of Medical Biochemistry and Chemistry and Centre for the Study of Mitochondria and Energy Metabolism, Bari, Italy and [2]Institute of Physiology, Czechoslovak Academy of Sciences, Praha, Czechoslovakia

INTRODUCTION

The H^+-ATP synthase (F_1-F_o) complex of mitochondria is an oligomeric protein of the inner mitochondrial membrane constituted by a hydrophylic sector (F_1), which contains the catalytic site(s) for ATP synthesis or hydrolysis, connected to a hydrophobic protein (F_o), inserted in the lipid membrane, whose function is the coupling of catalytic activities of F_1 to transmembrane proton translocation (1).

Molecular architecture of F_oF_1 ATP-synthase in all types of energy transducing membranes is very complex. The catalytic part (F_1) of the enzyme always consists of five types of protein subunits (1) with stoichiometry $3\alpha, 3\beta, 1\gamma, 1\delta, 1\epsilon$ and Mr ranging from 55 kDa (α) to 6 kDa (ϵ). In mitochondria, each F_1 is also relatively freely associated with one ATPase inhibitor protein (IF_1; 10 kDa) (2,3).

The polypeptide composition of the F_o sector varies, on the other hand, among species (1). In E. Coli, F_o consists of three types of subunits a,b and c encoded by genes of the unc or atp operon. Their structure has been described and they have been shown to be essential for proper assembly and function of F_o in the ATP synthase (4).

The subunit composition of F_o in eukaryotic ATP synthase is more complex and even preparations from bovine-heart mitochondria exhibit, besides the five F_1 subunits and the ATPase inhibitor protein, seven to nine proteins which may belong to F_o (5-8).

The whole mitochondrial H^+-ATPase complex could be isolated in a relatively pure form, and especially F_1 and IF_1 could be purified to homogeneity. Antibodies could be raised against these preparations, or even against individual H^+-ATPase subunits (as purified by preparative SDS-polyacrylamide electrophoresis of F_1 or F_o (7,8)). These antibodies (antisera) could be used in immunoblotting (see below) for characterization of H^+-ATPase structure (7,8) and for comparing quantities of individual H^+ATPase subunits in e.g. submitochondrial particles lacking different parts of H^+-ATPase (9,10), or for quantification of H^+-ATPase in different types of mitochondria (ontogenic and tissue-specific differences) (11).

The catalytic activity of membrane bound F_1 is compulsorily coupled to transmembrane H^+ translocation through the membrane sector of F_o (1). However, in certain vesicular preparations of inner mitochondrial membrane, obtained by exposure of mitochondria to ultrasonic energy in presence of EDTA and at alkaline pH, the proton conductivity of F_o results in passive proton diffusion that can be blocked by oligomycin and DCCD (12-14). From these particles IF_1 can be removed by Sephadex chromatography with increase of oligomycin-sensitive ATPase activity and proton conductivity (4,15).

After urea treatment of Sephadex particles, the membrane are stripped of F_1 so that no more ATPase activity can be observed but the oligomycin sensitive proton conductivity is increased (12-14).

Sequential addition of purified F_1 (16) and IF_1 (4) to USMP can restore the ATPase activity and the control of the proton conductivity by F_1 (4,15).

The whole mitochondrial H^+-ATP synthase complex or the F_o moiety can be isolated in a relatively pure form, by detergent extractions from the membrane (5,17). The isolated F_o-F_1 complex and the isolated F_o are not able to translocate protons; however reconstitutions in artificial phospholipid membrane can restore the oligomycin sensitive proton translocations (6). In addition the isolated protein subunits could be used for molecular analysis (aminoacid composition and sequence (8)). These structural and functional resolutions and reconstitutions of F_o-F_1 complex allow for studies on functional interaction between the two subunits in the integrated mechanism of the enzyme.

EQUIPMENT AND CHEMICALS

A. Equipment

- Sonicator Cell Disruptor
- Refrigerated high speed centrifuge and ultracentrifuge (Beckman)
- UV/Vis Spectrophotometer (Perkin Elmer)
- Amicon Diaflo apparatus with PM-10 filters
- Electrometer (Keithley)
- Combined pH electrodes
- Vortex mixer
- Apparatus for slab polyacrylamide gel electrophoresis (Bio Rad)
- Apparatus for electrotransfer (Bio Rad)
- Electrophoretic power supply (LKB)
- Densitometer (CAMAG)
- Shaking thermometer water bath (0° -100°)

B. Chemicals

- Bovine heart mitochondria
- Sephadex G-50 (Pharmacia)
- Zwitterionic CHAPS detergent (Sigma)
- Asolectin (Sigma)
- Potassium cholate (Sigma)
- Potassium deoxycholate (Sigma)
- NADH, ATP, ADP (Boehringer)
- Phosphoenolpyruvate (PEP) (Boehringer)
- Pyruvate kinase (PK), lactate dehydrogenase (LDH) and catalase (Boehringer)

- Valinomycin and oligomycin (Sigma)
- Rabbit antisera against F_1 and F_0 subunits (see ref.s 7,8) diluted with glycerol and stored at 20^oC until used
- Acrylamide and N,N'-methylene-bis-acrylamide (Serva)
- Sodium dodecylsulfate, Goat anti-rabbit IgC labeled by peroxidase (Bio Rad)
- Horseradish Peroxidase Color Development Reagent (HPR-reagent) (Bio Rad)
- Nitrocellulose membrane (0.45μm pore size) (Schleicher and Schull).

EXPERIMENTAL PROCEDURE

Preparations of submitochondrial particles with various degrees of resolution of F_0-F_1 complex and of F_0, F_1 and IF_1.

ESMP are obtained by exposure of 60s of beef heart mitochondria to ultrasonic Branson Sonifier (Model W 185) output 70 W (12). From these particles, the ATPase protein inhibitor is removed by passing ESMP through a Sephadex G-50 column (4). F_1 deprived particles are obtained by urea treatment of Sephadex particles (12). F_1 is obtained by chloroform extraction described by Beechey et al. (16). This procedure is particularly convenient being easy and rapid, although it shows a lower activity of ATPase as compared to other preparations (16).

For isolation of IF_1, Mg-ATP submitochondrial particles (4) are used as source materials. These particles are suspended in 0.25 M sucrose, at a protein concentration of 25-30 mg/ml and heated in a test tube for 4 min at 75°C with occasional shaking. The denaturated proteins are removed by centrifugation at 4°C for 10 min at 12.000 x g. The supernatant is removed carefully from the tube and, for each ml, 2.8 ml of ice cold 95% ethanol is added. The mixture is agitated, for 20-30 sec, on a vortex mixer and centrifuged 10 min at 12.000 x g. The pellet obtained represents the IF_1 and the polyacrylamide gel electrophoresis of the purified protein, performed or linear gradient (12-20%) slab gel, showed, generally only one band of apparent Mw 10.500 (4).

The F_0-F_1 complex from bovine heart mitochondria is purified using the following modification of the procedure developed by McEnery et al. (17): a) ESMP are used as starting materials; b) after CHAPS treatment, 4 ml of the membrane extract are layered on 32 ml of 20% sucrose in TA buffer (50 mM Tricine, 1 mM ATP, 25 mM EDTA, 0.5 mM DTT and 5% ethylene glycol, pH 7.9) containing 0.2% CHAPS and centrifuged for 10 h at 27.500 rpm in the Beckman SW 28 rotor at 2°C; c) the top 9 ml are carefully removed and next 21 ml are collected and concentrated 4-5 times (to about 200μg protein/ml) in an Amicon Diaflo apparatus, using a PM-10 filter.

For purification of F_0, USMP are suspended in 20 mM D-mannitol, 70 mM sucrose, 2 mM Hepes and 0.5 mg/ml of defatted bovine serum albumine, pH 7.4, at a concentration of 40-50 mg protein/ml and stored in liquid nitrogenum. The stock USMP are thawed at room temperature, diluted to 2 mg protein/ml with 0.15 M K_2HPO_4, 1 mM ATP, 25 mM EDTA, 0.5 mM DTT and 5% ethylene glycol, pH 7.9, and centrifuged at 105.000 x g for 45 min. The pellet is suspended in the same buffer and centrifuged again twice. The final pellet is suspended in TA buffer (see F_0-F_1 preparation) at a protein concentration of 4 mg/ml and incubated with 1% CHAPS for 10 min at 0°C. After 1 h of centrifugation at 105.000 x g the supernatant is layered on 20% sucrose in TA buffer (4 ml supernatant on 32 ml TA-sucrose) containing 0.2% CHAPS and centri-

fuged for 10 h at 27.500 rpm at 2°C. After centrifugation, the central 21
ml are concentrated (4-5 times) in an Amicon Diaflo apparatus with a PM-10
filter. The yeld of purified F_o sector is about 8-9 mg protein/100 mg protein of USMP.

Preparation of F_o liposomes

F_o liposomes are prepared by the dialysis method: 3 mg protein of isolated F_o are mixed with 30 mg of acetone-washed sonicated asolectin in 1 ml
0.1 M phospahte buffer, pH 7.2, containing 1.6% potassium cholate, 0.8% potassium deoxycholate and 0.2 mM EDTA. The mixture is dialyzed over-night
against 0.1 M potassium phosphate buffer pH 7.5, followed by a 3 h dialysis
against 100 mM sodium tricine buffer pH 7.5. Both dialysis media contained
0.25 mM EDTA and 2.5 mM $MgSO_4$.

Reconstitution of F_o-F_1 particles

To rebind F_1 to USMP (USMP-F_1); USMP are incubated in 0.25 M sucrose,
10 mM Tris/acetate, 1 mM EDTA, 6 mM $MgCl_2$, pH 7.5 (0.5 mg particles protein/
/ml), with bovine heart chloroform-released F_1 at a F_1/USMP protein ratio
of 0.2. After 30 min at 25°C, incubation is stopped by centrifugation (10^5
x g; 0°C).

To rebind IF_1 to Sephadex particles, or to USMP-F_1 submitochondrial
particles (5 mg protein/ml) are preincubated with IF_1 (4μg IF_1/mg particles
protein) for 15 min at room temperature and at pH 6.2. Then 50μg particles
protein are added to the reaction media.

ATPase activity assay

The ATPase activity assay is performed as described in (12) using an
ATP-regenerating system.

In the presence of pyruvate kinase each molecule of ADP, formed by the
ATPase activity of particles, F_1 and F_o-F_1 complex, is converted into ATP
by phosphoenolpyruvate. The product, pyruvate, is rapidly reduced by lactate dehydrogenase to lactate in the presence of NADH that is transformed in
NAD^+.

Mixture 1			In the cuvette	
Tris/acetate pH 7.5	25 mM		Mixture 1	1.94 ml
K-acetate	"	25 mM	NADH 0.1 M	0.002 ml
Sucrose		300 mM	PK (3.0 units)	0.01 ml
$MgCl_2$		2 mM	LDH (3.0 units)	0.005 ml
PEP		1 mM	Particles or isolated F_1 or F_o-F_1	
			complex added in the desired	
			amount of proteins.	
			Add H_2O to 2 ml of volumes.	

The absorbance decrease of NADH at 340 nm ($\Delta\epsilon=6.22$ mM^{-1}) is followed.
The reaction is started by the addition of ATP 2 mM (final concentration).

Measurement of proton translocation

Proton translocation in submitochondrial particles is analyzed by following potentiometrically either the anaerobic release of the respiratory
proton gradient (12-15) or the H^+ release induced by a diffusion potential
(positive inside) imposed by valinomycin-mediated potassium influx (6). The

latter procedure is also used with F_o liposomes (8).

For the kinetic analysis of the proton release from submitochondrial particles or F_o liposomes, the potentiometric traces are converted into proton equivalents by double titration with standard HCl and KOH.

For measurements of anaerobic H^+-release, submitochondrial particles (3 mg/ml) are incubated in a reaction mixture containing: 200 mM sucrose, 30 mM KCl, 0.5µg valinomycin/mg particles protein; 0.2 mg/ml catalase and 20 mM succinate (potassium salt) pH 7.5. Incubation is carried out in a stirred glass vessel under constant stream of N_2 at 25°C. After anaerobiosis, respiration driven proton translocation is activated by repetitive pulses (1-3%) H_2O_2 (5µl/ml) and the pH of the suspension monitored potentiometrically.

For measurements of H^+ release induced by diffusion potential (positive inside), submitochondrial particles (3 mg/ml) of F_o-liposomes (0.4 mg protein/ml) are incubated for 2 min in 0.15 M KCl then 2µg valinomycin/mg particles protein are added and proton release is measured potentiometrically.

Electrophoretic analysis

SDS electrophoresis is performed on slab (0.75x140x140 mm) on linear gradient of polyacrylamide (14-20%) gel according to Laemmli (18). Before electrophoresis, protein samples are suspended in 2.3% SDS, 5% 2-mercaptoethanol, 10 mM Tris-HCl, pH 6.8 and heated on boiling water bath for 3 min. Gels are stained for protein by 0.25% Coomassie brilliant blue R-250, in 45% methanol, 9% acetic acid (v/v) and destained in 25% ethanol, 8,2% acetic acid (v/v).

Transfer to nitrocellulose (blotting)

Proteins separated on the slab gel are transferred electrophoretically to nitrocellulose (NC) to obtain a replica of the protein separation pattern. Proteins remain bound to the membrane which is a firm support with large pores (in contrast to fragile and small pore size of polyacrylamide gel). Therefore steps of incubations and washing needed for immunodetection (see below) can be easily performed (7-9).
Solutions:
- Blotting buffer. Prepare stock solutions of 194.4 g glycine, 43.2 g Tris and 3.6 g SDS in 3 1 of distilled water (store in cold room). Before use mix 0.5 1 stock solution with 0.6 1 methanol and add water to 3 1 (final conc.: 150 mM glycine, 20 mM Tris and 0.02% SDS, 20% methanol, pH 8.3).
Note: Do not adjust pH. All the following steps are performed in clean glass dishes or test tubes, on laboratory rotator, at lab. temperature. NC must not be touched by a bare hand.
Procedure
- Incubate the gel immediately after electrophoresis in blotting buffer (300 ml) on a laboratory rotator (30 min).
Note: A strip of the slab gel containing e.g. separated Mr standards might be stained for proteins by Coomassie blue. Althought the size of polyacrylamide gel after the staining and that of the corresponding immunoblot (see below) will be slightly different, approximate Mr of the antigen can be roughly calculated. Alternatively, pre-stained standards for SDS-electrophoresis (from Bio Rad) can be used.

- Soak in the blotting buffer: filter papers (4 sheets), sponge pads (2 pieces) and NC (1 sheets). Filter papers and NC should be slightly larger than the slab gel.
- In a large dish, under the blotting buffer make a sandwich-like assembly containing: plexiglass frame (cathodic)   (1)

$$
\begin{array}{ll}
\text{sponge pad} & (2) \\
\text{two filter papers} & (3) \\
\text{polyacrylamide slab gel} & (4) \\
\text{NC sheet} & (5) \\
\text{two filter papers} & (6) \\
\text{sponge pad} & (7) \\
\text{plexiglass frame (anodic)} & (8)
\end{array}
$$

Note: No air bubbles must be present in between gel and NC.
- Insert the assembly in the Trans-Blot-Cell (NC facing anod) containing blotting buffer.
- Run the transfer (with mixing using magnetic stirrer) in a cold room, at 7 V/cm (60 V with Trans-Blot-Cell) for 3 h.
- After transfer use NC for immunodetection.

Note: NC can be stored at $-20°C$, on a clean glass plate covered by a plastic band. The slab gel can be stained for proteins to check the efficiency of transfer.

Immunodetection

Before probing with antibodies the membrane matrix must be first incubated with blocking agent. The blocking agent prevents unspecific binding of the probe to the matrix by saturation of the unspecific binding sites. In our protocol, detergent Tween-20 will be used, since a higher sensitivity of detection is usually observed than when various protein blocking agents are used. Rabbit antibodies will be used to detect the corresponding antigens specifically and the immune complexes will be visualized by using second antibodies labeled by peroxidase. For immunodetection the specific rabbit serum against F_1 (1/250 (7)) or against other F_o-F_1 protein can be used (7,8).

Solutions:
- PBS (Phosphate Buffer Saline)
  0.15 M NACl, 0.02 M Na-Pi, pH 7.4
  (Dissolve 26.3 g NaCl, 8.28 g NaH_2PO_4 or 9.36 g NaH_2PO_4 3 H_2O in 800 ml of water, adjust pH with NaOH to pH 7.4 and water to 3 1).
- PBS-Tween
  0.15 M NaCl, 0.02 M Na-Pi, 0.05% Tween-20, pH 7.4
- 3% H_2O (freshly diluted)
- 0.05% NaN_3
- 5 mM K-acetate, pH 5.0
  (Make 0.1 M stock solution: dissolve 9.8 g K-acetate in 100 ml water, adjust pH with KOH to pH 5.0).

Procedure
- After electrotransfer incubate NC in PBS-Tween, 30 min
- Transfer NC to a dish of corresponding size or cut to strips (1 cm) for incubation in test tubes (in the case of a single protein sample separated in the whole slab gel)
- Incubate for 2 h with specific rabbit antiserum diluted in PBS-Tween (1/200) - 1/1000, depending on the titre of specific antibodies in the

serum used). Total volume for incubation is 50 ml for NC sheet size 10x 14 cm, or 3 ml for a single strip in the test tube.
- Wash in PBS-Tween (at least 5 times during 1 h, better overnight).
- Incubate for 45 min with anti-rabbit IgG labeled by peroxidase, diluted in PBS-Tween (1/3000). Total volume is the same as in the case of incubation with the first antibodies.
- Wash by PBS-Tween, than by PBS (several times during 1 h).
- Prepare staining solution for peroxidase by using Horseradish Peroxidase Color Development Reagent (HRP-reagent).

HRP-reagent:
Dissolve 30 mg HRP reagent in 10 ml methanol, add 50 ml PBS and 0.3 ml 3% H_2O_2.

Note: HRP-reagent should be stored at $-20°C$.
- Incubate NC in staining solution (2-5 min) with little shaking untill staining is developed (grey color with HRP-reagent).
- Stop reaction in 200 ml 0.05% NaN_3 (5 min).
- Wash in water and dry in between filter patterns. Store in dark.
- Evaluate the result by densitometry.

RESULTS

    I. Resolution and reconstitution of mitochondrial H^+-ATPase complex. Effect on ATPase activity and on passive oligomycin sensitive proton conduction by H^+-ATP synthase complex.

    In the F_1-F_o complex, normally arranged in the membrane, the oligomycin sensitive proton conductivity by F_o is strictly connected to ATP synthesis or hydrolysis at F_1 level. Following treatment with EDTA, at alkaline pH, the F_1 is displaced from F_o, so that the passive oligomycin sensitive proton conduction by F_o is uncoupled from the catalytic activity of F_1 (Scheme 1b and Table 1). Subsequent removal of inhibitor protein (IF_1) by Sephadex chromatography favours uncoupling between passive proton conduction by F_o and the ATPase activity (Scheme 1c and Table 1). Urea treatment of these particles caused an oligomycin sensitive increase of proton conductivity (Table 1) due to removal of F_1 subunit (Scheme 1d). In fact the ATPase activity in these particles is practically 0; however an oligomycin sensitive ATPase activity can be reconstituted by rebinding of purified F_1 (Table 1) and this is accompanied by inhibition of oligomycin sensitive proton conduction by F_o (Table 1). This observation suggests that cooperative interaction of F_1 and F_o would control proton access (during ATP hydrolysis) and proton exit (during ATP synthesis) at the matrix side of the proton channel in F_o (19). Addition of purified IF_1 to Sephadex particles or to reconstituted USMP-F_1 particles caused inhibition of both ATPase activity and oligomycin sensitive proton conduction by F_o (Table 1). This observation indicates that, in addition to the specific inhibitor effect on ATPase activity, IF_1 can control proton conductivity favouring the binding of F_1 to F_o (4).

II. Polypeptide composition of F_1-F_o complex and F_o.

    SDS polyacrylamide gel electrophoresis of isolated F_o-F_1 complex (Fig.1) showed, in addition to the 5 polypeptides of F_1, 8 prominent protein bands of Mr lower than that of γ-F_1 (33 kDa).

Scheme I

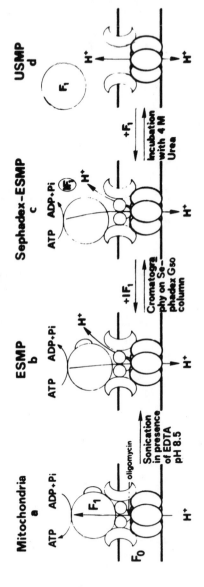

Table 1

Particles	ATPase activity (μmoles ATP hydrolyzed min⁻¹·mg protein⁻¹)		Anaerobic release of respiratory proton gradient (1/t₁/₂ sec⁻¹)	
	–	+oligomycin (1.5 μg/mg protein)	–	+oligomycin (1.5 μg/mg protein)
ESMP	1.08	0.30	1.00	0.32
Sephadex	2.02	0.50	1.81	0.30
Sephadex+IF₁ (4 μg/mg protein)	0.90	0.25	0.96	0.25
USMP	0.03	0.00	2.30	0.32
USMP+F₁	1.35	0.35	1.55	0.50
USMP+F₁+IF₁	0.60	0.15	1.00	0.30
F₁	25.00	25.00	–	–

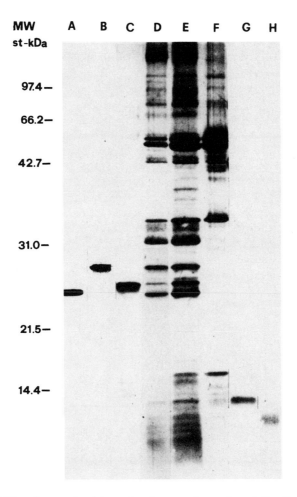

Fig.1 - SDS-polyacrylamide gel electrophoresis of F_o-F_1 complex, F_o, F_1, IF_1, F_oI, OSCP, F_oII and F_6.

F_O-F_1 complex, F_o, F_1 and IF_1 were obtained as described in the experimental procedure. F_oI, OSCP and F_oII were isolated as described in refs. (7,8). F_6 was isolated as described in ref.(20). On SDS-slab gel the following samples were analyzed: A) 1µg F_oII protein; B) 1µg F_oI protein; C) 1µg OSCP; D) 8µg protein of F_o; E) 15µg protein of F_o-F_1; F) 5µg protein of F_1; G) 2µg IF_1; H) 2µg F_6.

The F_o preparation contained low amounts of F_1 subunits (Fig.1) and a lower content of OSCP with respect to the F_o-F_1 complex (Fig.1) as observed for other preparations of F_o (see refs. 21,22). Among the other 7 polypeptides, with Mr lower than 33000, present in F_1-F_o preparation the following F_o subunits were recognized: F_oI (7,8), F_oII (7), F_6 and at least other two of apparent Mr 9000 and 7000 that could represent respectively the DCCD binding protein and the A6L, a product of the mitochondrial DNA (see ref. 23 for review). The inhibitor protein (IF_1) (apparent Mr of 11000)

205

was present in the F_o-F_1 complex but disappeared from the F_o preparation (Fig.1). In addition a polypeptide of Mr 31000 was retained in the F_o preparation, although its identity and pertinence to F_o is questionable as this protein is present only in some F_o-F_1 preparations (5,17).

Using a preparative gel electrophoresis the protein bands can be cut and electroeluted in glycerol-H_2O (6-8). Samples of the purified proteins, whose homogeneity was checked electrophoretically (Fig.1), were used for production of polyclonal antibodies (7,8), or subjected to automated Edman degradation for determiantion of the aminoacid sequence at the N-terminus after electrotransfer onto solid polyvinyldiene difluoride membrane (8).

The content of individual F_o-F_1 subunits in the F_o preparation can also be tested with specific antibodies. Immunoblot analysis of ESMP, USMP, F_o-F_1 and F_o with a polyvalent serum specific for F_1 subunits (7,8) showed a very low content of F_1 subunit in USMP and F_o with respect to ESMP and F_o-F_1 complex respectively (Fig.2).

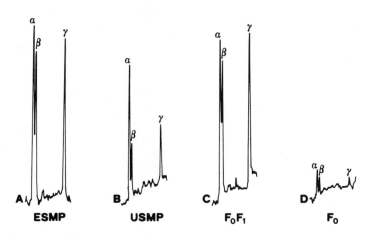

Fig.2 – Comparison of F_1 subunits content in ESMP, USMP, F_o-F_1 complex and F_o by immunoblot.

After SDS-electrophoresis of 15µg of ESMP (A), USMP (B) and 10µg of F_o-F_1 complex (C) and F_o (D), aliquots of proteins were transferred to nitrocellulose and immunodecoration was performed using the antiserum against F_1. Immunoblots were scanned at 590 nm in a CAMAG TLC II spectrophotometer. For other details see Experimental Procedure.

DISCUSSION

The experiments reported in this paper could offer an example how the resolution and reconstitution of membrane complexes (F_1-F_o ATP synthase in the present paper) can be used:

a) for analysis of integrated functions between the subunits.

b) For the study of the role of the single subunits in the functions of the membrane enzyme.

c) For the analysis of polypeptide composition of a single subunit.

d) To make polyclonal antibodies against isolated polypeptide, that can be also used for aminoacid composition analysis.

e) For qualitative and semiquantitative analysis of the single polypeptide in various conditions of the enzyme (i.e. in pathological conditions) by immunoblot analysis.

f) To study protein-protein and protein-lipid interactions.

REFERENCES

1. S. Papa, K. Altendorf, L. Ernster and L. Packer, "H^+-ATPase (ATP Synthase): Structure, Function, Biogenesis. The F_oF_1 complex in coupling membranes," ICSU Press, Miami/Adriatica Editrice, Bari (1984).

2. M.E. Pullman and G.C. Monroy, A naturally occurring inhibitor of mitochondrial adenosin triphosphatase, J. Biol. Chem. 238:3762 (1963).

3. F. Guerrieri, R. Scarfò, F. Zanotti, Y.W. Che and S. Papa, Regulatory role of the ATPase inhibitor protein on proton conduction by mitochondrial H^+-ATPase complex, FEBS Lett. 213:67 (1987).

4. J.P. Aris, D.J. Klionsky and R.D. Simoni, The F_o subunits of Escherichia Coli F_1F_o-ATP synthase are sufficient to form a functional proton pore, J. Biol. Chem. 260:11207 (1985).

5. D.L. Stigall, Y.M. Galante and Y. Hatefi, Preparation and properties of an ATP-Pi exchange complex (complex V) from bovine heart mitochondria, J. Biol. Chem. 259:956 (1978).

6. F. Zanotti, F. Guerrieri, Y.M. Che, R. Scarfò and S. Papa, Proton translocation by the H^+-ATPase of mitochondria. Effect of modification by monofunctional reagents of thiol residues in F_o polypeptides, Eur. J. Biochem. 164:517 (1987).

7. J. Houstek, J. Kopecky, F. Zanotti, F. Guerrieri, E. Jirillo, G. Capozza and S. Papa, Topological and functional characterization of the F_oI subunit of the membrane moiety of the mitochondrial H^+-ATP synthase, Eur. J. Biochem. 173:1 (1988).

8. F. Zanotti, F. Guerrieri, G. Capozza, J. Houstek, S. Ronchi and S. Papa, Identification of nucleus-encoded F_oI protein of bovine heart mitochondrial H^+-ATPase as a functional part of the moiety, FEBS Lett. 237:9 (1988).

9. M. Buckle, F. Guerrieri and S. Papa, Changes in activity and F_1 content of mitochondrial H^+-ATPase in regenerating rat liver, FEBS Lett. 188: 345 (1985).

10. F. Guerrieri, F. Capuano, M. Buckle and S. Papa, Alteration of mitochondrial H^+-ATPase complex in tissue regeneration and neoplasia, in: "The Molecular Biology of Human Diseases," J.W. Gorrod, O. Albano and S. Papa, eds., Ellis Horwood Limited, Cholchester,West Sussex, England (in press).

11. J. Houstek, J. Kopecky, Z. Rychter and T. Soukup, Uncoupling protein in embryonic brown adipose tissue-existence of nonthermogenic and thermogenic mitochondria, Biochim. Biophys. Acta 935:19 (1988).

12. A. Pansini, F. Guerrieri and S. Papa, Control of proton conduction by the H^+-ATPase in the inner mitochondrial membrane, Eur. J. Biochem. 92:545 (1978).

13. F. Guerrieri and S. Papa, Effect of chemical modifiers of amino acid residues on proton conduction by the H^+-ATPase of mitochondria, J. Bioenerg. Biomembr. 13:393 (1981).

14. J. Kopecky, F. Guerrieri and S. Papa, Interaction of dicyclohexylcarbo-diimide with the proton conducting pathway of mitochondrial H^+-ATP-ase, Eur. J. Biochem. 131:17 (1983).

15. F. Guerrieri, F. Zanotti, Y.W. Che, R. Scarfò and S. Papa, Inactivation of the mitochondrial ATPase inhibitor protein by chemical modifica-tion with diethylpyrocarbonate, Biochim. Biophys. Acta 892:284(1987).

16. R.B. Beechey, S.A. Hubbard, P.E. Linnett, A.D. Mitchell and E.A. Munn, A simple and rapid method for the preparation of adenosine tripho-sphatase from submitochondrial particles, Biochem. J. 148:533 (1975).

17. M.W. McEnery, E.L. Buhle Jr., U. Aebi and P.L. Pedersen, Proton ATPase of rat liver mitochondria, J. Biol. Chem. 259:4642 (1984).

18. J.H. Laemmli, Cleavage of structural proteins during the assembly of the head of Bacteriophage T4, Nature (Lond). 227:680 (1970).

19. A. Pansini, F. Guerrieri and S. Papa, Mechanism of proton conduction by the H^+-ATPase of mitochondria. Studies with chemical modifiers of amino acid residues and amphiphilic ions, in: "Membrane Bioenerge-tics," C.P. Lee, G. Schatz and L. Ernster, eds., Addison-Wesley Pu-blishing Company, Inc., USA (1979).

20. B.I. Kanner, R. Serrano, M.A. Kandrach and E. Racker, Preparation and characterization of homogeneous coupling factor 6 from bovine heart mitochondria, Biochem. Biophys. Res. Commun. 69:1050 (1976).

21. Y.M. Galante, S. Wong and Y. Hatefi, Resolution and reconstitution of complex V of the mitochondrial oxidative phosphorylation system: properties and composition of the membrane sector, Arch. Biochem. Biophys. 241:643 (1976).

22. J.A. Berden and M.M. Voorn-Brouwer, Studies on the ATPase complex from beef-heart mitochondria. I. Isolation and characterization of an oligomycin-sensitive and oligomycin-insensitive ATPase complex from beef-heart mitochondria, Biochim. Biophys. Acta 501:424 (1978).

23. S. Papa, F. Guerrieri, F. Zanotti, J. Houstek, J. Kopecky, G. Capozza and S. Ronchi, Mitochondrial F_o,F_1 H^+-ATP synthase. Characterization of functional components of the F_o-moiety, in: "Molecular Basis of Biomembrane Transport," F. Palmieri and E. Quagliariello, eds., Elsevier Science Publishers B.V., Amsterdam (1988).

THE MITOCHONDRIAL b-c$_1$ COMPLEX: POLYPEPTIDE COMPOSITION AND ENZYMATIC ACTIVITIES

M. Lorusso, T. Cocco and D. Boffoli

Institute of Medical Biochemistry and Chemistry, University

of Bari, Bari, Italy

INTRODUCTION

The second segment of the respiratory chain of mitochondria (b-c$_1$ complex, complex III or ubiquinol-cytochrome c reductase, E.C. 1.10.2.2) is an oligomeric integral protein of approx. 240 kDa in a monomeric form, which catalyses the transfer of electrons from ubiquinol or synthetic analogs to ferricytochrome c. The flow of reducing equivalents from dehydrogenase or quinols to cytochrome c, along the b-c$_1$ complex, results in the effective translocation of two protons from the matrix space (Negative phase) to the outer space (Positive phase); two additional protons, deriving from the scalar oxidation of a hydrogen donor, are released in the P phase, thus giving an overall H$^+$/2e$^-$ stoicheiometry of 4.

The mechanism of electron transfer and coupled proton translocation through the complex is still a matter of debate and different models have so far been proposed (1-3).

Purification of the b-c$_1$ complex from bovine mitochondria has been carried out by i) cholate solubilization and fractionation (4); ii) hydroxyapatite chromatography of the complex in Triton X-100 (5) and iii) affinity chromatography using a thiol-sepharose-bovine cytochrome c (6).

Reconstitution of isolated b-c$_1$ complex into phospholipid vesicles shows an uncoupler-stimulated rate of electron flow and uncoupler-sensitive proton translocation, similar to that observed in intact mitochondria (1,7,8).

I. SPECTROPHOTOMETRIC ANALYSIS OF THE b-c$_1$ COMPLEX

b-c$_1$ complex isolated from bovine heart mitochondria according to (4) is diluted with 100 mM phosphate buffer (pH 7.2) containing 1% Tween-80 to a final concentration of 0.2 mg/ml. Absolute oxidized and dithionite reduced spectra are obtained (Fig. 1A). Readings at 562-575 nm (ΔA_{red-ox}, 20mM^{-1}

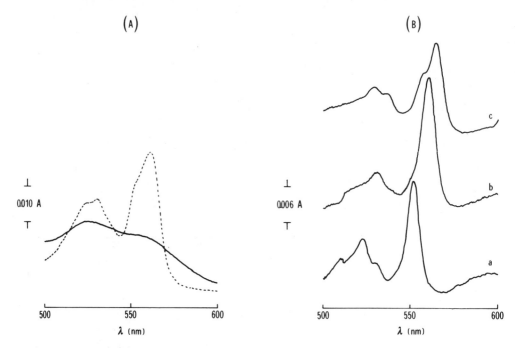

Fig. 1   Spectrophotometric analysis of the $b-c_1$ complex purified from
bovine heart mitochondria.
A) Absolute oxidized (solid line) and dithionite-reduced (dashed
line) spectra.
B) Difference spectra of the individual heme centers: a) heme c_1,
(ascorbate reduced minus oxidized); b) heme b-562, (ascorba-
te/TMPD minus ascorbate); c) heme b-566, (dithionite minus
ascorbate/TMPD).

for heme b) and at 552-540 (ΔA_{red-ox}, 17.5 mM^{-1} for heme c_1) allow an estimation of the heme content of the complex preparation. The standard preparation contains 8 nmol heme b and 4 nmol heme c_1 per mg of proteins.

By poising the $b-c_1$ complex suspension at different potentials, the
optical difference spectra of the individual heme centers (c_1, b-562 and
b-566) can be recorded (Fig. 1B). The $b-c_1$ complex at 0.3 mg protein/ml in
a mixture as above is placed in both reference and sample cuvette of a
split-beam spectrophotometer. The procedure is as follows:

- add 3 mM ascorbate in the sample cuvette. The resulting difference spectrum with a peak at 553 nm refers to the heme c_1 (trace a).

- add 3 mM ascorbate also in the reference cuvette so as to obtain a new
  base line after 5 min incubation. Add 100 µM TMPD in the sample cuvette.
  The difference spectrum refers now to the high potential heme b center
  (b-562), (trace b).

- add 100 µM TMPD also in the reference cuvette and after 5 min obtain a
  new base-line. Addition of few grains of solid sodium-dithionite allows
  the recording of the difference spectrum of the low potential heme b
  center (b-566), (trace c).

II. SDS-PAGE ANALYSIS OF THE b-c$_1$ COMPLEX

Whatever the method of purification used, the bovine b-c$_1$ complex shows invariably eleven polypeptide subunits, namely eight supranumerary subunits in addition to the catalytic subunits holding the two b hemes, the heme c$_1$ and the 2 Fe-2S cluster.

The procedure that will be described is essentially that developed by Schägger et al. (9).

A. Reagents and Buffers

	separating	stacking
Final acrylamide concentration	16%	4%
48% (w/v) acrylamide, 1.5% (w/v) bisacrylamide	10 ml	1 ml
3.0 M Tris, 1.0 N HCl, 0.3% (w/v) SDS	10 ml	3 ml
Glycerol	4 g	
Water to a final vol of	30 ml	12 ml
N,N,N',N'-tetramethylethylenediamine	10 µl	10 µl
10% (w/v) ammonium persulfate	100 µl	100 µl

The above quantities are given for a 24 x 15.5 x 0.075 cm slab gel:

- overlay solution: 1.0 M Tris, 0.33 N HCl, 0.1% (w/v) SDS
- anode buffer   : 0.2 M Tris/Cl, pH 8.9
- cathode buffer   : 0.1 M Tris, 0.1 M Tricine, 0.1% (w/v) SDS
- sample buffer   : 50 mM Tris/Cl, 5% (w/v) SDS, 15% (w/v) Glycerol,
                      2% Mercaptoethanol, 0.003% Bromophenol Blue, pH 6.8.
- Tetramethylbenzidine staining solution: warm saturated solution of
     tetramethylbenzidine, 10% (v/v) isopropanol, 7% (v/v) acetic acid.
- Fixing solution: 50% (v/v) Methanol, 10% (v/v) Acetic acid.
- Coomassie staining solution: 0.1% (w/v) Coomassie Brillant Blue G-250,
     10% (v/v) Acetic acid.
- Destaining solution: 10% (v/v) Acetic acid.

B. Equipment

- Complete vertical slab gel electrophoresis set with a power supply.
- Densitometer at variable wavelength connected with a chromato-integra-
  tor.

C. Procedure

A 16% poliacrylamide slab gel is prepared and assembly on a gel elec-trophoresis vessel and covered with the overlay solution. The polymeriza-tion takes place within 15 min. Some minutes afterwards, the overlay solu-

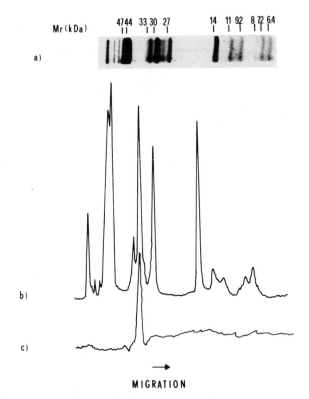

Fig. 2   SDS-Poliacrylamide gel electrophoresis.

a) Coomassie-blue stained gel. The constituent polypeptide subunits are: the two core proteins (47 and 44 kDa), the apopro-
tein of b cytochromes (apparent Mr of 33 kDa), the cytochrome
c_1 (30 kDa), the Rieske Fe-S protein (27 kDa) and six more low
molecular mass polypeptides whose function is still unknown.
Contaminant polypeptides are also present including a 66 kDa
band corresponding to succinate dehydrogenase.

b) densitometric profile at 590 nm of the Coomassie-blue stained
gel.

c) densitometric profile at 690 nm of the tetramethylbenzidine
stained gel. The only absorbing band refers to the heme c_1
containing polypeptide.

tion is decanted and replaced by the stacking gel. The lower end of the
inserted comb should be 1 cm spaced from the separating gel. 40 μl of the
sample buffer are mixed with 5 μl sample (containing 40-60 μg of proteins)
and after 30 min incubation at room temperature, applied to the gel. The
electrophoresis is performed at room temperature for 1-2 h at 30 V and 17
h at 90 V. After electrophoresis the gel is placed into a tetramethylbenzidine solution and incubated for 15 min. The gel is then transferred into
a 30% solution of H_2O_2; the greenish-blue bands of the heme containing peptides develop in a few minutes and the gel can immediately be scanned at
690 nm (Fig. 2c) (10). The gel is subsequently fixed for 1 h, stained with
Coomassie for 2-4 h and destained overnight. The destained gel (Fig. 2a)
is scanned at 590 nm (Fig. 2b).

III. RECONSTITUTION OF PURIFIED b-c$_1$ COMPLEX INTO LIPOSOMES

A. Materials and solutions

- Purified b-c$_1$ complex

- acetone-washed soybean phospholipids

- 20% (w/v) solution (pH 8.0) of recrystallized cholic acid

- 0.1 M potassium phosphate (pH 7.8)

- 0.125 M potassium chloride.

B. Procedure

It is essentially based on the "cholate dialysis method" developed by Leung and Hinkle (7). 120 mg of phospholipids are added to 4 ml of 0.1 M potassium-phosphate containing 2% cholate. The mixture, in an ice bath and under nitrogen atmosphere, is sonicated using a microtip of a Branson type sonicator until clear. Three pulses of 2 min sonication with 2-3 min intervening periods are usually given. b-c$_1$ complex (8 mg proteins) is then added and the mixture immediately dialyzed for 4 hours against 200 volumes of phosphate buffer. A second overnight dialysis is performed against 800 ml of 0.125 M KCl, followed by 2 hours final dialysis against 0.125 M KCl or 0.220 M sucrose/1 mM K-Hepes (pH 7.2) to decrease, where necessary, the activity of K$^+$ ions in the suspending medium of vesicles.

IV. MEASURE OF PASSIVE H$^+$ CONDUCTANCE IN THE b-c$_1$ VESICLES

When the potassium-ionophore valinomycin is added to K$^+$ containing vesicles (as prepared) suspended in a medium at low concentration (1 mM) of K$^+$, equilibration across the vesicles will be controlled by the rate of charge compensating H$^+$ diffusion into the vesicles. The measure of the initial rate of alkalinization of the external medium, in the presence of valinomycin, will thus give a measure of the passive proton conductance of the vesicles. The lower proton conductance of the vesicles, then the higher can be considered the quality of the proteoliposome preparation. This represent the basic requirement when dealing with reconstitution of respiratory chain complexes exhibiting a proton translocating activity.

A. Materials and solutions

- b-c$_1$ vesicles prepared as described (at 2 mg of proteins/ml of phospholipid suspension), and dialyzed during the final dialysis against 220 mM sucrose/1 mM K-Hepes (pH 7.2)

- Valinomycin, dissolved in ethanol at 1 mg/ml

- Carbonylcyanide-p-trifluoromethoxy-phenylhydrazone (FCCP), 3 mM in ethanol.

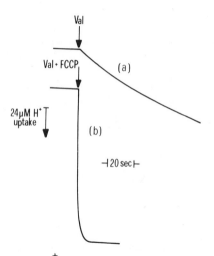

Fig. 3 Measure of passive H^+ conductance in the b-c_1 vesicles. Potentiometric measurement of passive H^+ conductance is measured in the absence (a) and in the presence (b) of FCCP.

B. Procedure

b-c_1 vesicles are suspended in 220 mM sucrose/1 mM K-Hepes (pH 7.2) at a concentration of 0.2 mg/ml. Final volume, 2 ml. The alkalinization of the medium is followed by a glass combination electrode (Beckmann n. 39305) connected to an electrometer (Keithley mod. 604). The signal is displayed on a strip-chart recorder set up at 20 mV full scale. The reaction starts by the addition of 1 μl of valinomycin solution (Fig. 3a). The observed proton permeability is compared with that exhibited by fully uncoupled vesicles (Fig. 3b), i.e. in the presence of the protonophore FCCP. The signal is calibrated with additions of a standard 10 mM solution of HCl.

V. MEASURE OF OXIDATION CONTROL RATIO

In respiring b-c_1 vesicles the electrochemical proton gradient, gene-rated by respiration, controls the rate of electron flow from substrate durohydroquinone to cytochrome c (controlled state, state 4 respiration). The electron flow can be activated by the addition of either valinomycin plus nigericin or by FCCP (activated state, state 3 respiration). The ratio between the rate of electron flow in the activated and controlled states is referred to as "oxidation control ratio" (O.C.R.) and can be taken as an index of the coupling of the b-c_1 vesicles preparation, i.e. their capability to maintain the transmembrane H^+ gradient.

A. Materials, reagents and solutions

- b-c$_1$ vesicles prepared as described above and dialyzed against 0.125 M KCl.

- Incubation mixture consisting of 125 mM Choline-Cl, 0.5 mM potassium phosphate (pH 7.2), 10 mM KCl, 0.050 mM EDTA, 1 mM Na-azide.

- Ferricytochrome c 3 mM, in water.

- Durohydroquinone solution 10 mM, in ethanol.

- Valinomycin 0.1 mg/ml, in ethanol.

- Nigericin 0.1 mg/ml, in ethanol.

B. Experimental procedure

2 ml of the incubation mixture are supplemented with 8 μl ferricyto-chrome c. Cytochrome c reduction is followed at 550-540 nm (ΔA_{red-ox} = 19.1 mM^{-1}), with the photomultiplayer signal displayed on a strip-chart recorder set up at 100 mV full scale. A non-enzymatic reduction of cytochrome c is obtained by the addition of 30 μM durohydroquinone, after which 3 μg//ml of b-c$_1$ vesicles are added. The difference between the two slopes gives the enzymatic reaction rate in the controlled state (Fig. 4a). The experiment is repeated in the presence of 0.05 μg/ml of both valinomycin and nigericin, thus giving the rate of electron flow in the activated state (Fig. 4b).

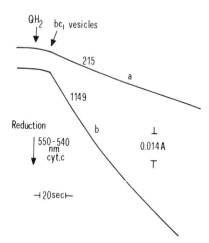

Fig. 4   Measure of oxidation control ratio in b-c$_1$ vesicles.
The rate of reduction of ferricytochrome c, followed at 550-540 nm, is measured in the absence (trace a) and in the presence (trace b) of valinomycin plus nigericin. Measured O.C.R. in this expt. is 5.3.

Oxidation control ratios ranging from 4 to 8 are usually obtained with DQH_2. Higher values (up to 15) are obtainable with more suitable substrates like Q_2H_2.

VI. MEASURE OF REDOX-LINKED PROTON TRANSLOCATION IN b-c$_1$ VESICLES

The measurement of redox linked proton translocation in b-c$_1$ vesicles can be carried out by an oxidant or reductant pulse type experiment (1). In the latter, that will be described, a quinol pulse is given to b-c$_1$ vesicles supplemented with ferricytochrome c. The H^+/e^- ratio may be calculate either by the extent or the initial rates of cytochrome c reduction and proton translocation (Fig. 5).

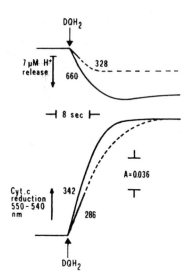

Fig. 5   Proton translocation and electron flow elicited by quinol pulses of oxidized b-c$_1$ complex.
Proton translocation is measured potentiometrically with a glass electrode. Electron transfer activity is measured at 550-540 nm. The experiment is carried out in the absence (solid line) and presence (dashed line) of FCCP. Figures on the traces are nmol H^+ and $e^- \cdot min^{-1} \cdot mg$ prot.$^{-1}$.

A. Materials, reagents and solution

- b-c$_1$ vesicles prepared as described and dialyzed (final dialysis) against 0.125 M KCl.

- Incubation medium consisting of 0.125 M KCl/5 mM $MgCl_2$ containing

216

1 mM Na-azide to inhibit cytochrome oxidase activity which may be
present in the purified b-c₁ complex.

- Ferricytochrome c 3 mM, in water.

- Durohydroquinone 10 mM, in ethanol.

- Valinomycin 1 mg/ml, in ethanol. Valinomycin has to be added to collapse the membrane potential ($\Delta\psi$) and maximize proton translocation.

- FCCP 3 mM, in ethanol.

- Standard solution of HCl, 5 mM.

B. Experimental procedure

In a spectrophotometric 3 ml cuvette, also housing a glass conbination
electrode, 0.2 ml of b-c₁ vesicles (corresponding to 0.4 mg of protein) are
added to 1.6 ml of incubation mixture containing 8 μM ferricytochrome and
2 μg of valinomycin, to a final concentration of 0.9 μM of cytochrome c_1.
The reaction is started by the addition of 10 μM durohydroquinone. Cytochrome c reduction is followed spectrophotometrically at 550-540 nm. Proton
translocation is followed potentiometrically in the same sample of vesicle
suspension and calibrated using additions of standard solutions of HCl.
Both signals are displayed on a double-channel strip-chart recorder.
A parallel experiment carried out in the presence of 3 μM FCCP allows for
the measurements of proton translocation deriving only from the chemical
oxidation of durohydroquinone by cytochrome c, with a stoicheiometry amoun-
ting to 1 H^+ per electron.

REFERENCES

1. S. Papa, M. Lorusso, D. Boffoli and E. Bellomo, Redox-linked proton
   translocation in the b-c₁ complex from beef-heart mitochondria reconstituted into phospholipid vesicles. General characteristics and control
   of electron flow by $\Delta\mu H^+$, Eur. J. Biochem. 137:405 (1983).
2. P. Mitchell, Possible molecular mechanism of the protonmotive function
   of cytochrome system, J. Theor. Biol. 62:327 (1976).
3. M. Wikström and K. Krab, The semiquinone cycle. A hypothesis of electron
   transfer and proton translocation in cytochrome bc-type complexes, J.
   Bioenerg. Biomembr., 18:181 (1980).
4. J.S. Rieske, Preparation and properties of reduced coenzyme Q-Cytochrome
   c reductase (complex III of the respiratory chain), Methods Enzymol.
   10:239 (1967).
5. W.D. Engel, H. Schägger and G. Von Jagow, Ubiquinol-cytochrome c reductase (EC 1.10.2.2). Isolation in Triton X-100 by hydroxyapatite and
   gel chromatography. Structural and functional properties. Biochim. Biophys. Acta, 592:211 (1980).
6. K. Bill, C. Broger and A. Azzi, Affinity chromatography purification
   of cytochrome c oxidase and b-c₁ complex from beef heart mitochondria.

Use of thiol-sepharose-bound saccharomyces cerevisiae cytochrome c. Biochim. Biophys. Acta, 679:28 (1982).

7. K.H. Leung and P. Hinkle, Reconstitution of ion transport and respiratory control in vesicles formed from reduced coenzyme Q-cytochrome c reductase and phospholipids. J. Biol. Chem., 250:8467 (1975).

8. F. Guerrieri and B.D. Nelson, Studies on the characteristics of a proton pump in phospholipid vesicles inlayed with purified complex III from beef-heart mitochondria. FEBS Lett. 54:339 (1975).

9. H. Schägger, U. Borchart, H. Aquila, T.A. Link and G. Von Jagow, Isolation and amino acid sequence of the smallest subunit of beef heart b-c$_1$ complex. FEBS Lett., 190:89 (1985).

10. C Broger, M.J. Nalecz and A. Azzi, Interaction of cytochrome c with cytochrome bc$_1$ complex of the mitochondrial respiratory chain. Biochim. Biophys. Acta, 592:519 (1980).

IN VITRO SYNTHESIS AND IMPORT OF MITOCHONDRIAL PROTEINS

Ersilia Marra, Sergio Giannattasio and Ernesto Quagliariello

Department of Biochemistry and Molecular Biology University of Bari, C.N.R. Centro di Studio sui Mitocondri e Metabolismo Energetico, Bari, Italy

INTRODUCTION

The majority of the mitochondrial proteins are encoded by nuclear DNA, synthesized in the cytosol on free cytoplasmic ribosomes and subsequently taken up by mitochondria where they are rendered functional. Most nuclear-coded mitochondrial proteins are synthesized as larger molecular weight precursors, which contain an amino-terminal extension. They are cleaved by a specific matrix protease upon translocation of the precursor into mitochondria (for Ref. see 1).

To date several approaches have been used to study the transport of nuclear-coded mitochondrial proteins. They concern both in vitro and in vivo experiments (for Ref. see 1). The huge number of in vitro experiments allow a basic understanding of mitochondrial protein import process. In a typical in vitro experiment total poly A[+]-RNA is isolated and used to direct protein synthesis in cell-free systems. The translation products are subsequently incubated with isolated mitochondria and the protein of interest analysed by immunoprecipitation with specific antibody followed by SDS/polyacrylamide gel electrophoresis and fluorography (for Ref. see 1,2).

On the other hand import of proteins into mitochondria has also been extensively studied by using a model system in which the import process of the purified mature enzyme is investigated by measuring the increase of its intramitochondrial activity[3-5].

Through these experimental approaches many general features of the mitochondrial import of many nuclear-coded mitochondrial proteins have been understood. The translocation of proteins across mitochondrial membranes is strictly dependent on the electrochemical potential across the inner mitochondrial membrane and is mediated by specific receptors on the mitochondrial membranes[6-8].

Thus, although many features of the transport process are already clear, the nature of the targeting signals and of the receptors which mediate this process still need further investigation. In this regard the use of recombinant DNA tecniques would appear to be a valuable tool in studying the role played by specific protein domains in the import process. Suitable expression vectors, containing the relevant control signals for the transcription and translation of cloned DNA sequences,allow for the expression of mitochondrial precursor proteins. Moreover, mutagenesis and gene fusion experiments could also be used to identify that sequence contained within the amino-acid framework of nuclear-coded mitochondrial precursor proteins which directs the specific segregation of these proteins into their mitochondrial compartments. Thus either chimeric protein products or mutated polypeptides could be obtained and their capacity to enter mitochondria studied (for review see 9).

In this paper we describe the experimental procedures which allow the monitoring of the import into isolated mitochondria of an _in vitro_ synthesized nuclear-coded mitochondrial protein whose coding nucleotide sequence has been inserted into an expression plasmid. The first section concerns the experimental procedure for synthesizing in a cell-free transcription-translation system a mitochondrial precursor protein, whose coding nucleotide sequence has been inserted into the expression plasmid pOTS. The second deals with the experimental methods for studying the import into isolated mitochondria of the _in vitro_ expression products of the recombinant plasmid pOTS along with some experiment to characterize the translocation process.

IN VITRO MITOCHONDRIAL PROTEIN SYNTHESIS IN A CELL-FREE COUPLED TRANSCRIPTION-TRANSLATION SYSTEM DIRECTED BY A RECOMBINANT EXPRESSION PLASMID

The ability to synthesize a protein in a cell-free expression system is strictly dependent on the availability of a suitable, high level expression vector in which the cDNA encoding that protein may be cloned. The expression plasmid pOTS has proved to be a very efficient vector for synthesizing a recombinant polypeptide both _in vivo_ , in properly transformed host bacterial cells[10], and _in vitro_ , in a cell-free coupled transcription-translation system directed by a recombinant expression plasmid[11].

pOTS is a derivative of pAS1 which carries regulatory signals derived from the bacteriophage λ genome[12]. Phage regulatory information proves to be more efficient than its host-derived counterpart. Transcription of cDNA cloned in pOTS starts at the λ P_L promoter which is followed by two utilization sites for the antitermination factor N, the ribosome binding site and the initation codon of λ cII protein. Plasmids carrying P_L are often unstable, presumably owing to the high level of P_L-directed transcription. To overcome such instability, the P_L transcription may be repressed by using bacterial hosts that contain an integrated copy of the genome (i.e., bacterial lysogens). In these cells, P_L transcription is controlled by the phage λ repressor protein (cI), which is continuously synthesized and autogenously regulated in the lysogen. The cells can be stably transformed and the vector maintained in these lysogenic hosts. Moreover P_L-directed transcription can be activated at any time by using a lysogen carrying a temperature-sensitive mutation in cI gene (cI857). Induction is accomplished by simply raising the temperature of the cell culture from 32°C to 42°C. Thus, cells carrying the vector can initially be grown to high density without the expression of the cloned gene at 32°C, and subsequently induced to synthesize the product at 42°C. The cDNA of a mitochondrial precursor protein may be inserted immediately downstream of the cII initiator codon thus obtaining the recombinant expression plasmid pOTS[12].

In the following part of this paper the experimental procedure to synthesize _in vitro_ a mitochondrial precursor protein whose coding DNA sequence is inserted into pOTS, is described. This is a three phase procedure:
1) amplification of the recombinant plasmid in a bacterial host;
2) extraction of the recombinant plasmid DNA;
3) _in vitro_ synthesis of the recombinant polypeptide in a cell-free transcription-translation system supplied with the recombinant DNA.

The recombinant plasmid DNA is used to transform competent E. coli cells lysogenic for λcI repressor (MM294/cI$^+$ strain), according to[13]. The transformed cells are grown overnight at 37°C in 1% (w/v) bacto-tryptone, 0,5% (w/v) bacto-yeast extract, 1% (w/v) NaCl (LB-broth) containing 40 µg/ml ampicillin, in a 250 ml flask under continuous shaking. The cells are harvested and the plasmid DNA is extracted. The method used for the large scale preparation of the recombinant expression plasmid is the extraction of the plasmid DNA through lysis by sodium dodecylsulfate (SDS)/alkali followed or not by centrifugation to equilibrium in caesium chloride-ethidium bromide density gradient[13]. The DNA that is used to direct a cell-free expression system must be RNase-free. Thus, methods of DNA preparation involving the use of ribonucleases should be avoided, unless these enzymes can be completely inactivated. The DNA should also be free of phenol, caesium chloride and ethidium bromide.

In order to synthesize _in vitro_ mitochondrial precursor proteins the recombinant plasmid DNA pOTS may be used to direct a procaryotic coupled transcription-translation system. The "prokaryotic DNA-directed translation kit" supplied by Amersham Int. is a suitable _in vitro_ system for expressing either wild type or recombinant pOTS DNA. This cell-free coupled transcription-translation system derives from E. coli and allows for the expression _in vitro_ of genes contained in bacterial plasmids, provided that the relevant control signals are present. The following set of reagents which enables the reproducible expression of protein-coding information contained in a wide variety of exogenous circular DNA template is supplied:
solution 1: an S-30 extract prepared from E. coli. Strain MR600, RNaseI$^-$;
solution 2: supplement solution. This contains sufficient nucleotides for transcription, tRNA for translation, an energy generating system and inorganic salts;
solution 3: amino acids minus methionine. An equimolar mixture of amino acids compatible with the use of Amersham's L-[^{35}S]-methionine;
solution 4: dilution buffer;
solution 5: standard DNA. Plasmid pAT153;
solution 6: methionine chase solution.

The experimental procedures are carried out avoiding contamination with ribonucleases and using RNase-free plasticware and glassware. In a typical recombinant pOTS expression assay, the following reaction mixture is prepared in 1.5 ml Eppendorf tubes kept on ice:

solution 1	5.0 µl	
solution 2	7.5 µl	
solution 3	3.0 µl	
L-[^{35}S]-met (1300 Ci/mmol)	2.0 µl	(30 µCi)
solution 4	7.5 µl	
recombinant pOTS DNA (0.5 µg/µl)	5.0 µl	

Both the supplement solution and the amino acid mixture minus methionine should be vortexed before being added to the reaction mixture. The reaction mixture is gently mixed and incubated at 37°C for one hour. Then 5 µl of the methionine chase solution are added to the tube and 5 min incubation at 37°C is allowed. The reaction is stopped by rapidly cooling with ice. A control expression assay is performed under the same conditions, with the exception that no recombinant pOTS DNA is added.

L-[^{35}S]-methionine incorporation into newly synthesized poly-peptides may be measured by a trichloroacetic acid (TCA) precipitation assay. 2 µl of the _in vitro_ protein syntesis mixture are put onto squared cut Whatman 3MM paper. Precipitation of polypeptides is allowed by soaking the filter paper first in ice cold 10% TCA for 10 min and subsequently in boiling 5% TCA for 10 min. The filter is washed twice with ethanol and twice with acetone, and is then dried in an oven. Incorporated radioactivity is determined by liquid scintillation counting in a toluene solution of diphenyl oxazole (5 g/l).

A successful _in vitro_ polypeptide synthesis is achieved when the ratio between the incorporated radioactivity measured in the pOTS expression assay and that measured in the control assay is greater than 6. The _in vitro_ synthesized protein products are subsequently analysed by SDS/polyacrylamide gel electrophoresis and fluorography.

A typical expression assay was made up as previously described, using different circular plasmid DNA templates, such as pAT153, wild type pOTS, and two recombinant pOTS, pOTS/pre-mAspAT and pOTS/mAspAT, which contain the DNA sequence encoding for the precursor and the mature form of mitochondrial aspartate aminotransferase, respectively[10]. A small aliquot of the total volume of each _in vitro_ expression mixture was then analysed by SDS/polyacrylamide gel electrophoresis and fluorography. Fig. 1 shows the electrophoretic pattern of the _in vitro_ expression products of pAT153, pOTS, pOTS/pre-mAspAT and pOTS/mAspAT (Fig. 1, lanes A, B, C and D). The two recombinant pOTS direct the _in vitro_ synthesis of two polypeptides which have the same molecular weight of the precursor (Fig. 1, lane C) and the mature form (Fig. 1, lane D) of mitochondrial aspartate aminotransferase.

IMPORT INTO MITOCHONDRIA AND BINDING TO MITOCHONDRIAL MEMBRANES OF RECOMBINANT pOTS EXPRESSION PRODUCTS

To carry out both the import and the binding assay, coupled mitochondria are needed. Thus, much care must be used in the isolation of the organelles and their ability to generate electrochemical proton gradient and to give oxidative phosphorylation must be strictly controlled before each assay. Methods of preparing mitochondria from various sources have been widely reported[16], however, the general features of these preparations and purity criteria vary according to the experiment in which the mitochondrial suspension is used.

In the reported experiments mitochondria were prepared with a solution containing 150 mM sucrose, 20 mM Tris-Cl, 50 mM KCl (TKS-medium). Where necessary, polarographic measurements were made of oxygen uptake by mitochondria due to succinate (plus rotenone) stimulated by either ADP (1 mM) or carbonyl cyanide m-cholrophenylhydrazone (1 µM). Mitochondria showing a respiratory control index lower than three were discarded.

A fast assay of the mitochondrial proteins was taken according to Waddel and Hill[17]. Briefly two samples were prepared containing 3.3 mM NaOH (3 ml each), one of which used as a blank, the other was added with 20 μl of 1:10 (v/v) mitochondrial stock suspension. The absorbance of the test sample at 215 and 225 nm was measured. The amount of the mitochondrial protein/ml stock suspension was given by the following formula:

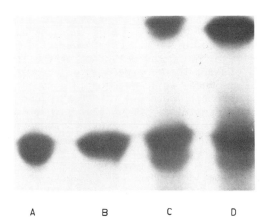

A   B   C   D

Fig. 1. Fluorography of 10% SDS/polyacrylamide gel after electro-phoresis analysis of the _in vitro_ expression products of various circular plasmid DNA. The expression assay is performed using the desired plasmid DNA template in the conditions described in the text. L-$[^{35}S]$-methionine incorporation is determined by TCA precipitation followed by liquid scintilla-tion counting. If polypeptide synthesis is successful, part of the protein synthesis mixture, usually 2 μl, is analysed by SDS/polyacrylamide gel electrophoresis according to Laemmli[14]. After electrophoresis, the gel is treated for fluorography using sodium salyciate[15]. For this purpose the gel is soaked first for 10 min in boiling 5% TCA an then for 30 min in ice cold 5% TCA. After neutralization with 1 M Tris-Cl pH= 7.4, the gel is saturated with 1 M sodium salycilate for 30 min. The gel is then dried and a fluorograph is made at -70°C using Kodak X-Omat AR film. Lane A, pAT153 expression product; lane B, wild type pOTS expression product; lanes C, D, respect-ively pOTS/pre-mAspAT and pOTS/mAspAT expression product.

$$mg\ prot/ml = \frac{A_{215} - A_{225}}{3.74} \times 500$$

The prepared mitochondrial suspension may thus be used to test the capacity of the recombinant pOTS expression products to be taken up by mitochondria.

In order to assess whether the _in vitro_ synthesized recombinant mitochondrial proteins can enter mitochondria, the coupled transcription-translation system containing [^{35}S] -methionine labeled protein is incubated for 15 min at 37°C with the mitochondrial suspension. The composition of the incubation mix is the following:

TKS-medium	60 µl
250 mM succinate	2 µl
100 mM ADP	1 µl
mitochondria (37.5 mg/ml)	20 µl
S-30 expression system	8 µl

After incubation the mixture is centrifuged and the mitochondrial pellet is resuspended in TKS-medium and incubated for 20 min at 23°C in the presence of pronase in order to digest externally bound protein thus allowing the detection of the incorporated protein (IMPORT ASSAY). Under these experimental conditions the pronase concentration is as much as is necessary to completely digest the _in vitro_ synthesized protein both in the absence and in the presence of mitochondrial lysate proteins. Soon after pronase digestion a mixture of protease inhibitors (antipain, pepstatin, leupeptin, chymostatin) is added (300 µM each) in order to avoid degradation of the imported protein after mitochondrial lysis. Mitochondria are separated by centrifugation and the mitochondrial pellet is immediately lysed in 1% SDS in which protease inhibitors (300 µM each) are also present. When binding of the recombinant pOTS expression products to mitochondria is measured pronase treatment is omitted. In this case both bound and imported protein are detected. To reveal either import into mitochondria alone, or import plus binding to the mitochondrial membranes, SDS/polyacrylamide gel electrophoresis analysis and fluorography are performed.

The result of a typical import and binding assay is shown in Fig.2. When mitochondria are treated with pronase a single band, corresponding to the import and processed protein (i.e. mature protein) is detected (Fig. 2, lane C), whereas when pronase treatment is omitted two bands showing the externally bound precursor protein and the imported mature protein are observed (Fig. 2, lane B).

A further distinction between the two processes can be obtained by testing the energy dependence of binding and import. Since import into mitochondria requires an electrochemical potential across the inner membrane either electron transfer inhibitors or uncouplers can prevent it. Thus, in the experiment carried out to study the energy dependence of the import process, isolated mitochondria are incubated at 37°C for 2 min with 2.5 µg carbonyl cyanide m-chlorophenylhydrazone (CCCP) before the addition of the products of the _in vitro_ synthesis. When mitochondria are added with CCCP and the pronase treatment is absent, only the binding process is allowed (Fig 2, lane D), whereas when CCCP is added followed by pronase treatment neither binding nor import processes are detectable (Fig. 2, lane E).

MITOCHONDRIAL SUBFRACTIONATION

To identify the submitochondrial localization of the imported protein, mitochondria subfractionation can be carried out as summarized in Table 1.

In a typical experiment the recombinant pOTS expression products are incubated at 37°C with a mitochondrial suspension (15 mg proteins) under import assay conditions. Incubation is stopped by rapidly centrifuging. The mitochondrial pellet is resuspended in TKS-medium (500 µl). An aliquot of this suspension (750 µg of mitochondrial proteins) is analysed by SDS/polyacrylamide gel electrophoresis and fluorography in order to detect both the protein molecules bound to and those imported into mitochondria. In order to digest externally bound proteins the remaining suspension is treated with pronase and subsequently, protease inhibitor mixture (see above) is added. Import assay is carried out with a 750 µg mitochondrial protein aliquot. Mitochondrial suspension is centrifuged and the pellet resuspended with 20 mM Tris-HCl buffer (pH=7.8) to give a final concentration of about 4 mg mit.prot./ml, homogenized 5 times with a Dounce homogenizer and kept on ice for 10-15 min. Following centrifugation (10 min, 12,000xg) outer mitochondrial membrane and

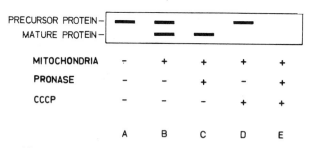

Fig. 2. Schematic representation of the fluorography of an SDS/polyacrylamide gel after electrophoresis analysis, in a typical experiment in which mitochondrial import of an _in vitro_ synthesized mitochondrial precursor protein, encoded by a recombinant expression plasmid pOTS, has been studied. Lane A shows the _in vitro_ synthesized precursor protein. Isolated mitochondria are incubated with the recombinant pOTS expression products as described in the text. After centrifugation the mitochondrial pellet is analysed without (lane B) and following pronase treatment (lane C). Lanes D and E show the result of the same experiment as illustrated in lanes B and C, using mitochondria that were preincubated with CCCP.

intermembrane space fractions are found in the supernatant, whereas mitoplasts have settled in the pellet.

Mitoplasts are used to obtain membrane and matrix fractions as follows. 2 ml of mitoplast suspension (5 mg mit.prot./ml) in TKS-medium are sonicated five times for 10 s at 50 W, with 2 min intervals, and centrifuged at 105,000xg for 1 h. Matrix fraction is released in the supernatant, whereas membranes are found in the pellet. The intramitochondrial localization of the imported protein is determined by SDS/polyacrylamide gel electrophoresis and fluorography of each isolated fraction. The amount of each fraction analysed is that resulting from 750 µg of proteins of the starting mitochondrial suspension. In the subfractionation procedure absence of contamination in each isolated fraction was checked by testing the activity of specific mitochondrial marker enzymes. In Table 2 the activities of

Table 1. Flow Diagram Summarizing The Experimental Procedure For
Subfractionation Of Isolated Mitochondria

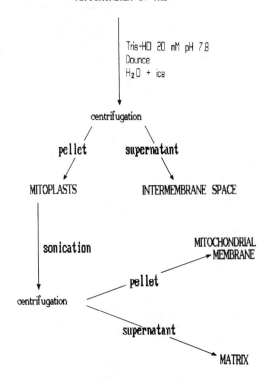

MITOCHONDRIA IN TKS

Tris-HCl 20 mM pH 7.8
Dounce
H₂O + ice

centrifugation

pellet   supernatant

MITOPLASTS   INTERMEMBRANE SPACE

sonication   MITOCHONDRIAL
                              MEMBRANE

                     pellet

centrifugation

                supernatant

                              MATRIX

Table 2. Enzyme Activity Measurement of Glutamate Dehydrogenase
and Adenylate Kinase in Isolated Mitochondria and in
Submitochondrial Compartments

	GDH enzyme act. ΔA/minxmg mit.prot.	%	AK enzyme act. ΔA/minxmg mit.prot.	%
MITOCHONDRIA	0.69	-	1.97	-
MITOPLASTS	0.65	95	0.18	10
INTERMEMBRANE SPACE	0.034	5	1.68	90
MITOCHONDRIAL MEMBRANE	0.033	5	-	-
MATRIX	0.62	95	-	-

adenylate kinase and glutamate dehydrogenase, marker enzymes of the intermembrane space and matrix respectively, are reported as measured in the subfractionation procedure[18].

CONCLUDING REMARKS

An efficient method of studying the mitochondrial import of an _in vitro_ synthesized nuclear-coded mitochondrial precursor protein is to clone the cDNA encoding the mitochondrial protein into the expression vector pOTS and bring it to expression in a coupled transcription-translation system supplied with $[^{35}S]$-methionine. The mixture containing the labeled polypeptide product of the recombinant pOTS is incubated with isolated mitochondria and the protein import and binding processes are assayed.

By allowing for the expression of cloned genes under the control of a phage promoter, this experimental system can be used to obtain a precursor protein with an altered amino acid sequence either by mutagenizing the cDNA sequence using recombinant DNA techniques[10], or by constructing fused genes. The recombinant plasmid pOTS, containing the manipulated DNA sequence, can then be brought to expression and thus the capacity of the new expression products to enter isolated mitochondria may be studied.

ACKNOWLEDGMENT

The authors wish to thank Dr. Maria Filomena Abruzzese for her help in the development of these procedures. This work was partially supported by a grant from Progetto Finalizzato Ingegneria Genetica e Basi Molecolari delle Malattie Ereditarie of the C.N.R.

REFERENCES

1. S. Doonan, E. Marra, S. Passarella, C. Saccone and E. Quagliariello, Transport of proteins into mitochondria, _Int. Rev. Cyt._ 91:141 (1984)
2. B. Hay, P. Bohni and S. Gasser, How mitochondria import proteins, _Biochim. Biophys. Acta_ 779:65 (1984)
3. E. Marra, S. Doonan, C. Saccone and E. Quagliariello, Selective permeability of rat liver mitochondria to purified aspartate aminotransferases _in vitro_, _Biochem. J._ 164:685 (1977)
4. S. Passarella, E. Marra, S. Doonan and E. Quagliariello, Selective permeability of rat liver mitochondria to purified malate dehydrogenase isozymes _in vitro_, _Biochem. J._ 192:649 (1980)
5. E. Marra, S. Passarella, S. Doonan, E. Quagliariello and C. Saccone, Protease resistence of aspartate aminotransferase imported into mitochondria, _FEBS lett._ 122:33 (1980)
6. S. Passarella, E. Marra, S. Doonan, L. R. Languino, C. Saccone and E. Quagliariello, Uptake of aspartate aminotransferase into mitochondria _in vitro_ depends on the transmembrane pH gradient, _Biochem. J._ 202:353 (1982)
7. S. Passarella, E. Marra, S. Doonan and E. Quagliariello, Uptake of malate dehydrogenase into mitochondria _in vitro_. Some characteristics of the process, _Biochem. J._ 210:207 (1983)
8. E. Marra, S. Passarella, E. Casamassima, E. Perlino, S. Doonan and E. Quagliariello, Kinetic studies of the uptake of aspartate aminotransferase and malate dehydrogenase into mitochondria _in vitro_, _Biochem. J._ 228:493 (1985)

9. M. G. Douglas, M. T. McCammon and A. Vassarotti, Targeting proteins into mitochondia, <u>Microbiol. Rev.</u> 50:166 (1986)
10. R. Jaussi, R. Behra, S. Giannattasio, T. Flura and P. Christen, Expression of cDNA encoding the precursor and the mature form of chicken mitochondrial aspartate aminotransferase in <u>Escherichia coli</u>, <u>J. Biol. Chem.</u> 262:12434 (1987)
11. S. Giannattasio, E. Marra, M. F. Abruzzese, M. Greco and E. Quagliariello, Synthesis of the mature form of mitochondrial aspartate aminotranserase in a cell-free coupled transcription-translation system, <u>It. J. Biochem.</u> 36:163 (1987)
12. M. Rosenberg, Y.-S. Ho and A. Shatzmann, The use of pKC30 and its derivatives for controlled expression of genes, <u>Methods Enzymol.</u> 101:123 (1983)
13. T. Maniatis, E. F. Fritsch and J. Sambrook, "Molecular cloning, a laboratory manual", Cold Spring Harbour (1982)
14. U. K. Laemmli, Cleavage of structural proteins during the assembly of the head of bacteriophage T4, <u>Nature</u> 227:680 (1970)
15. J. P. Chamberlain, Fluorographic detection of radioactivity in polyacrylamide gels with the water-soluble fluor, sodium salicylate, <u>Analyt</u>. <u>Biochem</u>. 98:132 (1979)
16. Methods Enzymol. vol. X, Estabrook R. W. and Pullman M.E., eds., Academic Press, New York and London (1967)
17. W. J. Waddel and C. Hill, A simple ultraviolet spectrophotometric method for the determination of proteins, <u>J. Lab. Clin. Med.</u> 48: 311 (1956)
18. H. U. Bergmeyer, "Methods of Enzymatic Analysis", Verlag Chemie, Academic Press, New York (1963)

CONTROL OF LIPOGENESIS IN MAINTENANCE CULTURES OF ISOLATED HEPATOCYTES

Gabriele V. Gnoni

Laboratory of Biochemistry
University of Lecce
Italy

INTRODUCTION

Various hormones such as insulin,glucagon,thyroid hormones and others participate in the regulation of lipid metabolism in hepatic tissue.This has been well documented in intact animals, perfused liver or liver slices.While these systems have provided extensive and complementary information,each possesses some limitations. One is that intact liver consists of hepatocytes, Kuppfer cells,bile duct and connective tissue which often restricts interpretation of the data since one cannot be sure which cell type is contributing to and/or modifying the metabolic end point being studied.This disadvantage can be resolved by using primary cultured adult rat hepatocytes,which also offer the advantage of cellular homogeneity,the opportunity to conduct experiments with homologous cells from a single rat,and especially the ability to study the effect of various nutrients or hormones without the secondary hormonal responses from other organs, as can occur in whole animal studies.In the following experiments the short-term regulation of both fatty acid and cholesterol synthesis by triiodothyronine in primary cultures of rat hepatocytes will be determined.

EQUIPMENT AND SOLUTIONS

A.Equipment
  -intravenous flask
  -perfusion apparatus
  -peristaltic pump
  -Carbogen (95% O_2 + 5% CO_2) tank
  -CO_2 incubator

B.Solutions
  -Buffer A ("Collagenase buffer")
   Stock sol. : 4.0 g NaCl

```
       0.5 g KCl
      24.0 " Hepes
       0.7 " CaCl₂.2 H₂0
       0.5 " Collagenase
         x ml NaOH 1M up to pH 7.4
```

0.5 g KCl
24.0 " Hepes
0.7 " $CaCl_2.2 H_2O$
0.5 " Collagenase
x ml NaOH 1M up to pH 7.4

Bring to 100 ml with redistilled water.Centrifuge for 30 min at 4500 x g.
Freeze in 10 ml portions.On the day of use : dilute a 10 ml portion up to 100 ml,add 0.36 g glucose.
Addition of glucose to the perfusion buffers has been recommended for minimizing glycogenolysis (Harris R.A.,1975).

-Buffer B ("Perfusion buffer")
Stock sol. : 83 g NaCl
             5 " KCl
            24 " Hepes
             x ml NaOH 1M up to pH 7.4

Freeze in 20 ml portions.On the day of use : dilute a 20 ml portion with H_2O up to a volume of 500 ml (pH 7.4) containing 1.8 g glucose.

-Buffer C (modified Krebs-Henseleit bicarbonate buffer)
Stock sol. : 68.97 g NaCl
              3.53 " KCl
              1.61 " KH_2PO_4
              2.96 " $MgSO_4$. 7 H_2O
              3.68 " $CaCl_2$. 2 H_2O

Bring to 1 l with redistilled H_2O and freeze in 50 ml portions.For daily use : dilute a 50 ml portion + 1.05 g $NaHCO_3$ in a final volume of 500 ml containing 1.8 g glucose; let CO_2/O_2 bubble through for 15 min.The pH should be 7.3-7.4.

-Culture medium

HAMS F 12	10.7 g
TES (N-tris(hydroxymethyl)methyl--2-aminoethanesulfonic acid)	2.86 g
MES (2-(N-morpholino(ethanesulfonic acid)	2.44 g
$NaHCO_3$	1.18 g

The solution is sterilized and protected with 100 units of penicillin and 100 µg streptomycin per ml.
Bring to a volume of 800-900 ml, let CO_2/O_2 bubble for 30 min,add 10 g of defatted bovine serum albumin fraction V, and then bring to a final volume of 1 l (pH 7.4)

ISOLATION OF RAT-LIVER PARENCHIMAL CELLS (Hepatocytes)

     The procedure is essentially based on the original method of Berry and Friend with the modification of Seglen.The method consists of two steps :
i) In situ perfusion of the liver with non recirculating medium

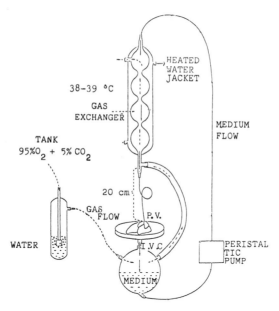

Fig. 1 Apparatus for rat liver perfusion

and ii) enzymatic digestion (by collagenase) with recirculating medium.

1. First of all set up all necessary apparatus (Fig.1). Put 100 ml of "Collagenase buffer"(buffer A) in the reservoir of the in vitro perfusion apparatus,which is gassed with 95% O_2 + 5% CO_2.The water bath of the latter is set at 38°C so that the fluid which will perfuse the liver will hold at 37°C.Start the peristaltic pump.Make sure that the perfusion fluid coats the gas exchanger evenly.

2. A rat (approximately 200-300 g) is anesthetized with ether in a plastic box.Apply ether to filter paper overlying the box. Place a shield over the filter paper to contain the fumes.

3. The anesthetized rat is pinned to a small slab that lies in a metal tray. The head is kept in a small jar with a pad to which ether is applied.

4. The peritoneal cavity is opened by making two lateral incisions starting at the midline of the lower abdomen and proceeding anteriorly up both sides of the rat until the diaphragm is reached (Fig.2). Use care to avoid cutting or puncture of the diaphragm. The resulting flap of skin and muscle is then lifted, exposing the peritoneal cavity.Isotonic saline (0.9% NaCl) containing 5,000 units heparin per 200·ml is used to wash away blood in the abdominal cavity.

5. Two ligatures are loosely placed around the portal vein, one anterior to the entrance of the superior mesenteric vein,and the other (preferably another colour and at 1 cm from the first) just posterior to the entrance of the splenic vein.

6. In the meantime,an intravenous flask filled with 500 ml of preheated (38 °C)Ca^{++}-free "Perfusion buffer" (buffer B) is

hung about 80 cm above the rat.To it is attached conventional
I.V. tubing with an adjustable clamp.
7. Place a flattened matchstick under the portal vein caudal to
the last ligature and with the point of a N° 11 scalpel make a
slit (4-5 mm long) in the vein in a longitudinal direction only
on its ventral aspect.
8. With the clamp on the I.V. tubing slightly open, insert the
tip of the cannula into the slit;push so that the tip just passes the first portal ligature and pull the distal ligature tight.
Open the clamp wide.As the liver distends ,cut the inferior vena
cava caudal to its suture.Secure the two portal ligatures.The
clamp on the I.V. tubing is now adjusted to provide a flow rate
of two drops per second (as seen in the flow indicator).The
liver blanches as its blood is displaced.
9. The thorax is opened by extending the incisions on either
side through the rib cage.Cut the inferior vena cava between
diaphragm and heart.
10.The liver is then dissected free for transfer to the <u>in vitro</u>
perfusion apparatus;taking care not to cut lobes of the liver.
11.Lay the liver on the round glass plate (platform) and wait
till 250-300 ml of the perfusion buffer has been passed.
12.Quickly remove the portal needle attachement and insert the
tube leading from the gas exchanger on the perfusion apparatus
(Fig.1).Avoid entrapping gas bubbles which can block the flow.
Place the glass plate over the lower round bottom flask which
is the reservoir for the medium (100 ml).The portal inlet can
be held upright by forming a simple loop in the tubing and by
taping it in place.The flow of effluent into the reservoir will

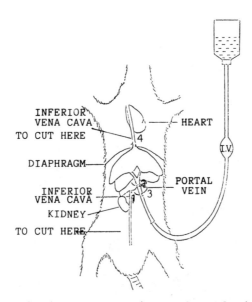

Fig.2. Diagram depicting cannulation and perfusion procedures.

be a continuous stream.From the beginning,keep two plastic
shields over the liver.

13.Let this perfusion continue in recirculating perfusion apparatus for 8-10 minutes.Note that the liver becomes limp and that
a serous fluid begins to ooze onto the glass plate as the tissue
structure deteriorates.At this point the liver perfusion is ter-
minated.

14.Transfer the liver from the perfusion apparatus to a wide
low beaker containing about 100 ml Krebs-Henseleit buffer (buffer C).By a forceps the Glisson membrane is taken off and the
liver vigorously stirred.The dispersed cells are filtered through
250 μ Nytal mesh and the cell suspension kept in continuous
shaking under an atmosphere of CO_2/O_2 for 15 min.After that,cell
suspension is refiltered through 100 μ Nytal mesh and centrifuged at 50 x g per 2 min.The pelletted cells are suspended in modified Krebs buffer containing 1% fatty acid free albumin. The
washing procedure is repeated twice in the same conditions and
finally the isolated hepatocytes are suspended in about 50 ml
Hams F 12 medium.

To check cell viability,a droplet of the cell suspension
is examined under the microscope in the presence of a vital dye
(trypan blue : 0.5% in 0.9% NaCl) .A preparation is acceptable
for study if 85% or more of the cells reject the dye.

Then the cells are counted in a Burker chamber and the suspension diluted with Hams F 12 medium,containing 10% fetal calf
serum,to give approximately $5x10^5$ cells per ml.Inclusion of the
serum is necessary for the formation of monolayer.

Four ml of the cell suspension are added to vented plastic Petri
dishes (60 mm) for plating and kept at 37 °C in a humidified
incubator in equilibrium with a 95% air and 5% CO_2 gas mixture.
In 2 h the majority of the cells becomes firmly attached to the
bottom of the dishes so that the medium can be changed without
loss of hepatocytes.After this 2 h plating period 4 ml fresh
medium without fetal calf serum is added together with the hormone,which remains in contact with the cells for additional
four hours.

ASSAYS

Synthesis of fatty acids and cholesterol is monitored by
measuring the rate of 1-^{14}C acetate (2.5 μCi,specific activity
0.96 mCi/mmole) incorporation into these lipid fractions.Fresh
medium,with or without hormone,is added together with labelled
substrate 1 h before ending the incubations.To terminate the
lipogenic assay,medium is aspirated from the plated cultures
and the adherent cells washed three times with 4 ml of ice-cold
0.15 M KCl and the reaction is stopped with 1.5 ml of 0.5 N KOH.
Subsequently the cells are scraped off with a rubber policeman,
aliquot are taken for protein determination,and then transferred
to extraction tubes for lipid extraction.After addition of 2 ml
of ethanol,the mixtures are saponified at 90 °C for 2 h,after
which non saponifiable lipids (NSL) are extracted three times

with 5 ml of petroleum ether (b.p. 40-60).
The pooled extracts are evaporated to dryness under a N_2 stream and NSL fraction radioactivity counted.To measure the incorporation of labelled precursor into digitonin precipitable-cholesterol (which is a part of the NSL fraction),a modification of the Sperry and Webb method can be used :
1. The ether extracts (NSL fraction) are transferred to conical glass centrifuge tubes.Add 0.1 mg cold cholesterol (dissolved in chloroform).
2. Add 1 ml acetone-95% ethanol (1:1,v:v).Mix on vortex and add 1 ml of about 1% digitonin (heat till 60° 1 g of digitonin in 100 ml 50% ethanol).Mix on vortex and leave at room temperature for 10 min.
3. Spin in table top clinical centrifuge full speed for 10 min.
4. Discard supernatant and allow to drain for 5 min.
5. Suspend residue in 4 ml acetone.Mix on vortex.Leave to room temperature for 10 min.
6. Spin as under step 3.
7. Discard supernatant and allow to drain for 10 min.
8. Remove last traces of solvent under N_2.Do not overdo it, otherwise it is extremely difficult to dissolve.
9. Dissolve residue in 1 ml toluene/methanol (2:1).
10.Transfer to scintillation vial with Pasteur pipette and rinse centrifuge tube twice with 5.0 ml scintillation fluid each time. Transfer the rinses to the scintillation vial and count.

Table 1.Effect of triiodothyronine on the rate of fatty acid and cholesterol synthesis in hepatocyte cultures from eu-and thyroidectomized (Tx) rats.

| T_3 | (nmoles acetyl units inc./mg protein per h) | | | |
| | Fatty acid sinthesis | | Cholesterol synthesis | |
(M)	Euthyroid	Tx	Euthyroid	Tx
0	1.38	0.73	0.17	0.06
1.48×10^{-5}	1.42	0.75	0.17	0.06
7.40×10^{-7}	1.68	0.95	0.19	0.07
1.48×10^{-6}	1.69	1.10	0.19	0.08
1.48×10^{-5}	1.89	2.05	0.24	0.15

Plated cells were in contact with the indicated T_3 concentration for a total period of 4 h .After 3 h,fresh medium containing the appropriate hormone concentration together with labelled acetate was added.Reactions were stopped after incubation for an additional hour and lipid extracted.

FATTY ACID EXTRACTION

Following extraction of NSL,the aqueous phase is acidified by adding 1.5 ml HCl 7 N,the pH is checked (it should be 1-2) and then fatty acid are extracted with 5 successive extractions (5 ml each) of petroleum ether (40-60°).The pooled extracts are evaporated to dryness under a stream of N_2.Residua are dissolved in scintillation fluid and the radioactivity counted.

RESULTS AND COMMENTS

In order to maximize the response to added triiodothyronine (T_3),hepatocytes from thyroidectomized (Tx) rats have been utilized.For comparison,liver cells from euthyroid rats were also included in this study.

Hepatocyte cultures from thyroidectomized rats show a smaller rate of lipogenesis than cultures from euthyroid animals (Table 1).Acetate incorporation into fatty acid is noticeably enhanced in comparison to controls that are incubated in the absence of added T_3 (Table 1).This stimulation of fatty acid synthesis is most evident in hepatocytes from thyroidectomized rats.In this case a stimulation is already observed at a concentration of T_3 of about 10^{-7} M.A three-fold enhancement is observed at around 10^{-5} M. The stimulatory effect by T_3 is less evident in hepatocyte cultures from euthyroid rats.

The rate of incorporation of labelled acetate into cholesterol parallels the rate of fatty acid synthesis in response to T_3 although changes in the rate of cholesterol synthesis are often less pronounced.

SELECTED REFERENCES

1. M.N.Berry and D.S.Friend,High-yield preparation of isolated rat liver parenchymal cells,J. Cell.Biol. 43:506-520 (1969).
2. M.N.Berry,High-yield preparation of morphologically intact isolated parenchymal cells from rat liver,Methods in enzymology 32,Part B, 625-632 (1974).
3. M.J.H.Geelen and D.M.Gibson , Control of lipogenesis in maintenance cultures of freshly isolated hepatocytes, In:Use of isolated liver cells and kidney tubules in metabolic studies,Tager J.M.,Soling H.D. and Williamson J.R., Eds,North-Holland Publ.Co,Amsterdam,219-230 (1976).
4. G.V.Gnoni,C.Landriscina and E.Quagliariello,Thyroid hormone stimulation of lipogenesis in isolated rat hepatocytes, Biochem.Med. 24 : 336-347 (1980).
5. G.V.Gnoni,M.J.H.Geelen,C.Bijleved,E.Quagliariello and S.G. van den Bergh,Short-term stimulation of lipogenesis by triiodothyronine in maintenance cultures of rat hepatocytes, Biochem.Biophys.Res.Comm. 128:525-530 (1985).
6. R.A.Harris,Studies on the inhibition of hepatic lipogenesis by Dibutyryl-Adenosine 3',5',-monophosphate,Arch.Biochem. Biophys.169 : 168-180 (1975).

7. P.O.Seglen,Preparation of rat liver cells.I.Effect of Ca^{++}
   on Enzymatic dispersion of isolated,perfused liver,Expt.Cell
   Res. 74 : 450-454 (1972).
8. W.M.Sperry and M.Webb, A revision of the Schoenheimer-Sperry
   method for cholesterol determination, J.Biol.Chem.187: 97-
   -110 (1950).
9 S.R.Wagle and W.R.Jr.Ingebretsen, Isolation,purification,and
   metabolic characteristics of rat liver hepatocytes,Methods
   in Enzymology.35,Part B, 579-594 (1975).

ENZYMIC REACTIONS IN UREOGENESIS: ANALYSIS OF THE CONTROL OF CITRULLINE

SYNTHESIS IN ISOLATED RAT-LIVER MITOCHONDRIA

A.J. Meijer

Laboratory of Biochemistry, University of Amsterdam

INTRODUCTION

In the experiment to be described the flux control coefficient of some steps in the pathway from ammonia to citrulline in isolated rat-liver mitochondria will be determined[1]. The pathway is shown in Fig.1.

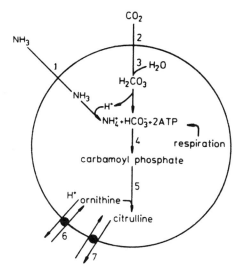

1. Ammonia transport

2. CO_2 transport

3. Carbonic anhydrase

4. Carbamoyl-phosphate synthetase

5. Ornithine carbamoyl-transferase

6,7. Mechanisms of ornithine transport

Fig. 1 Synthesis of citrulline in liver mitochondria.

In addition to NH_4^+, HCO_3^- and ornithine as precursors for citrulline synthesis, succinate (+ rotenone) will be used as the respiratory substrate to provide ATP. Although in the experiment some extramitochondrial ATP is added it cannot be used for intramitochondrial ATP-dependent processes since ATP cannot enter energized mitochondria; its function, however, is to prevent leakage of intramitochondrial ATP to the extramitochondrial fluid. The transaminase inhibitor aminooxyacetate (AOA) is present in order to prevent degradation of ornithine in the mitochondria.

The concentrations of substrates (NH_4^+, HCO_3^-, ornithine) are such that during the course of the experiment (10 min) their concentration can be considered to be constant. Although the concentration of the product of

the pathway, citrulline, is not constant but increases with time this
is not a problem because citrulline has no feed-back effects on its own
production; thus, variation in its concentration does not affect flux
through the pathway.

In the experiment the flux control coefficient of ornithine carbamoyl-
transferase (OCT) in the pathway of citrulline synthesis (defined as:
$C_{oct} = \dfrac{dJ/J}{d(oct)/oct}$) will be determined by titration of the flux through
the pathway (the rate of citrulline production) with norvaline, a competitive inhibitor of the enzyme with respect to ornithine. The following equation can be derived[2]:

$$C_{oct} = \frac{-K_i(1 + \frac{S}{K_m})}{J} \cdot \left(\frac{dJ}{dI}\right)_{I \to o}$$

in which:

$\quad K_i$ = K_i for norvaline (70μM)
$\quad S$ = intramitochondrial ornithine (1.5 mM)[1].
$\quad K_m$ = K_m for ornithine (0.4 mM)
$\quad J$ = flux through the pathway in the absence of norvaline (the
$\quad\quad$ rate of citrulline formation)
$\quad I$ = the intramitochondrial concentration of norvaline (equal to
$\quad\quad$ the extramitochondrial concentration[1].
$\quad \dfrac{dJ}{dI}$ = slope of the curve of J vs. I when I approaches zero.

The flux control coefficient of carbamoyl-phosphate synthetase (CPS)
cannot be determined directly since specific inhibitors of this enzyme
are not available. Its value, however, can be calculated on the basis
of the connectivity principle which states that the flux control coefficients of two neighbouring enzymes in a pathway are related to the
elasticity coefficients of the two enzymes for their common intermediate
(carbamoyl phosphate, CP, in our experiment) as follows:

$$C_{cps} \cdot \varepsilon_{cp}^{cps} + C_{oct} \cdot \varepsilon_{cp}^{oct} = o$$

$$\left(\varepsilon_{CP}^{CPS} = \frac{\partial CPS/CPS}{\partial CP/CP}; \quad \varepsilon_{CP}^{oct} = \frac{\partial OCT/OCT}{\partial CP/CP}\right)$$

The elasticity coefficients of CPS and OCT for carbamoyl phosphate can
be determined in situ, in the intact mitochondrion, in the following
way:

a. By adding small amounts of malonate, a competitive inhibitor of succinate dehydrogenase, the supply of ATP will be diminished. A new
steady state will be reached in which both the rate of citrulline
synthesis (J) and the intramitochondrial concentration of carbamoyl
phosphate are decreased. Since at steady state the flux through the
entire pathway is equal to the flux through each of the individual
enzymes of the pathway this experiment gives information on the
activity of OCT at varying intramitochondrial concentrations of
carbamoyl phosphate. From this ε_{cp}^{oct} can be calculated[1].
b. Similarly, by adding small amounts of norvaline (instead of malonate) to decrease the activity of OCT, one is able to study the relationship between the activity of CPS at raised intramitochondrial
carbamoyl phosphate concentration. From this ε_{cp}^{cps} can be calculated[1].

The flux control coefficient of mitochondrial carbonic anhydrase in the pathway of citrulline synthesis will be determined in a rather straight forward way. Firstly, activity of carbonic anhydrase can be estimated from the rate of swelling of the mitochondria in an iso-osmotic solution of NH_4HCO_3. Since the mitochondrial inner membrane is not permeable to HCO_3^- (in contrast to CO_2) net uptake of NH_4HCO_3 by the mitochondria requires participation of mitochondrial carbonic anhydrase. By testing the effect of increasing concentrations of acetazolamide, a specific inhibitor of carbonic anhydrase, on the rate of swelling one can calculate the percentage inhibition of enzyme activity obtained at a particular concentration of acetazolamide. Secondly, in a parallel experiment, the effect of acetazolamide on the rate of citrulline synthesis will be studied. From a comparison of the effect of a particular low concentration of acetazolamide on the activity of carbonic anhydrase (as measured by mitochondrial swelling in NH_4HCO_3) and its effect on citrulline synthesis the flux control coefficient of this enzyme can be directly calculated.

EXPERIMENTAL

Mitochondrial incubations

The following, two times concentrated, incubation mixture (mix) must be used:
200 mM KCl; 100 mM Tris/HCl; 20 mM NH_4Cl; 2 mM EGTA; 2 mM ATP; 40 mM Tris- succinate; 10 mM K-phosphate; 6 mM ornithine; 33.3 mM $KHCO_3$; 2 mM aminooxyacetate; 4 µg/ml rotenone. Final pH, 7.4.

A. Norvaline titration

Add to incubation flasks (counting vials):

Addition	Flask no.							
	1	2	3	4	5	6	7	8 (zero time)
Mix	0.50	0.50	0.50	0.50	0.50	0.50	0.50	0.50 ml
H_2O	0.40	0.39	0.38	0.34	0.30	0.25	0.40	0.40
Norvaline 0.2M	–	0.01	0.02	0.06	0.10	0.15	–	–

Flush with 95% O_2 + 5% CO_2; close flasks; equilibrate 5 min at 25°.

Mitochondria at t = 0 min								
	0.10	0.10	0.10	0.10	0.10	0.10	0.10	0.10

After 10 min at 25° 0.7 ml of the suspension must be carefully layered on top of a precooled Eppendorf tube filled with 0.15 ml (14% $HClO_4$ + 5 mM EDTA) and 0.6 ml silicone oil. Centrifuge in a microcentrifuge for 45 sec: the mitochondria are now at the bottom of the tube. AFTER CENTRIFUGATION, KEEP SAMPLES ON ICE AS CARBAMOYL PHOSPHATE IS UNSTABLE AT ROOM TEMPERATURE! Transfer 0.5 ml of the extramitochondrial fluid (upper layer) to an Eppendorf tube filled with 0.2 ml 14% $HClO_4$; centrifuge 1 min. These samples are ready for citrulline determination.

<u>Neutralization of mitochondrial extract</u> (must be carried out within 5 min after centrifugation of the mitochondria through silicone oil). Remove remainder of upper layer and most of the oil by suction. Carefully resuspend the mitochondrial pellet with a small stirring rod and centrifuge again. Take 0.125 ml of the extract and transfer to an empty Eppendorf tube (on ice). Add 0.25 ml <u>cold</u> H_2O and neutralize with a mixture of 2 M KOH + 0.3 M Mops to pH 7. (First try a blank with 0.125 ml 14% $HClO_4$ to see how much KOH-Mops is approximately needed! Avoid pH values above 7.5: carbamoyl phosphate will then hydrolyze).

B. <u>Malonate titration</u>

Add to incubation flasks (counting vials):

Addition	Flask no.							
	1	2	3	4	5	6	7	8 (zero time)
Mix	0.50	0.50	0.50	0.50	0.50	0.50	0.50	0.50 ml
H_2O	0.40	0.39	0.38	0.37	0.36	0.35	0.40	0.40
malonate 0.1M	–	0.01	0.02	0.03	0.04	0.05	–	–

Flush with 95% O_2 + 5% CO_2; close flasks; equilibrate 5 min at 25°

Mitochondria at t = 0 min								
	0.10	0.10	0.10	0.10	0.10	0.10	0.10	0.10

After 10 min at 25° 0.7 ml of the suspension must be carefully layered on top of a precooled Eppendorf tube filled with 0.15 ml (14% $HClO_4$ + 5 mM EDTA) and 0.6 ml silicone oil. Centrifuge for 45 sec in a microcentrifuge. For further processing of the samples, see (A).

C. <u>Acetazolamide titration</u>

Add to incubation flasks (counting vials):

Addition	Flask no.							
	1	2	3	4	5	6	7	8 (zero time)
Mix	0.50	0.50	0.50	0.50	0.50	0.50	0.50	0.50 ml
H_2O	0.40	0.39	0.38	0.37	0.30	0.20	0.40	0.40
Acetazola-mide 2 mM	–	0.01	0.02	0.05	0.10	0.20	–	–

Flush with 95% O_2 + 5% CO_2; close flasks; equilibrate 5 min at 25°.

Mitochondria at t = 0 min								
	0.10	0.10	0.10	0.10	0.10	0.10	0.10	0.10

After 10 min at 25°, transfer 0.5 ml of the suspension to an Eppendorf tube filled with 0.2 ml 14% $HClO_4$. After removal of the protein by centrifugation (microcentrifuge 45 sec) the samples are ready for determination of citrulline.

D. Measurement of mitochondrial swelling in iso-osmotic NH_4HCO_3 as an indicator of carbonic anhydrase activity; titration with acetazolamide

Mitochondria (20 µl) are preincubated for 3 min at room temperature in an Eppendorf tube with 80 µl 250 mM mannitol in the absence or presence of 5 µl 200 µM, 5 µl 400 µM and 5 µl 1000 µM acetazolamide, respectively. After this period, these mitochondria are added to a medium (2 ml, room temperature) containing the following components: 100 mM NH_4HCO_3 (saturated with 50% CO_2), 50 mM Tris/HCl, 1 mM EGTA and 1 µg/ml rotenone; pH, 7.4. Monitor mitochondrial swelling as the decrease in absorbance of the suspension at 520 nm.

Assay of citrulline

Add the following components to glass tubes:

Addition	Tube no.					
	1	2	3	4	5	samples
H_2O	0.20	0.19	0.18	0.17	0.16	–
Sample	–	–	–	–	–	0.20 ml
Citrulline standard (2mM)	–	0.01	0.02	0.03	0.04	–
Reagent A+B (1:1)	1.80	1.80	1.80	1.80	1.80	1.80

MIX WELL (BE CAREFUL FOR EYES AND CLOTHES! REAGENT CONTAINS CONCENTRATED H_2SO_4!). Incubate 20 min at 90-100°C.
Afterwards, cool samples in a waterbath. Read absorbance at 466 nm. Numbers 1-5 are standards.

Solution A: 0.85 gram antipyrin dissolved in 100 ml 40% H_2SO_4.
Solution B: 0.625 gram diacetylmonoxim dissolved in 100 ml 5% acetic acid (in dark bottle).
Reagent is prepared by mixing equal volumes of A and B (CAREFUL!).

Assay of carbamoyl phosphate (fluorimeter; excitation 340 nm; emission, 470 nm)

Add to a cuvet: 2 ml assay mix
            0.15 ml sample (or 0.15 ml H_2O as a blank)
First remove all ATP present in the sample by addition of 10 µl hexokinase (1:10). Then, after the base line is stable again, start the assay of carbamoyl phosphate by addition of 5 µl carbamate kinase. After the determination of carbamoyl phosphate is finished, add 10 µl 0.2 mM ATP as an internal standard.

Principle:

Glucose + ATP ———>Glucose 6-phosphate + ADP (hexokinase)

Glucose 6-phosphate + $NADP^+$———>6-P-gluconolactone + NADPH + H^+
(glucose-6-phosphate dehydrogenase)

Carbamoyl phosphate + ADP———>CO_2 + NH_3 + ATP (carbamate kinase)

Composition assay mix:
KCl, 75 mM
EDTA, 3.7 mM
Triethanolamine, 50 mM
$MgCl_2$, 7.5 mM
(final pH 7.5)

Per 100 ml of this mixture is added:
0.5 ml 0.5 M glucose, 5 mg $NADP^+$,
0.1 ml 0.1 M ADP, 0.1 ml glucose
6-phosphate dehydrogenase

CALCULATIONS

After having finished all measurements:

1. Construct a graph in which the rate of citrulline production is plotted against the intramitochondrial concentration of carbamoyl phosphate (experiments A and B).
2. Calculate from these curves the elasticity coefficients of CPS and OCT for their common intermediate, carbamoyl phosphate.
3. From experiment A, calculate the flux control coefficient of OCT in the pathway of citrulline synthesis using the equation mentioned in the Introduction.
4. Calculate the flux control coefficient of CPS using the connectivity equation (see Introduction).
5. From experiments C and D, calculate the flux control coefficient of mitochondrial carbonic anhydrase.

REFERENCES

1. R.J.A. Wanders, C.W.T. van Roermund, and A.J. Meijer, Analysis of the control of citrulline synthesis in isolated rat-liver mitochondria. Eur. J. Biochem. 142: 247-254 (1984)
2. A.K. Groen, R. van der Meer, H.V. Westerhoff, R.J.A. Wanders, T.P.M. Akerboom and J.M. Tager, Control of metabolic fluxes, in: "Metabolic Compartmentation", H. Sies, ed., pp.9-37, Academic Press, London (1982).

CHROMAFFIN GRANULES AND GHOSTS: ISOLATION AND MEASUREMENT OF ELECTROCHEMICAL PROTON GRADIENT AND AMINE UPTAKE

Antonio Scarpa

Dept. of Physiology and Biophysics
School of Medicine
Case Western Reserve University
Cleveland, OH 44106

OVERVIEW

The chromaffin granule is that subcellular organelle within the chromaffin cell of the adrenal medulla in which most of the body catecholamines are localized, and where these agents are stored prior to release. During the last 15 years, the mechanism of the accumulation and storage of the large quantities of catecholamines found within the intragranular matrix space has been elucidated. The basic elements of the mechanism, the coupling to gradient coupling (Johnson et al., 1978; Njus and Radda, 1978; Johnson and Scarpa, 1979) are now generally accepted. It has been shown that a proton-translocating ATPase in the chromaffin granule membrane, in conjunction with the extremely low permeability of the membrane to protons and cations (Johnson et al., 1978) is responsible for the generation and maintenance of an acidic granule interior and positive transmembrane potential. Both the pH gradient (ΔpH) and potential gradient ($\Delta\Psi$) across the chromaffin granule membrane, which constitute the electrochemical proton gradient (Δu_H^+) in accordance with the chemiosmotic hypothesis of Mitchell (1968), are thought to provide the driving force for the carrier-mediated accumulation of catecholamines against a concentration gradient (see also Johnson et al., 1981; Phillips, 1978; Apps. 1978).

COMPOSITION

Several excellent reviews are available on the membrane composition and content of the chromaffin (Winkler 1976; Winkler and Carmichael, 1982). Fig. 1 diagrammatically illustrates the major components.

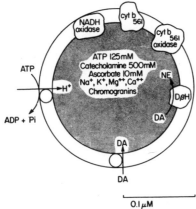

Composition of the Chromaffin Granules

Fig. 1

Chromaffin granules contain one of the lowest protein-lipid ratio of any isolated organelle. Of the total phospholipid content 13-18% is constituted by lysophosphatidylcholine.

Among the major proteins in the membrane are: dopamine-β hydroxylase, a mixed-function oxidase responsible for the β hydroxylation of dopamine to form norepinephrine: a H^+-translocating ATPase of the V class which hydrolizes cytosolic ATP and translocates H^+ in the intravesicular space; cytochrome b_{561} and possibly an NADH oxidoreductase.

In the intravesicular space, the proteins are dopamine-β-hydroxylase, enkefalin, chromagranin A and small peptides. Also stored within the intragranular spaces are large amounts of catecholamines, ATP and ions. If all the catecholamines and ATP were free in solution, their concentration would approach 0.5 and 0.2 M, respectively.

PERMEABILITY

The membrane of the chromaffin granule has been measured as the most impermeable to cations of any previously isolated subcellular organelle (Johnson and Scarpa, 1976; Johnson et al., 1978).

The conductance to protons may be one order of magnitude less than that of a mitochondrion, formerly thought to maintain the lowest permeability to protons. Cations such as Na^+ and K^+ also have very limited permeability across the granule membrane. In addition, the intravesicular buffering capacity of the chromaffin granules is quite high, approaching 300 μmol H^+/pH unit/g dry wt. These properties help to explain the experimental observation that isolated granules suspended at 4°C can maintain an internal pH of 5.5 for over 48 hours (Johnson et al., 1978). Other physiologically relevant cations and anions also diffuse very slowly, with the exception of Cl.

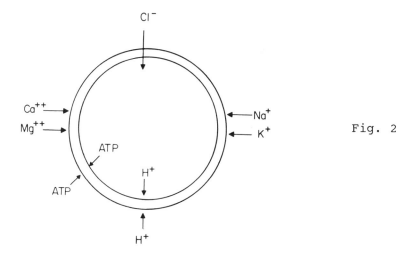

Fig. 2

The H^+ ATPase

The majority of ATPase activity appears to be due to a H^+-translocating ATPase, a V class of ion motive ATPase. Verification of H^+-translocating ATPase activity is based upon fulfillment of the minimal criteria that: 1) the addition of ATP to an isolated ghost preparation in which the internal and external pH values are initially identical results in generation of a ΔpH, inside acidic (Johnson et al., 1979); 2) there is a fixed stoichiometry between ATP hydrolysis and H^+ translocation (Flatmark and Ingebretsen, 1977; Njus et al., 1978); and 3) under appropriate conditions ATP can be synthesized at the expense of a proton gradient (Roisen et al., 1980). In addition, recent attempts at isolation of the ATPase have resulted in measurements of physico-chemical properties which are consistent with the presence of a H^+ translocating ATPase which is sensitive to the mitochondrial proton pump inhibitors DCCD and the alkyl tin compounds. The kinetic properties of this ATPase are illustrated in Table 1 (from Johnson et al., 1982).

TABLE 1
Properties of the Chromaffin Granule $H^{\pm}$-ATPase

Specificity	ATP = ITP > GTP > UTP > CTP
Km	69 µM
Vmax	10-50 nmol/min/mg
pH optimum	7.3
Stoichiometry (H+/ATP)	1.6-2
Inhibitors	Alkyn tin
	DCCD

MEASUREMENT OF THE $\Delta\mu_H^+$:

Activity of the H^+-translocating ATPase within the chromaffin granule membrane is responsible for the development of an electrochemical proton gradient ($\Delta\mu_H^+$) across the membrane, composed of intraconvertible electrical and concentration components according to the chemiosmotic hypothesis of Mitchell (1968):

$$\Delta\mu_H^+ = \Delta\Psi - Z\Delta pH,$$

where $Z = 2.3\ RT/F$. Unfortunately, the small size of the granules precludes direct measurement of the internal pH or membrane potential with microelectrodes. Thus, for measurement of the ΔpH, $[^{14}C]$-methylamine distribution, ^{31}P NMR, and fluorescent dye distribution are commonly utilized (Johnson and Scarpa, 1976; Njus et al., 1978; Salama et al., 1980), while for measurement of the $\Delta\Psi$, the independent methods of $[^{14}C]$ SCN^- and fluorescent dye distribution have yielded equivalent results (Holtz, 1978; Johnson and Scarpa, 1979; Salama et al., 1980).

In intact chromaffin granules suspended in isotonic sucrose at physiologic pH, the ΔpH across the membrane approaches 2 pH units, indicating an internal pH of 5.5 (Bashford et al., 1976; Johnson and Scarpa, 1976). (Fig. 3). This acidic internal pH is independent of the ionic composition of the media (NaCl, KCl, choline Cl) and of the presence of Mg ATP, indicating that it is not due to the establishment of a Donnan equilibrium. The transmembrane potential can reach large positive or negative values, depending upon the experimental conditions (Johnson and Scarpa, 1979). In the presence of Mg ATP, a of 80 mV, inside positive (as measured by $[^{14}C]$ SCN^- distribution) is observed.

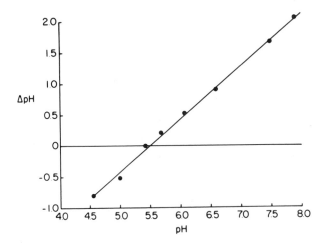

Fig. 3

This large $\Delta\Psi$ most likely exists under physiological conditions. When FCCP, an ionophore that allows protons to distribute across membranes down their electrochemical gradient, is present, a large negative potential (measured by [^{14}C] TPMP) approaching the Nernst potential for protons is measurable. On the other hand, incubation of the granules with NH_4^+ causes dose-dependent alkalinization of the interior and consequent collapse of the ΔpH without an effect on the transmembrane potential.

AMINE UPTAKE

The properties of amine accumulation are consistent with a carrier-mediated active process based upon the following empirical evidence: uptake exhibits structural specificity and stereospecificity, a Q_{10} of 4.6, dependence upon the presence of ATP and Mg, and specific inhibition by reserpine. The most important recent contribution to the study of amine uptake has been to elucidate and quantitate the contribution of the $\Delta\mu_H+$ as the driving force for amine accumulation.

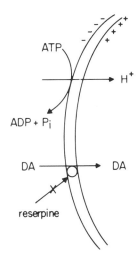

Protonmotive Force and Biogenic Amino Accumulation

Fig. 4

The chromaffin ghost preparation has provided the means by which to approach these quantitations. Utilizing the selective permeability properties of the membrane to Cl^-, ghosts can be formed in differing ionic media so as to generate a ΔpH alone (Fig. 5A), a $\Delta\Psi$ alone (Fig. 5B) or both upon Mg ATP addition.

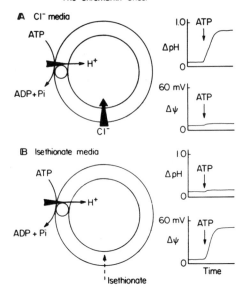

The Chromaffin Ghost

A Cl⁻ media

Fig. 5A & 5B

Recent extensive investigation has clearly demonstrated that either the ΔpH, ΔΨ, or both can drive amine accumulation, with the maximum rate and extent of accumulation occurring in the presence of both gradients together (Johnson et al., 1981).

The improvement of methodology for preparation of highly purified ghosts has permitted the investigation of the precise quantitative relationship between the $\Delta\mu_H^+$ and $\Delta\mu_A$ as measured under identical conditions. Their relationship is predicted by the chemiosmotic hypothesis, which states that the energy for substrate transport systems is derived from the $\Delta\mu_H^+$ generated by the proton pumping ATPase, and that the coupling between the transported species (amine and proton, in this case) would approach equilibrium, i.e., net transport vanish, whenever:

$$\Delta\mu_A - n\Delta\mu_H^+ = 0,$$

where "n" is the number of protons translocated in the opposite direction (antiport), i.e., the stoichiometry of the reaction (Mitchell, 1970; Rottenberg, 1979). Thus, H+ amine antiport is defined by nonspontaneous obligatory coupling through the existence of a putative translocator.

GENERAL MECHANISM FOR UPTAKE OF VARIOUS AMINES

Other subcellular organelles accumulate and store biogenic amines, including the serotonin dense granules of platelets, the histamine granules of mast cells, anterior pituitary granules, insulin granules of pancreatic β cells, and intestinal peptide granules. The evidence to date strongly suggests that each of these amine containing subcellular organelles maintains a similar mechanism for amine accumulation based upon the force provided by the electrochemical proton gradient.

EXPERIMENT #1

Isolation of Intact Adrenal Chromaffin Granules Using Isotonic Percoll Density Gradients

1. Introduction

The granules are fragile and susceptible to changes in external osmotic conditions, so that only isotonic gradient centrifugation minimizes membrane properties and redistribution of intravesicular contents.

This experiment takes advantage of the properties of Percoll, a high density, low viscosity suspension of minute silica particles coated with polyvinylpyrrolidine. Percoll can be diluted with sucrose solution and centrifuged to form a biologically inert, isotonic continuous density gradient.

A full description of this method and the results obtained can be found in Carty et al. (1980).

2. Solutions

Percoll Gradient	Sucrose-Tris
30.0 ml Percoll	Vol. = 6 liters
9.25 g sucrose	555.42 g sucrose (0.27 M)
1 ml Tris-Maleate, 1 M	60 ml Tris-Maleate, 1 M
pH 7.0 (w. H_2SO_4)	pH 7.0

(volumes must be 100 ml exactly)
N.B. All solutions made with deionized H_2O.

3. Procedure

1. Obtain fresh glands (approx. 40) from slaughterhouse, and place on ice immediately.

2. Trim off fat and place glands in plastic 600 ml beaker containing 300 ml S-T until full. Then cut around edge to medulla, slice out medulla and place in 250 ml plastic beaker containing 125 ml S-T (Remove all outer membranes). Drain excess S-T and mince with scissors into fine pieces (5-10 min.). Avoid orange and/or black-specked tissue.

3. Fill size "C" homogenizing tube with 1 1/2 minced medulla, add 4 parts S-T, and homogenize 3 X. Keep homogenizing vessel in ice. Continue in batches until glands are used up, then pour into S-T-rinsed centrifuge bottles (in ice), balance bottles, and spin @ 2750 rpm for 5 min. Discard pellet, which contains unbroken cells and cell nuclei.

4. Filter supernatant through 4 layers cheesecloth into 6 S-T rinsed large centrifuge bottles in ice, spin @ 7250 rpm for 30 min.

5. Swirl to dislodge brown fluffy mitochondrial layer, then discard supernatant. Resuspend pellet in S-T with BB size pestle and BB homogenizing tube 2 X. Avoid resuspending erythrocytes. Yield about 60 ml, approx. 3 spins. Cool suspension in ice.

6.   Percoll Gradient: Prespin 50 ml plastic tubes
containing approx. 40 ml 30% Percoll for 5 min. at 13,000 rpm
(use timer). Keep cold. Calibrate with Density Marker Beads
#3 (green), #4 (red), and #10 (blue) so as to obtain maximum
separation between red and green beads. Then layer 2-2.5 ml
homogenate carefully on top of gradient, spin @ 9,000 rpm for
40 min. Aspirate down to red beads, dilute to 20 ml with cold
S-T, and wash twice as in step #8; no need to homogenize.

7.   Using cold ultracentrifuge head as a test-tube rack,
aspirate supernatant from tubes, clean mitochondria from sides
of tube with Q-tip, and add 2 ml of S-T. Then aspirate S-T
and resuspend granules with 1 ml S-T and fine homogenizing
pestle. Do six tubes, then remaining six tubes. Swirl and
pour into size A homogenizing vessel, homogenize 2 X.

8.   Dilute to 150 ml S-T, stir, divide into S-T rinsed
plastic centrifuge tubes, and spin @ 9,000 for 30 min.

9.   Discard supernatant, resuspend granules to desired
final volume, and refrigerate until use, at which time
granules should be washed again.

10.   Percoll silica beads tend to coat all glass and
plastic tubes, so glassware should be washed ASAP.

EXPERIMENT #2

Isolation of Chromaffin Ghosts

1.   Introduction

This method yields closed ghosts devoid of ATP and
catecholamines which can be resealed in different media.   A
reference for this method and results may be obtained in
Johnson et al. (1981).

2.   Purification of Dialysis Tubing

Purpose: To remove sulfur and organic by-products
present in dialysis tubing.

Stock Solutions:

a) 1 M Tris Acetate prepared by adjusting pH of 1 M Tris
with acetic acid to pH 8.0.

b) 1 M EDTA.

Procedure:

1) Prepare 3 liters of 10 mM Tris Acetate, 1 mM EDTA.

2) Bring 1.5 liters of (1) to boil. Add tubing and allow
to boil for 15 min. with continual stirring. Repeat with
fresh remaining 1.5 liters of (1).

3) Soak in several liters of room temp. distilled water with continual stirring.

4) Soak in 1 liter of your dialysis medium before use.

3.   Preparation

1.   After resuspension of one prep's worth (approx. 15 ml of 20 mg/ml) of chromaffin granules in a small volume of S-T, add it to 400 ml of room temperature lysing medium. Incubate 5 min. at room temperature. Spin in rinsed plastic tubes @ 15,000 for 20 min.

2.   Aspirate all supernatant, and resuspend pellets in 20 ml total of the appropriate dialysis buffer (cold); revesicularization is occurring.

3.   Soak 10-12" dialysis tubing in deionized water to loosen, then twist apart, and tie one end in a knot. Transfer vesicle suspension into tubing; leave a small air bubble and tie off tube. Dialyze in 3.5 liter ice-cold buffer overnight in cold room with stirrer on; cover to inhibit evaporation of the buffer.

4.   Break dialysis bag into 30 ml of fresh buffer. Spin @ 14,000 for 20 min. Resuspend pellet to desired final volume for experiments, using same buffer as for dialysis.

5.   Gradients will be generated upon the addition of 25 mM Mg ATP.

4.   Solutions

Lysing Medium

5 mM EDTA
5 mM Tris-Maleate
0.06% BSA

Dialysis Buffers

To generate ΔpH:	IN:	185 mM KCl 2 mM HEPES pH 7.0
	OUT:	185 mM KCl 30 mM HEPES pH 7.0
To generate $\Delta\Psi$:	IN:	185 mM Na Isethionate 30 mM HEPES pH 7.0
	OUT:	Same

To generate both ΔpH and ΔΨ:

                    IN: 20 mM ascorbate
                           285 mM Na Isethionate
                           5 mM HEPES
                           pH 7.0

                    OUT: 185 mM Na Isethionate
                           30 mM HEPES
                           pH 7.0

EXPERIMENT #3

Measurement of ΔpH or ΔΨ or Catecholamine Uptake Through Isotope Distribution Ratios

1. ## Introduction

    This experiment permits the determination of the intravesicular pH of chromaffin granules (see Fig. 3). More details in Johnson and Scarpa (1976; 1979).

2. ## Solutions:

Mg ATP Stock	FCCP Stock
0.2 M ATP	1 mM FCCP
0.2 M $MgSO_4$	1 ml absolute ethanol
pH 7.0 w/NaOH	keep in freezer
keep in freezer	

$(NH_4)_2SO_4$ Stock	KSCN Stock
2 M $(NH_4)_2 SO_4$	1 M KSCN
2 mM Tris-Maleate	20 mM Tris-Maleate
pH 7.0	pH 7.0
Refrigerate in amber bottle	Refrigerate in amber bottle

3. ## Procedure

    ## Wear gloves and observe radiation precautions throughout

    1. Draw a map of the various experimental samples ordered by time in groups of four. Lay out, according to map, a 1.5 ml Eppendorf capped tube for each sample, and pipette into each tube 1 ml of the appropriate experimental medium (S-T for granules, various buffers for vesicles).

    2. Pipette 5 μl of ^3H_2O into inner north wall of each tube. Pipette 5 μl of appropriate ^{14}C isotope onto inner south wall (refer to map). Do not contaminate isotope vials, and use Oxford pipettes.

    3. To measure extravesicular space, include (and on map) a Dextrose sample, containing 1 ml medium, 5 μl ^3H_2O (north), and 10 μl HMW Dextran (south). If desired include a $^{14}C_1$ standard (to measure ^{14}C spill) containing 1 ml medium, 5 ml ^{14}C-isotope, and no ^3H_2O.

4.   Pipette appropriate volumes of cold stock solutions
onto inner east wall of tubes, according to map. Examples:

    60 μl of 1 M NaSCN for 5 mM NaSCN.
    30 μl of 1 M $(NH_4)_2 SO_4$ for 50 mM $(NH_4)_2 SO_4$.
    10 μl FCCP.
    50 μl Mg ATP for 8 mM Mg ATP.

    Start incubation period (use stopwatch) with addition of
200 μl granules/ 100 μl vesicles onto inner west wall of each
sample tube. Cap tubes and invert to mix ingredients.

    5.   Remove and discard caps from flexible (Markson) 200
μl centrifuge tubes. Transfer contents of 1 sample tube to 3
Markson tubes using Pasteur pipette. Place Markson tubes in
an Eppendorf desk centrifuge (use adaptors) and end the
incubation periods with 7 min. centrifugation for vesicles, 4
min. for granules.

    6.   Pipette 100 μl supernatant into numbered 14% PCA
tubes (pre-prepared with 200 μl PCA) and cap. On a rubber
cutting surface, slice pellet in half with scalpel and
transfer bottom half (uncontaminated by supernatant) to the
correspondingly numbered PCA tubes. Wipe scalpel clean and
change pipette tip after each sample to avoid contamination.

    7.   Preparation for scintillation counter:

        a) Vortex pellets until completely dissolved.

        b) Spin down supernatant and pellet protein for
        approx. 4 min.

        c) Into vials place 5 ml scintillation fluid, 0.2
        ml formic acid, and 0.2 ml sample, and cap. Shake
        vials thoroughly and label.

EXPERIMENT #4

Accumulation of Dopamine in Chromaffin Ghosts

1.   Introduction

    The purpose of this experiment is to generate a ΔpH
gradient in isolated chromaffin ghosts with ATP and to follow
the transport of dopamine. The conditions are similar to that
of Fig. 4A.

    Catecholamine uptake can be measured a) by labelled
intra-extravesicuar distribution, as described in Experiment
#3; b) by glassy-carbonyl electrodes; c) by HPLC. The latter
two procedures are described in detail by (Hayflick, et al.,
1981).

    An example of this transport is shown in Fig. 6, where
the time course of catecholamine uptake by the ghosts is
followed through both electrodes and HPLC.

2. Solutions

a) 100 ml 185 mM KCl
   100 ml 20 mm HEPES
   pH 6.8

b) 1 liter 0.5 M Ammonium phosphate (pH 4.3), HPLC grade.

c) 1 ml ATP, 0.5 M, pH 6.8

d) 1 ml Dopamine, 10 mM

e) approx. 10 mg protein ghosts prepared in 185 mM KCl, 2 mM HEPES.

f) standard containing 50 μM each: ATP, ADP, AMP, dopamine.

3. Instrumentation

a) HPLC pump, HPLC injector, HPLC detector at 214 nm, column C_{18} bondapak 10 μ and precolumn.

b) Millipore manifold filtration to collect filtrate and 0.45 μ filters.

c) Thermostated (37°) stirred cuvet (5 ml).

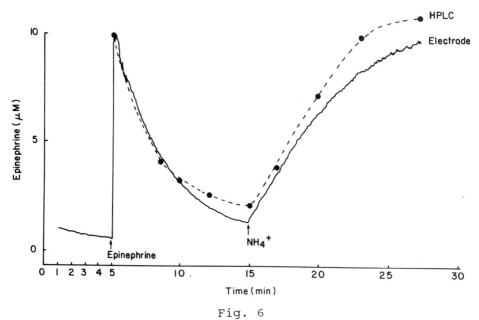

Fig. 6

4. The Experiment

1. Add 5 ml of medium (185 mM KCl, 20 mM HEPES) to the thermostated cuvet.

2. Add approx. 2.5 mg of chromaffin ghosts dialyzed and suspended in 185 mM KCl, 2 mM HEPES.

3. Add 25 µl dopamine.

4. Add 50 µl ATP.

5. Before addition of ATP and 1-2-3-5-10 min. after, withdraw 500 µl sample, filter, collect supernatant, and measure by HPLC.

The separation of the various nucleotides and catecholamines in the column is shown in Fig. 7. In this experiment a higher peak due to ATP should appear.

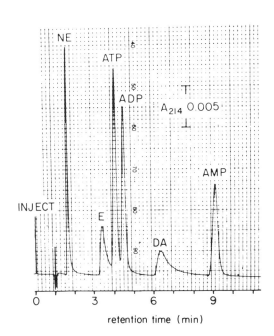

Fig. 7

REFERENCES

Apps, D.K.., and Glover, S.A, 1978, FEBS Lett., 85: 254-258.
Apps, D.K., and Schatz, G., 1979, Eur. J. Biochem., 100: 411-419.
Bashford, C.L., Casey, R.P., Radda, G.K., and Ritchie, G.A., 1976, Neuroscience, 1: 399-412.
Buckland, R.M., Radda, G.K., and Wakefield, L.M., 1979, FEBS Lett., 103: 323-327.
Carty, S.E., Johnson, R.G., and Scarpa, A., 1980, Anal. Biochem., 106: 438.
Flatmark, R., and Ingebretsen, O.C., 1977, FEBS Lett., 78: 53-56.

Hayflick, S., Johnson, R.G., Carty, S.E., and Scarpa, A., 1981, _Anal. Biochem._, 126: 58-66.

Holtz, R.W., 1978, _Proc. Natl. Acad. Sci. USA._, 75: 5190-5194.

Johnson, R.G., and Scarpa, A., 1976, _J. Gen. Physiol._, 68: 601-631.

Johnson, R.G., Beers, M.F., and Scarpa, A., 1982, _J. Biol. Chem._, 257: 10701-10707.

Johnson, R.G., Carlson, N.J., and Scarpa, A., 1978, _J. Biol. Chem._, 253: 1512-1521.

Johnson, R.G., Pfister, D., Carty, S.E., and Scarpa, A., 1979, _J. Biol. Chem._, 254: 10963-10972.

Johnson, R.G., and Scarpa, A., 1979, _J. Biol. Chem._, 254: 3750-3760.

Johnson, R.G., Carty, S.E., and Scarpa, A., 1981, _J. Biol. Chem._, 256: 5773-5780.

Johnson, R.G., Hayflick, S.J., Carty, S.E., and Scarpa, A., 1981, _FEBS Lett._, submitted.

Mitchell, P., 1968, _Chemiosmotic Coupling and Energy Transduction_, Glynn Research, Bodmin, Cornwall.

Njus, D., and Radda, G.K., 1979, _Biochem. J._, 180: 579-585.

Phillips, J.H., 1978, _Biochem. J._ 170: 673-674.

Roisin, M.P., Scherman, D., and Henry, J.P., 1980, _FEBS Lett._, 115: 143-146.

Rottenberg, H., 1979, _Biochim. Biophys. Acta_, 549: 225-253.

Salama, G., Johnson, R.G., and Scarpa, A., 1980, _J. Gen. Physiol._, 75: 109-140.

Winkler, H., 1976, _Neuroscience_, 1: 65-80.

Winkler, H., and Carmichael, S.W., _in_: "The Secretory Granule" E. Poisner and J. Trifaro, eds., Elsevier Press, Amsterdam (1982).

CONTRIBUTORS

Claude Alban
Laboratoire de Physiologie Cellulaire Végétale
URA CNRS no. 194
Départmnet de Recherche Fondamentale
Centre d'Etudes Nucléaires et Université Joseph Fourier
F-38041 Grenoble-Cedex
France

Angelo Azzi
Institut für Biochemie und Molekularbiologie
Universität Bern
Bühlstrasse 28
CH-3012 Bern
Switzerland

Rob Benne
E.C. Slater Institute for Biochemical Research
University of Amsterdam
Academic Medical Centre
Meibergdreef 15
1105 AZ Amsterdam
The Netherlands

Maryse A. Block
Laboratoire de Physiologie Cellulaire Végétale
URA CNRS no. 194
Département de Recherche Fondamentale
Centre d'Etudes Nucléaires et Université Joseph Fournier
F-38041 Grenoble-Cedex
France

D. Boffoli
Istituto Policattedra di Biochimica Medica e Chimica Medica
Università degli Studi di Bari
Piazza G. Cesare
70124 Bari
Italy

Abraham Bout
E.C. Slater Institute for Biochemical Research
University of Amsterdam
Academic Medical Centre
Meibergdreef 15
1105 AZ Amsterdam
The Netherlands

Stanley Brul
E.C. Slater Institute for Biochemical Research
University of Amsterdam
Academic Medical Centre
Meibergdreef 15
1105 AZ Amsterdam
The Netherlands

Xina Cao
Institut für Molekularbiologie und Tumorforschung
Philipps-Unvirsität
D-3550 Marburg
Federal Republic of Germany

Titiana Cocco
Istituto Policattedra di Biochimica Medica e Chimica Medica
Universitá degli Studi di Bari
Piazza G. Cesare
70124 Bari
Italy

John Davey
Department of Biochemistry
University of Birmingham
Birmingham B15 2TT
United Kingdom

Ben Distel
E.C. Slater Institute for Biochemical Research
University of Amsterdam
Academic Medical Centre
Meibergdreef 15
1105 AZ Amsterdam
The Netherlands

Roland Douce
Laboratoire de Physiologie Cellulaire Végétale
URA CNRS no. 194
Département de Recherche Fondamentale
Centre d'Etudes Nucléaires et Université Joseph Fournier
F-38041 Grenoble-Cedex
France

C. Eckerskorn
Max-Planck-Institut für Biochemie
Genzentrum
D-8033 Martinsried
Federal Republic of Germany

Sinikka Eskelinen
Laboratory of Physiological Chemistry
University of Groningen
Bloemsingel 10
9712 KZ Groningen
The Netherlands

Sergio Giannattasio
Dipartimento di Biochimica e Biologia Moleculari e
C.N.R. Centro di Studi sui Mitocondri e Metabolismo Energetico

Università degli Studi di Bari
70126 Bari
Italy

Gabriele V. Gnoni
Dipartimento di Biochimica
Universita di Lecce
Lecce
Italy

Albert K. Groen
Department of Gastroenterology
University of Amsterdam
Academic Medical Centre
Meibergdreef 9
1105 AZ Amsterdam
The Netherlands

Ferruccio Guerrieri
Istituto Policattedra di Biochimica Medica e Chimica Medica
Università degli Studi de Bari
Piazza G. Cesare
70124 Bari
Italy

Dieter Häussinger
Medizinische Universitätsklinik
Universität Freiburg
Hugstetterstrasse 55
D-7800 Freiburg
Federal Republic of Germany

Judith C. Heikoop
E.C. Slater Institute for Biochemical Research
University of Amsterdam
Academic Medical Centre
Meibergdreef 15
1105 AZ Amsterdam
The Netherlands

Lutger Hengst
Max-Planck-Institut für Biofysikalische Chemie
Abteilung Molekulare Genetik
D-3400 Göttingen
Federal Republic of Germany

Dick Hoekstra
Laboratory of Physiological Chemistry
University of Groningen
Bloemsingel 10
9712 KZ Groningen
The Netherlands

Jacques Joyard
Laboratoire de Physiologie Cellulaire Végétale
URA CNRS no. 194
Département de Recherche Fondamentale
Centre d'Etudes Nudcléaires et Université Joseph Fournier
F-38041 Grenoble-Cedex
France

Bernard Kadenbach
Fachbereich Chemie
Abteilung Biochemie
Philipps-Universität
D-3550 Marburg
Federal Republic of Germany

Jan Willem Kok
Laboratory of Physiological Chemistry
University of Groningen
Bloemsingel 10
9712 KZ Groningen
The Netherlands

Jan Kopecky
Institute of Physiology
Czechoslovak Academy of Sciences
142 20 Praha 4
Czechoslavakia

Michele Lorusso
Istituto Policattedra di Biochimica Medica e Chimica Medica
Università degli Studi di Bari
Piazza G. Cesare
70124 Bari
Italy

L. Lottspeich
Max-Planck-Institut für Biochemie
Genzentrum
D-8033 Martinsried
Federal Republic of Germany

Ersilia Marra
Dipartimento di Biochimica e Biologia Moleculari e
C.N.R. Centro di Studi sui Mitocondri e Metabolismo Energetico
Università degli Studi di Bari
70126 Bari
Italy

Alfred J. Meijer
E.C. Slater Institute for Biochemical Research
University of Amsterdam
Academical Medical Centre
Meibergdreef 15
1105 AZ Amsterdam
The Netherlands

Thomas Mengel
Fachbereich Chemie
Abteilung Biochemie
Philipps-Universität
D-3550 Marburg
Federal Republic of Germany

Paul A.M. Michels
Research Unit for Tropical Diseases
International Institute of Cellular and Molecular Pathology
Avenue Hioppocrate, 74
B-1200 Brussels
Belgium

Cesare Montecucco
Centro C.N.R. Biomembrane e Istituto di Patalogia Generale
Università di Padova
Via Loredan 16
35131 Padova
Italy

Michele Müller
Institut für Biochemie und Molekularbiologie
Universität Bern
Bühlstrasse 28
CH-3012 Bern
Switzerland

Fred R. Opperdoes
Research Unit for Tropical Diseases
International Institute of Cellular and Molecular Pathology
Avenue Hippocrate, 74
B-1200 Brussels
Belgium

Sergio Papa
Istituto Policattedra di Biochimica Medica e Chimica Medica
Università degli Studi di Bari
Piazza G. Cesare
70124 Bari
Italy

Emanuele Papini
Centro C.N.R. Biomembrane e Istituto di Patalogia Generale
Università di Padova
Via Loredan 16
35131 Padova
Italy

Ernesto Quagliariello
Dipartimento di Biochimica e Biologia Moleculare e
C.N.R. Centro di Studi sui Mitocondri e Metabolismo Energetico
Università degli Studi di Bari
70126 Bari
Italy

Rino Rappuoli
Centro Ricerche SCLAVO S.p.a.
Via Fiorentina 1
53100 Sienna
Italy

Cecilia Saccone
C.N.R. Centro di Studi sui Mitocondri e Metabolismo Energetica
e Dipartimento di Biochimica e Biologia Moleculare
Università degli Studi di Bari
70126 Bari
Italy

Dorianna Sandona
Centro C.N.R. Biomembrane e Istituto di Patalogia Generale
Università di Padova
Via Loredan 16
35131 Padova
Italy

Elisabetta Sbisá
C.N.R. Centro di Studio sui Mitocondri e Metabolismo Energetico
e Dipartimento di Biochimica e Biologica Molecolare
Universitá degli Studi di Bari
70126 Bari
Italy

Antonia Scarpa
Department of Physiology and Biophysics
School of Medicine
Case Western Reserve University
Cleveland, OH 44106
U.S.A.

Giampetro Schiavo
Centro C.N.R. Biomembrano e Istituto di Patalogia Generale
Universitá di Padova
Via Loredan 16
35131 Padova
Italy

Andrea Schlerf
Fachbereich Chemie
Abteilung Biochemie
Philipps-Universität
D-3550 Marburg
Federal Republic of Germany

Anneke Strijland
E.C. Slater Institute for Biochemical Research
University of Amsterdam
Academic Medical Centre
Meibergdreef 15
1105 AZ Amsterdam
The Netherlands

Guntram Suske
Institut für Molekularbiologie und Tumorforschung
Philipps-Universität
D-3550 Marburg
Federal Republic of Germany

Henk F. Tabak
E.C. Slater Institute for Biochemical Research
University of Amsterdam
Academic Medical Centre
Meibergdreef 15
1105 AZ Amsterdam
The Netherlands

Joseph M. Tager
E.C. Slater Institute for Biochemical Research
University of Amsterdam
Academic Medical Centre
Meidbergdreef 15
1105 AZ Amsterdam
The Netherlands

Ronald J.A. Wanders
Department of Pediatrics
University of Amsterdam
Academic Medical Centre
Meibergdreef 9
1105 AZ Amsterdam
The Netherlands

Andries Westerveld
Department of Human Genetics
University of Amsterdam
Academic Medical Centre
Meibergdreef 15
1105 AZ Amsterdam

Erik A.C. Wiemer
E.C. Slater Institute for Biochemical Research
University of Amsterdam
Academic Medical Centre
Meibergdreef 15
1105 AZ Amsterdam
The Netherlands

Franco Zannoti
Istituto Policattedra di Biochimica Medica e Chimica Medica
Unversitá degli Studi di Bari
Piazza G. Cesare
70124 Bari
Italy

Receptors (continued)
  for epidermal growth factor, 63,
      75
  ligand-receptor complex, 51
  acidic pH and, 51
  dissociation of, 51, 56
  for LDL, 51, 51, 62, 63
  and lipid flow, 61-77
  for mannose-6-phosphate, 53, 66,
      67
  and membrane flow, 47-57, 61
      77
  recycling of, 51, 56, 62-64
  for Semliki forest virus, 50
  for transferrin, 63-65, 73
  tyrosine residues in, 52, 53
  for viruses, 50, 67
Refsum disease
  adult form, 32
  infantile form, 32, 38-41
Regulation of metabolism (see also
      Control Analysis)
  Biochemical Systems Theory and,
      85, 86
  Metabolic Control Theory and, 85,
      86
Replication origin
  in mitochondrial DNA, 137
Recycling
  during lipid flow, 61-77
  during membrane flow, 47-57
Respiratory chain, 1-7
  genes for complexes of, 129
Rhizomelic Chondrodysplasia
      Punctata, see
      Chondrodysplasia Punctata
Rhodamine
  lipid analogues labelled with,
      69-74

Saccharomyces cerevisiae
  peroxisomal protein topogenesis
      in, 178-183
Secretion, 47, 61
Semliki Forest Virus (see also
      Receptors)
  endocytosis of, 50
Sphingomyelin, 76
Sialidase, 75
Sterol carrier protein, 74
[^{3}H]Sucrose
  as fluid-phase endocytosis
      marker, 50
Sulpholipids
  biosynthesis of, 172
  in plastid envelope membranes,
      167
Summation theorem, see Control
      Analysis

Toxin, see Diphteria toxin

Trans-Golgi network, 47-57, 61-77
Transferrin, see Receptors
Trypanosoma brucei, 188
  glycosomes in, 187, 188

Ubiquinone, 1,3
Ubiquinol: cytochrome c
      oxidodreductase, see
      Complex III
Urea cycle, 101-111
  and acid-base homeostasis, 108-
      110
  and ammonia elimination, 103-108
  and bicarbonate elimination,
      108-110
  control of flux in, 110, 111
  by N-acetylglutamate, 111

Vesicles
  coated, 47-57, 62
  fusion of, 54, 55
  and lipid flow, 61-77
  and membrane flow, 47-57, 61-77
  perinuclear, 56
  secretory, 48
  transport, 48, 64-69, 76
  uncoated, 50, 51, 62
  uncoating, 53
Viruses (see also Receptors)
  and membrane flow, 67

Zellweger syndrome, see Cerebro-
      hepato-renal syndrome
Zellweger-like syndrome, 32, 36